3rd International Conference on Aerospace Technology, Communications and Energy Systems (ATCES 2019)

IOP Conference Series: Materials Science and Engineering
Volume 685

Nanjing, China
26 - 29 September 2019

ISBN: 978-1-7138-0937-1
ISSN: 1757-8981

Printed from e-media with permission by:

Curran Associates, Inc.
57 Morehouse Lane
Red Hook, NY 12571

Some format issues inherent in the e-media version may also appear in this print version.

Institute of Physics
Dirac House, Temple Back
Bristol BS1 6BE UK

Phone: 44 1 17 929 7481
Fax: 44 1 17 920 0979

techtracking@iop.org

Additional copies of this publication are available from:

Curran Associates, Inc.
57 Morehouse Lane
Red Hook, NY 12571 USA
Phone: 845-758-0400
Fax: 845-758-2633
Email: curran@proceedings.com
Web: www.proceedings.com

TABLE OF CONTENTS

AEROSPACE TECHNOLOGY (SPACE)

ON ORBIT CAPTURE AND CONTROL OF SPACE MANIPULATOR BASED ON MAGNETO RHEOLOGICAL 1
Chao Wang, Jianwei Wang, Dong Fang, Donglin Li, Ying Zhang, Dalin Kong

RESEARCH ON MULTI-SATELLITE OBSERVATION MULTI-REGION TASK PLANNING BASED ON GENETIC ALGORITHM 7
Ying Zhang, Jianwei Wang, Bo Yuan, Chao Wang, Lei Shi

ORBIT-ATTITUDE COUPLED TRACKING AND LANDING CONTROL FOR AN ASTEROID 17
Lingchao Zhu, Shuquan Wang

MAGNETIC SIMULATOR TESTBED FOR APSCO STUDENT SMALL SATELLITE PROJECT 23
J Uscategui, W Xinsheng

AEROSPACE TECHNOLOGY (STRENGTH, DESIGN)

RELIABILITY DESIGN FOR DATA STORAGE IN AN INTEGRATED TEST SYSTEM 35
Zhang Bing, Sun Xiao, Shi Jianwei, Deng Rongjun

MECHANICAL SYSTEM PERFORMANCE ANALYSIS OF COMMUTER AIRCRAFT'S REAR LANDING GEAR DESIGN FOR UNPAVED RUNWAY 41
N A Rammon, W Nirbito

DERATED THRUST: METHOD ANALYSIS FOR OPTIMIZING TURBOFAN ENGINE TAKEOFF PERFORMANCES (SFC, EGT) DUE TO LOWER MAXIMUM TAKEOFF WEIGHT (MTOW) REQUIREMENT 48
R F Irsyadi, W Nirbito

DESIGN ANALYSIS OF DUCTED PROPELLER FOR BICOPTER DRONE PROPULSION 54
M R Priatmoko, W Nirbito

A DESIGN METHOD FOR TWIST DISTRIBUTION ALONG WING SPAN 61
Xiaoyan Liu, Pan Cheng, Yang Zhao

OPTIMIZATION OF HEATED STRUCTURES FOR SEVERAL LOAD CASES 67
Anatolii Kretov

AERODYNAMIC, ENGINE SYSTEM, IT TECHNOLOGY, COMMUNICATION

ADAPTIVE REFINEMENT BASED INTEGRAL METHOD FOR LOAD TRANSFER IN FLUID/STRUCTURE INTERACTION PROBLEMS 73
Pan Cheng, Zekai Ding, Wu Zhang

DISTANCE-AWARE DATA FORWARDING PROTOCOL FOR NAMED DATA MOBILE AD-HOC NETWORK (NDMANET) 83
Junyu Lai, Zhongwei Chen, Chengkang Li, Yingbing Sun, Han Xiao

ENABLING EDGE COMPUTING ABILITY IN MOBILE SATELLITE COMMUNICATION NETWORKS 91
Junyu Lai, Yudi Zhang, Lei Zhong, Ying Qu, Rui Liu

STUDY ON LINEAR AND NONLINEAR COUPLING DAMAGE MODEL OF BGA SOLDER JOINT 98
Dongyang Lei, Yufeng Sun, Chaojie Qi, Weiwei Hu

DEEP LEARNING FOR AIRCRAFT WAKE VORTEX IDENTIFICATION 104
Pan Weijun, Duan Yingjie, Zhang Qiang, Tang Jiahao, Zhou Jun

STUDY ON THE INFLUENCE OF SWEPT ANGLE ON THE AERODYNAMIC CHARACTERISTICS OF THE CROSS-SECTION AIRFOIL OF A VARIABLE SWEPT-WING AIRCRAFT 112
Su Xinbing, Jiang Wen, Zhao Xiwei

MODELING AND SIMULATION OF BOW WAVE IN AERIAL REFUELING 118
Kunxian Zhai, Suiping Zhou, Xiaoping Li, Yanming Liu

RESEARCH ON METHOD OF PHANTOM TRACK GENERATION BASED ON COOPERATIVE CONTROL OF UAVS IN UNIFORM LINEAR MOTION 127
Zhiping Ouyang, Chen Xu, Jiqiang Feng, Bo Wang

BASED ON MATLAB SIMULATION TO ANALYZE THE ONLINE SOURCE INTERACTION OF THE PHOTOVOLTAIC POWER GRID 133
Lin Qing Luo, Zhao Xing Ma

STUDY ON THE EFFECT OF EXCITATION CIRCUIT OF SPECTRAL LAMP ON THE TEMPERATURE COEFFICIENT OF PHYSICS PACKAGE OF RUBIDIUM ATOMIC CLOCK 141
Yonggang Guo, Jun Zhang, Shiwei Wang, Jianxiang Wang, Lili Zhu, Jianhui Tu, Zhidong Liu

CONCEPTUAL MULTIDISCIPLINARY DESIGN OPTIMIZATION OF COMMERCIAL AERO-ENGINE SYSTEM 148
Fushui Guo, Shiqi Zhao, Xiaodong Zhou, Jun Fan, Dong Mi

A TRANSMISSION SYSTEM PLANNING METHOD CONSIDERING DEMAND-SIDE RESPONSE AND CAPABILITY FOR ACCOMMODATING WIND POWER 156
Bo Sun, Bo Zou, Jubin Yan, Xiaoming Wang, Yuxuan Yang, Yan Huang

ANALYSIS AND APPLICATION OF DISTRIBUTED NAVIGATION SPOOFERS 162
Chao Ma, Jun Yang, Chao Zhou

AERODYNAMIC THERMODYNAMIC MODELING AND SIMULATION OF TURBOFAN ENGINE 170
Renjun Qian, Benwei Li, Hanqiang Song, Qing Dong, Yonghua Wang

STUDY ON THE FORECASTING OF THE HOT CORROSION RESISTANCE OF TYPICAL SUPERALLOYS FOR AEROENGINES 176
Guo Bingxiu, Xiaohui Wang, Yanyan Wang, Yue Shao

SINGLE-STATION TDOA PASSIVE LOCATION UNDER MULTIPATH PROPAGATION CONDITIONS 183
Luxi Zhang, Yuanyuan Song, Yijun Li, Qun Wan

PERFORMANCE ANALYSIS FOR MIMO SWIPT SYSTEMS WITH SPACE-TIME CODING 189
Tao Cui, Jun Li, Jun Shen, Shiwei Dong

A CENTRALIZED SCHEDULING ALGORITHM IN ON-BOARD HYBRID OPTICAL/ELECTRIC SWITCHING 196
Yi Zhang, Jun Li, Quan Zhou, Jingling Li, Wei Liang, Tao Cui

ANALYSIS OF IGS DATA BASED ON MILITARY ACTION PREDICTION 204
Qiang Fu, Guowei Shi, Wentao Fu, Yuanfa Ji, Xiyan Sun

APPLICATION OF DEEP CONVOLUTIONAL NEURAL NETWORK IN REMOTE PROTECTION OF TRANSMISSION LINES 211
Wenjie Zheng, Chengqi Li, Bo Yang, Xuebin Lv

CFD SIMULATION METHODS FOR ROTOR HOVERING BASED ON N-S EQUATION 217
Zhou Cheng-Long, Chen Ming, Xu An-An, Wang Fang

Author Index

IOP Conf. Series: Materials Science and Engineering **685** (2019) 011001 doi:10.1088/1757-899X/685/1/011001

PREFACE

ATCES 2019 was held in Nanjing, China, September 26-29, 2019. ATCES 2019 was organized by Asia Pacific Institute of Science and Engineering. This conference aims to provide a platform for researchers and engineers to share their ideas, recent developments and successful practices in Aerospace Technology, Communications and Energy Systems. The participants of the conference were from almost every part of the world, with background of either academia or industry, even well-known enterprise. The success and prosperity of the conference is reflected high level of the papers received.

The proceedings are a compilation of the accepted papers and represent an interesting outcome of the conference. This book covers 3 chapters: 1. Aerospase technology (Space); 2. Aerospase technology (Strength, Design); 3. Aerodynamic, Engine system, IT technology, Communication.

We would like to acknowledge all of those who supported ATCES 2019. Each individual and institutional help were very important for the success of this conference. Especially we would like to thank the organizing committee for their valuable advices in the organization and helpful peer review of the papers.

We sincerely hope that ATCES 2019 will be a forum for excellent discussions that will put forward new ideas and promote collaborative researches. We are sure that the proceedings will serve as an important research source of references and the knowledge, which will lead to not only scientific and engineering progress but also other new products and processes.

Prof. Kretov Anatolii
Conference Committee Chair

IOP Conf. Series: Materials Science and Engineering **685** (2019) 011002 doi:10.1088/1757-899X/685/1/011002

COMMITTEES

Honorable Chair

Alexander M. Mebel, Florida International University, USA
Nabil Ayoub, American University of Madaba AUM in Jordan, Jordan
Wagdi G. Habashi, McGill University, Canada

Conference Committee Co-Chair

Andrey B. Prokofiev, Samara University, Russia
Evgeniy V. Shakhmatov, Samara University, Russia
Kretov Anatolii, Nanjing University of Aeronautics and Astronautics, China
Yury A. Nozhnitskiy, Central Institute of Aviation Motors (Moscow), Russia

Program Committee

Mikhail Kolotnikov, Research Institute of Mechanics , Moscow State University, Russia
S. A. Channiwala, S. V. National Institute of Technology,Surat, India
Yaser Qudaih, American University of Madaba (AUM) - Jordan

Financial Chair

Jinping Lei, The Hong Kong University of Science and Technology, Hong Kong

General Conference Secretary

Dmitry A. Uglanov, Samara University, Russia

Local Committee

Oleksiy Chernykh, Nanjing University of Aeronautics and Astronautics, China
Tiniakov Dmytro, Nanjing University of Aeronautics and Astronautics, China

Technical Committee Member

Adel A. Elbaset, Minia University, EGYPT
Aleksandr A. Igolkin, Samara University, Russia
Abdelghani CHAHMI, University of Sciences and Technology of Oran, Algeria
Aleksandr S.Vinogradov, Samara University, Russia
Dmitry K. Novikov, Samara University, Russia
GUELAILIA Ahmed, Satellite Developement Center, Algerian Space Agency, Oran, Algeria
Hassan Migdadi, American university of Madaba, Jordan
Ifigeneia Antoniadou, The University of Sheffield, UK
Ilya N. Krupenich, Samara University, Russia
John Page,The University of New South Wales, Australia
Kouzou Abdellah, Djelfa University, Algeria
Mohammed Mahdi Ali, Philadelphia University, Jordan
Mohamed Yahia Edries, National Authority for Remote Sensing and Space Sciences, NARSS, Egypt
OKUYAMA Keiichi, Kyushu Institute of Technology, Japan
Pooya Sareh, Imperial College London,UK
Tarek Hassan Mohamed, Aswan University, Egypt
Thongchart Kerdphol, Kyushu Institute of Technology, Japan

IOP Conf. Series: Materials Science and Engineering **685** (2019) 011002 doi:10.1088/1757-899X/685/1/011002

Tieying Wu, Nanjing University of Aeronautics and Astronautics, China
Viktor Ya. Sverbilov, Samara University, Russia
Vladimir A. Zrelov, Samara University, Russia
Yehia Abdel-Aziz, Head of Space Research Lab, NRIAG, Egypt

IOP Conf. Series: Materials Science and Engineering **685** (2019) 011003 doi:10.1088/1757-899X/685/1/011003

Peer review statement

All papers published in this volume of *IOP Conference Series: Materials Science and Engineering* have been peer reviewed through processes administered by the proceedings Editors. Reviews were conducted by expert referees to the professional and scientific standards expected of a proceedings journal published by IOP Publishing.

IOP Conf. Series: Materials Science and Engineering **685** (2019) 012001 doi:10.1088/1757-899X/685/1/012001

On Orbit Capture and Control of Space Manipulator Based on Magneto Rheological

Chao Wang[1,2], Jianwei Wang[1,2],Dong Fang, Donglin Li[1,2], Ying Zhang[1,2]and Dalin Kong[1,2]

[1]Key Laboratory for Spacecraft In-Orbit Fault Diagnosis and Maintenance, Xi'an, 710043

[2]China Xi'an Satellite Control Centre, Xi'an, 710043, China

Email: chao860219@126.com

Abstract. With the rapid development of space technology, the performance and technology of spacecraft is in the continuous improvement. The key technology for the non-cooperative target satellite on-orbit capture and control becomes very important. For the question of on orbit capturing a non-cooperative target, a method based on Magneto Rheological (MR) damper is proposed. There are many successful applications of MR damper in ground-base and aviation systems. However, the application of this device on a space manipulator has not been addressed yet. In this paper, A 6DOFs space robot is built in Maplesim, and analyses the kinematic and dynamic model, and the component of MR damper is also created, after simulation, the results shows that the angular velocities of joints and base has been smoothly dampen. Although the control algorithm is simple, it's very valuable to use the MR damper in capture non-cooperative target.

1. Introduction

With the rapid development of space technology, the performance and technology of spacecraft is in the continuous improvement[1]. At the same time, its organizational structure is also increasing complexity. In this case, to ensure the spacecraft can work more lasting and stable in complex space environment, which is need to develop the technology of on-orbit service, therefore, the key technology for the non-cooperative target satellite on-orbit capture and control becomes very important.

There exist several methods for on-orbit capture of non-cooperative target. Yoshikawa et al[2] proposed the use of impedance control algorithm to capture non-cooperative target, and describes the conditions to ensure that the target is not pushed away after collision. Nenchev[3-4] proposed the use of "Reaction null space" control algorithm to process the collision of coupled angular momentum, and this method can effectively decoupling the dynamic of manipulators and the base, limiting the base attitude in a small value within the scope, reducing the angular velocity of joints. Cong Peichao proposed a method for planning the configuration of space manipulator before arrest so that the angular momentum of system caused by collision force is zero. Although these methods are different in detail, most of them use the same fundamental approach, which is based on the dynamic model and different control algorithms. Because of the complex environment in space, the boundary conditions when capturing the target are very demanding; using control algorithms mentioned above may result to fail.

To overcome the problem of capturing a non-cooperative target, each joint of the space manipulator can can be install an active damping device. According to condition at the contacting moment, and the rotational velocity of the target, the MR damper can generate the damping torque to protect the manipulator and base from damage during and after impact.

2. Kinematic And Dynamic Model ANALYSIS

2.1 *Kinematic Model*

In order to establish the kinematics model, we take the universal joint as two degrees of freedom joints and a zero-length link which is shown in Figure 1. The joints J_i are a single revolute for pitch direction; $J_i^{'}$ for yaw direction; and L_i equal to the length of link i; $L_i^{'}$ equal to zero.

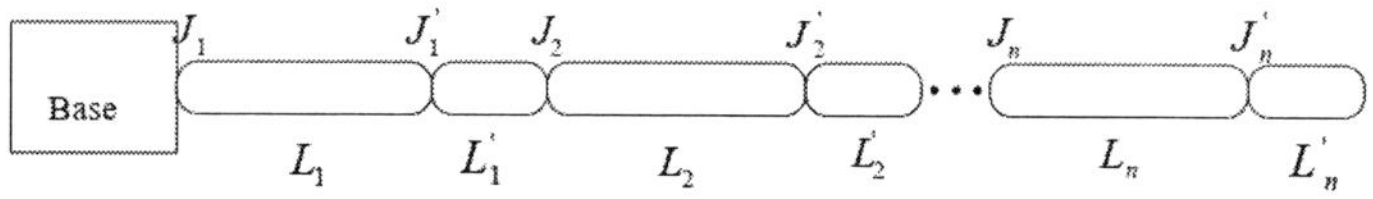

Figure 1. the equivalent configuration of space manipulator

We assume the space manipulator demonstrated in this paper is the free-floating system, and then a simplified model of space manipulator system is shown in Figure 2. From this, it can be seen that the vector from the inertial fixed origin O to i body's centre of mass (CM).

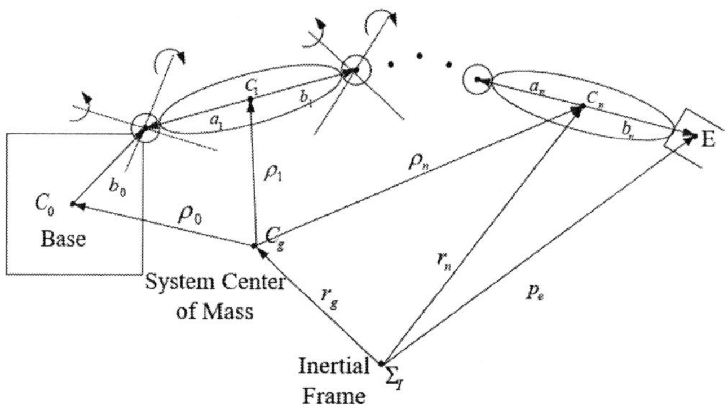

Figure 2. A simplified models of space manipulator and its end-effect

Because there is no external force action on the system, the mass centre of the system is not changed, the linear momentum of system is conservational, and then we have the following formulation.

$$\sum_{i=0}^{N} m_i r_i = M r_g \tag{1}$$

From (1), we can derive position vector for the mass centre of base:

$$r_0 = r_g - \frac{(m_1 + \cdots + m_n)(b_0 + a_1)}{M} - \cdots - \frac{m_n (b_{n-1} + a_n)}{M} \tag{2}$$

Then we can get the end-effector position vector:

$$P_e = r_g + \frac{m_0 b_0}{M} + \left(\frac{m_0 a_1}{M} + \frac{(m_0 + m_1) b_1}{M} \right) + \cdots + \left(\frac{(m_0 + \cdots m_{n-1}) a_n}{M} + b_n \right) \tag{3}$$

While the VM link $\hat{b}_i$ and $\hat{a}_i$ have the following formulation:

$$\hat{b}_i = \frac{\sum_{k=0}^{i} m_k}{M} b_i , \hat{a}_i = \frac{\sum_{k=0}^{i-1} m_k}{M} a_i \tag{4}$$

2.2 *Dynamic Model*

There are several mature methods for kinematics model of traditional space robot, such as the Lagrange method and Newton-Euler method. The paper uses the Lagrange method to describe the 5-links space manipulator.

The dynamic model of space manipulator should also be considered as the two degree of freedom joints and a zero-length link. According to the kinematic formulation in section 2.1, we have the following equations:

$$\omega_i = \omega_0 + \sum_{k=1}^{i} z_k \dot{\theta}_k \tag{5}$$

$$v_i = \dot{r}_i = v_0 + \omega_0 \times (r_i - r_0) + \sum_{k=1}^{i} \{ z_k \times (r_i - p_k) \} \dot{\theta}_k \tag{6}$$

Then, the kinetic energy of whole system can be described as follow:

$$T = \frac{1}{2} \sum_{i=0}^{n} \left(\omega_i^T I_i \omega_i + m_i v_i^T v_i \right) \tag{7}$$

Combining (5), (6), (7) and making the transfer, the following equations are obtained:

$$T = \frac{1}{2} \left[v_0^T, \omega_0^T, \dot{\Theta}^T \right] \begin{pmatrix} ME & M\tilde{r}_{0g}^T & J_{Tw} \\ M\tilde{r}_{0g} & H_w & H_{w\phi} \\ J_{Tw}^T & H_{w\phi}^T & H_m \end{pmatrix} \begin{bmatrix} v_0 \\ \omega_0 \\ \dot{\Theta} \end{bmatrix} \tag{8}$$

where $H_m = \sum_{i=1}^{n} \left(J_{Ri}^T I_i J_{Ri} + m_i J_{Ti}^T J_{Ti} \right)$

Combining (8) into the Lagrange equation as:

$$\begin{bmatrix} H_b & H_{bm} \\ H_{bm}^{\mathrm{T}} & H_m \end{bmatrix} \begin{bmatrix} \ddot{x}_0 \\ \ddot{\Theta} \end{bmatrix} + \begin{bmatrix} c_b \\ c_m \end{bmatrix} = \begin{bmatrix} F_b \\ \tau_m \end{bmatrix} + \begin{bmatrix} J_b^{\mathrm{T}} \\ J_m^{\mathrm{T}} \end{bmatrix} F_e \tag{9}$$

where the F_b are the Forces and moments for the base; F_e are the external forces and moments acting on the manipulator; τ_m are the driving torques; c_b、 c_m are the nonlinear force, including centripetal force and Coriolis force.

In this section, the dynamic model is analysed in MATLAB for mathematical modelling, and simulated in Maplesim for verification respectively. The results from MATLAB and Maplesim are discussed. The subsystem of base and manipulator built in Maplesim is shown in Figure 3.

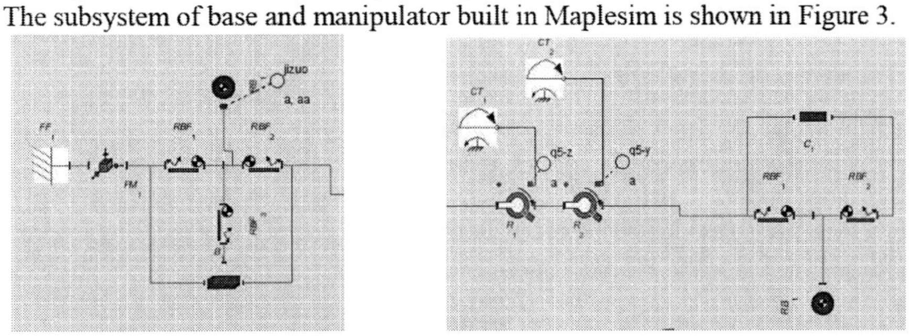

Figure 3. Subsystem of base and manipulator in Maplesim

In this simulation, the each torque of joints is 0.01N, and the range of simulation time is 10s. Then analyse the dynamic model of space model in MATLAB and Maplesim respectively, and can obtain the angular velocity of base in three axes. Figure 4 shown that, the result of numerical calculation in MATLAB is almost same with the result of simulation in Maplesim. So the mathematical model of the space manipulator has a high accuracy.

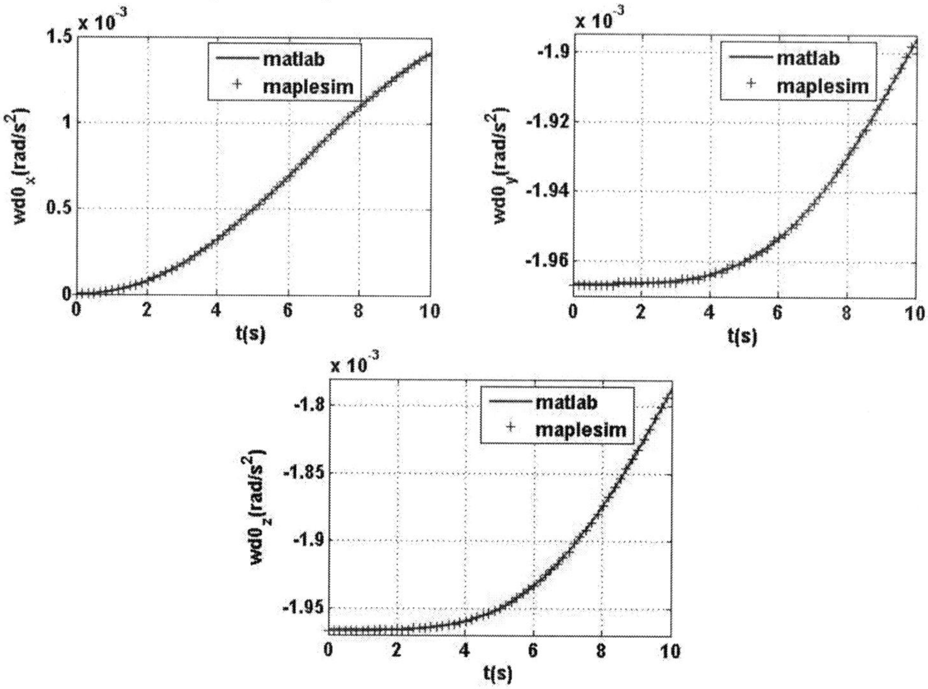

Figure 4. Angular velocity of base in X,Y,Z axis

3. On orbit Capture and Control Based on MR damper

3.1Characteristics and Model of MR damper

Magneto Rheological Fluid (MRF) is a smart material which can be controlled by the external magnetic field. Through the effect of external magnetic field , the MRF can change from a free-flowing state into a solid-like; after removing the magnetic field, the MRF can return to a free-flowing state, just as shown in Figure 5,

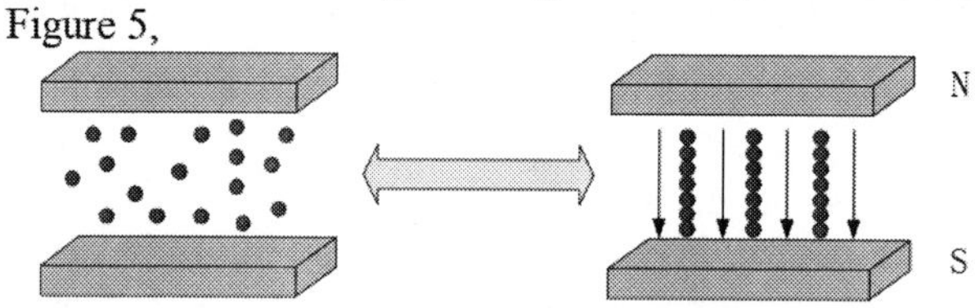

Figure 5. Effect of MRF

There are many researchers which have investigated the modelling of MRF, such as the Bingham model, Herschel-Bulkley model, Bouc-Wen model, the properties, modelling were demonstrated in details by Jolly. The paper mainly focuses on the implementation of MR damper to solve the problem of on-orbit capture by space manipulator. According to the result of Wei Zhou[5-6], the modified mathematical dynamic model is formulated as follows which is based on the Bingham model.

$$\begin{aligned}\tau = {} & f_0 \cdot \left(1-\exp\left(-\alpha \operatorname{abs}\left(\omega_{rel}\right)\right)\right) \cdot \operatorname{sgn}\left(\omega_{rel}\right) \\ & + c_0 \cdot \omega_{rel} + \beta\end{aligned} \tag{10}$$

$$f_0 = f_a + f_b \cdot P \tag{11}$$

$$c_0 = c_a + c_b \cdot P \tag{12}$$

$$\dot{P} = \gamma \cdot (I - P) \tag{13}$$

3.2On orbit Capture Based on Magneto Rheological

In this section, the simulation model and the results from the simulation study are discussed. The space robot is consisted of the base; the manipulators;and the target satellite. As shown in Figure 6, the target is connected rigidly by the gripper to the end-effector at the instant of capture, and the initial torque generated by contact is set to 2Nm at each axis. The joints have been equipped with the MR damper, while the base isn't controlled by the base mounted actuators.

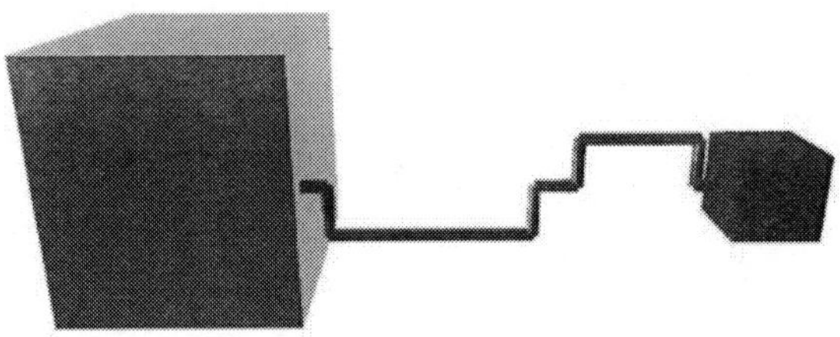

Figure 6. Configuration of the space robot

In this paper, the space robot models are created in Maplesim, and the parameter values are given in Table 1.

Table 1. Inertia parameters of space robot

Para	Mass(Kg)	Inertia tensor(Kgm^2)
Base	320	[25 18 24]
link1	11.5	[0.06 0.13 0.11]
link2	7.0	[0.02 1.28 1.28]
link3	10.3	[0.12 0.15 0.05]
link4	7.5	[0.53 0.46 0.08]
link5	10.5	[0.11 0.14 0.04]
target	100	[7.24 8.17 7.38]

To illustrate the performance of MR damper, the numerical simulation results on capture a non-cooperative target with a free-floating space manipulator installed the MR damper of the proportional feedback control law are shown in Figure 7.

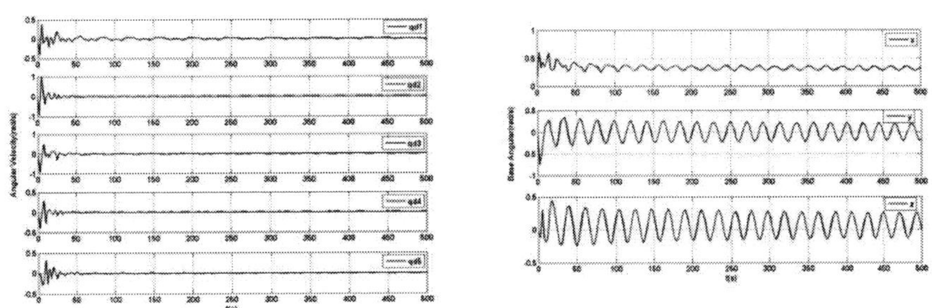

Figure 7. Relative angular velocity of joint and the base

The results presented that using MR damper to solve the on-orbit capture problem can really work. Under the effect of torque generated by MR damper, the relative angular velocity of joint has been smoothly dampen to zero. The angular velocity of base also has been decay from impulse to harmonic.

4. Conclusion

For the question of on-orbit capturing a non-cooperative target, a method based on MR damper is proposed. A 6DOF model of space model and component of MR damper are created in Maplesim. Although the control algorithm is simple, it's very valuable to use the MR damper in capture non-cooperative target.

References

[1] Boumans R, Heemskerk C. The European robotic arm for the international space station[J]. Robotics and Autonomous systems, 1998,23(1-2):17-27.
[2] Oda M. Space robot experiment on NASDA's ETS-VII satellite[C]. Proc. Of IEEE international conf. on robotics and automation, 1999:1390-1395.
[3] S.Hirose. Biologically inspired robots --- snake-like locomotors and manipulators[M]. Oxford: Oxford university press, 1993.
[4] Shigeo Hirose and Hiroya Yamada. Snake-Like Robots Machine Design of Biologically Inspired Robots[J]. IEEE Robotics & Automation Magazine. 2009, 3.
[5] Wei Zhou, Chee-Meng Chew, Geok-Soon Hong. Inverse Dynamics Control for Series Damper Actuator Based on MR Fluid Damper. IEEE/ASME International Conference on Advanced Intelligent Mechatronics, Monterey, California, USA, 2005. 473-478.
[6] Chee-Meng Chew, Geok-Soon Hong, Wei Zhou. Series damper actuator system based on MR fluid damper. Robotica, Vol. 24, 2006, pp. 699-710.

Research on Multi-satellite Observation Multi-region Task Planning based on Genetic Algorithm

Ying Zhang[1,2], Jianwei Wang[1,2], Bo Yuan[2], Chao Wang[1,2] and Lei Shi[1,2]

[1]Key Laboratory for Spacecraft In-Orbit Fault Diagnosis and Maintenance, Xi'an,710043, China
[2]China Xi'an Satellite Control Center, Xi'an,710043, China

Email:eva_zhang1331@163.com

Abstract. Space-based resources play an important role in the observation and handling of emergency events, and are widely used in national economic activities. In recent years, the large increase in the number of orbital spacecraft and the strategic transformation of the aerospace equipment application model have made the practical task planning for multi-type satellites of practical significance. The research and application of space-based regional reconnaissance algorithms are of great significance for improving the comprehensive application efficiency of spacecraft, forming uniform coverage of key areas, and improving the uniformity of satellite resource utilization and situational awareness. Based on the actual background of engineering application, this paper describes the scheduling problem of Earth observation satellites. Under the constraints of satellite resource capability, observable time, conversion time between reconnaissance missions, energy, etc., a multi-satellite multi-target reconnaissance mission is established. Planning model. With the most resource-saving and high situational awareness uniformity as the optimization index, the optimized satellite reconnaissance mission planning scheme is obtained based on genetic algorithm.

1. Introduction

In recent years, the aerospace industry has developed at a high speed, and the types and quantities of orbiting spacecraft have increased dramatically. The spacecraft has a wide coverage area, all weather, all-day time, is not restricted by air borders, and does not involve personnel safety[1]. These observation requests usually have higher timeliness, information acquisition accuracy, and resource usage synergy requirements. They often require multiple types of satellites to cooperate, and diverse spatial tasks urgently need to be played in a unified mission plan[2]. The greatest benefit of spacecraft.

Similar to the French SPOT-5 satellite system[3], the US Landsat-7 satellite system[4], the EO-1 satellite system[5], etc., in the past, China's satellites were managed separately and independently, according to professional vertical protection, various satellites and their The application system is self-contained, and these departments only perform mission planning for their respective satellites, which is difficult to integrate. There is a difficulty in emergency response, it is unable to efficiently meet the needs of emergency observation, and data from different sources cannot be effectively integrated and mutually verified and guided. After the strategic transformation of the aerospace equipment application model, the front-end proposes satellite observation requirements planning, the back-end satellites' measurement and control, and the unified management of operation and control. The satellite joint planning can optimize the integration of resources, realize integrated and multi-method integration of various systems, and improve application efficiency.

The existing satellite mission planning software is for the research of business process processing, model construction and solving algorithms for each series of satellites[6];[7];[8];[9];[10]. The main satellite mission planning models established by scholars at home and abroad include mathematical programming models, constraint satisfaction models and general models. The algorithm for solving the model algorithm is divided into an exact algorithm and an approximation algorithm. The approximation algorithm is further divided into an intelligent search algorithm and a rule-based heuristic algorithm. The satellite mission planning problem has been proved to be an NP-hard problem due to its multi-resource, multi-tasking and multi-time window characteristics. Considering the time and cost problems, most of the approximation algorithms are used to seek the optimal solution of mission planning[11].

From the perspective of engineering application, this paper studies and models multi-star and multi-objective mission planning problems, with the minimum total fuel consumption, average single-point reconnaissance coverage gap, and maximum reconnaissance time. It satisfies the satellite observable time window and task. The satellite attitude adjustment time, the longest continuous working time of the satellite, the satellite energy and other constraints are required, and the genetic algorithm is used to solve the problem.

2. Modeling the Earth Observation Satellite Scheduling Problem

During the on-orbit operation of the Earth observation satellite, the user first requests the product distribution center, and the distribution center converts it into a satellite observation order and submits it to the satellite control center. The control center sends it to the measurement and control center in the form of business data, and takes the form of remote control through the measurement and control station. The satellite is placed on the satellite, and then the satellite starts working. The original data is passed through the telemetry and sent to the user through data processing and product processing[12].

We study the geostationary satellite scheduling problem of point targets, and the regional targets can be converted into point targets. First, unify the point target into a "meta task". And construct the income function to adapt to the differences in income calculation of various targets. Finally, the Earth observation satellite scheduling model is established.

2.1. Atomic Task construction

The work of the Earth observation satellite requires the camera, radar and other payloads to adjust the attitude to the target[13]. In an effective time window, the satellite observes a point target. That is to say, a Atomic Task refers to a time window satellite observing a point target. The subtasks obtained after the point target decomposition are collectively referred to as Atomic Task. o_{ijkv} indicates that the satellite s_j decomposes the target task t_i in the kth time window, and the obtained v subtask, o_{ijkv} represents an observation activity of the satellite to the mission.

The Atomic Task is represented by a six-element array: $o_{ijkv} = \{AtomicID, TaskID, SatID, Win, Angle, Coordinate\}$. Among them, AtomicID is the Atomic Task identifier; TaskID is the target observation task identifier; SatID is the satellite resource identifier used by the observation activity; Win indicates the start and end time of the observation activity; Angle is the side view angle adopted by the satellite in the observation activity, that is, imaging Attitude; Coordinate identifies the coordinate information of the observation activity to the ground coverage area. The Atomic Task coordinate information of the point target decomposition is the latitude and longitude coordinates of the point target.

After the point target is decomposed, information about the observed resources, time and angle is determined in each Atomic Task. These Atomic Tasks are irreducible, and can be completed by a single satellite. Therefore, this paper uses each Atomic Task as the basic element of scheduling, which facilitates the unified scheduling of point targets.

Set in the scheduling period, the number of available time windows for the satellite s_j to the task t_i is N_{ij}, and o_{ijkv} indicates that the satellite s_j performs a reconnaissance task on the task t_i in the kth time window, which is the vth subtask of the task t_i. Therefore, there are: task t_i based on the satellite s_j decomposition of the Atomic Task set $\mathbf{O}_{ij} = \{O_{ij1}, O_{ij2}, \cdots, O_{ijN_{ij}}\}$. The Atomic Task set of task t_i decomposition is $\mathbf{O}_i = \{O_{i1}, O_{i2}, \cdots, O_{iN_S}\}$.

The Atomic Task set obtained after each task is decomposed is called the Atomic Task Group of the task.

$$O_i = \bigcup_{j=1}^{N_s} \bigcup_{k=1}^{N_{ij}} o_{ijkv}, i \in [1, \cdots, N_T] \quad (1)$$

Where N_{ij} is the number of time windows of the satellite s_j to the task t_i; N_{ijk} is the number of atomic task constructed within the k-th time window of the satellite s_j and the task t_i. v indicates that the atomic task is the vth subtask of the atomic task group. The time window of the atomic task o_{ijkv} is $[ws_{ijkv}, we_{ijkv}]$ and the observation angle of s_j to t_i in this time window is g_{ijkv}.

In summary, the atomic task concept can completely describe the observation tasks of satellite-to-point targets. One observation task can contain multiple atomic tasks, and they are a one-to-many logical relationship. Therefore, the process of satellite resource scheduling is essentially the process of selecting atomic tasks.

2.2. Revenue function construction

2.2.1. Multi-objective evaluation system. The earth observation satellite resource scheduling is to obtain the maximum comprehensive benefit under the condition of satisfying various constraints within a certain period of time, with the least resource consumption and the best imaging effect[14]. In summary, the main optimization indicators considered in this paper are:

1) Under the conditions of meeting various resource constraints, arrange as many observation tasks as possible, and prioritize tasks with high importance levels;

2) The satellite resources consumed for completing all observing tasks are the least. The resource consumption of this paper considers the number of satellites used and the number of observation windows, and is only used as a constraint for satellite energy consumption.

3) The observation coverage gap is average and the observable time is the largest.

Next, the corresponding income function is constructed separately for the above optimization indicators.

2.2.2. Income function. 1) The number of observation tasks completed the most

A point target can be considered as a completed task by arranging only one atomic task. There are only two states, scheduled and unscheduled. Therefore, the revenue of the task can be calculated based on the completion status of its atomic task. When the task is completed, the revenue of the task is , and the completion status of its atomic task is defined as:

$$x_{ijkv} = \begin{cases} 1, & \textit{If the atomic task } o_{ijkv} \textit{ is executed} \\ 0, & \text{otherwise} \end{cases} \quad (2)$$

According to the arrangement state of the atomic task of task t_i, the gain of the satellite observation task t_i can be obtained, and the income function of the point target is:

$$CSpot(t_i) = \sum_{j=1}^{N_S} \sum_{k=1}^{N_{ij}} x_{ijkv} \times \omega_i \tag{3}$$

The point target has a unique constraint that only one imaging can be scheduled. During the scheduling process, only one meta-task that arranges the point target is selected, so the revenue of the point target is not repeated here.

2) Minimal resource consumption

Suppose the satellite's resource set is $S = \{s_1, s_2, \cdots, s_{N_S}\}$, and whether the satellite s_j participates in the observation task is defined as:

$$y_j = \begin{cases} 1, & \text{If the satellite } s_j \text{ is involved in the observing mission} \\ 0, & \text{otherwise} \end{cases} \tag{4}$$

According to the state of arrangement of all observation tasks, the satellite resource use income function of participating in the observation task can be obtained as follows:

$$CSat(s_j) = \sum_{j=1}^{N_S} y_j \tag{5}$$

Assuming that the window resource set of task t_i is $Windows_i = \{v_{i1}, v_{i2}, \cdots, v_{N_{windows\_i}}\}$, according to the arrangement state of all observation tasks, the satellite window resource usage benefit function of the participating observation tasks can be obtained as follows:

$$CWindows(i) = \sum_{i=1}^{N_T} N_{windows\_i} \tag{6}$$

3) Optimal observation time

When some specific targets are being reconnaissance, the user has corresponding requirements for time. For example, the length of each observation is not less than $Time_{see}$; the maximum invisible duration cannot exceed $Time_{blind}$; the average coverage gap is observed. Among them, the maximum invisible duration can be regarded as the constraint constraint of the observation coverage gap evaluation, so it can be represented by a gain function. Considering the observation task time optimization index contains multiple aspects, which can be established separately when constructing the time benefit function, as follows:

(a) The longest total observation time

Under the various constraints, the total observation time of the atomic task set O_i is as follows:

$$CTime_{see}(t_i) = \sum_{j=1}^{N_S} \sum_{k=1}^{N_{ij}} (we_{ijk} - ws_{ijk}) \tag{7}$$

(b) Observation coverage gap average

Assume that the starting time of the total observation mission planning is $Time_{start}$ and the ending point is $Time_{end}$. In order to meet the maximum invisible duration requirement, at least N_{all_i} atomic tasks are required to complete the task O_i, namely:

$$N_{all\_i} = floor\left(\frac{Time_{end} - Time_{start}}{Time_{blind}} \right) \tag{8}$$

Assuming that the task O_i that satisfies the constraint actually has N_{actual_i}, the average coverage of the observation gap is expressed by the standard deviation of the time interval between the atomic tasks o_{ijkv}, namely:

$$\sigma_i = \sqrt{\frac{\sum_{l=1}^{N_{actual\_i}-1}\left[(ws_{i(l+1)} - we_{il}) - \mu_i\right]^2 + \left[Time_{blind} - \mu_i\right]^2}{N_{actual\_i}}} \tag{9}$$

$$\mu_i = \frac{\sum_{l=1}^{N_{actual\_i}-1}(ws_{i(l+1)} - we_{il}) + Time_{blind}}{N_{actual\_i}} \tag{10}$$

In summary, under the conditions of the observing task, the income function of the coverage average of the reaction observations is as follows:

$$CGap(\sigma_i) = \begin{cases} \sum_{i=1}^{N_T}\sigma_i, & N_{actual\_i} \geq N_{all\_i} \\ \sum_{i=1}^{N_T}\sigma_i + k_{gap}(N_{all\_i} - N_{actual\_i}), & N_{actual\_i} < N_{all\_i} \end{cases} \tag{11}$$

2.2.3. Satellite scheduling model. After the point target is uniformly represented as a atomic task, the earth observation satellite scheduling model can be established based on the actual problem constraints[15]. Considering the constraints in the earth observation satellite scheduling problem, if the model is modeled by using the planning model, the knapsack problem model and the graph theory, it is difficult to express and deal with various complex constraints. The Constraint Satisfaction Problem (CSP) can describe various variables and constraints in a problem with a more natural language with certain semantics[16]. The modeling process is relatively straightforward and simple. Therefore, this paper uses the constraint satisfaction problem to establish a geostationary satellite scheduling model.

The decision variables for the Earth observation scheduling problem are expressed as:

$$\boldsymbol{X} = \left\{x_{ijkv} \middle| i = 1,\cdots,N_T; j = 1,\cdots,N_S; k = 1,\cdots,N_{ij}; v = 1,\cdots,N_{ijk}\right\}$$

$$\boldsymbol{Y} = \left\{y_j \middle| j = 1,\cdots,N_S\right\}$$

among them

$$x_{ijkv} = \begin{cases} 1, & \textit{If the atomic task } o_{ijkv} \textit{ is executed} \\ 0, & \text{otherwise} \end{cases} \tag{12}$$

$$y_j = \begin{cases} 1, & \text{If the satellite } s_j \text{ is involved in the observing mission} \\ 0, & \text{otherwise} \end{cases} \tag{13}$$

Since the earth observation task corresponds to the observation activity of the satellite and is the main inspection object when dealing with various constraints, first, the start time set SWS_j, the end time set SWE_j, and the observation angle set SG_j corresponding to the task set of the decision variable are derived:

$$\begin{aligned} SWS_j &= \left\{ws_{ijkv} \middle| x_{ijkv} = 1, i \in [1, N_T]\right\} \\ SWE_j &= \left\{we_{ijkv} \middle| x_{ijkv} = 1, i \in [1, N_T]\right\} \\ SG_j &= \left\{g_{ijkv} \middle| x_{ijkv} = 1, i \in [1, N_T]\right\} \end{aligned} \tag{14}$$

The following is an optimization model for the geostationary satellite scheduling problem. According to formula (3-19), the total return of the satellite observation point target can be obtained as follows:

$$C_{Spot}=\sum_{i=1}^{N_T} CSpot(t_i)=\sum_{i=1}^{N_T}\sum_{j=1}^{N_S}\sum_{k=1}^{N_{ij}} x_{ijk}\times\omega_i \tag{15}$$

According to formula (3-21), the total revenue of satellite resources can be obtained as follows:

$$C_{resource}=CSat(s_j)+CWindows(i)=\sum_{j=1}^{N_S} y_j+\sum_{i=1}^{N_T} N_{windows\_i} \tag{16}$$

According to formulas (3-22) and (3-24), the total return of observation time can be obtained as follows:

$$\begin{aligned}C_{Time}&=\sum_{i=1}^{N_T}\left[CTime_{see}(t_i)+CGap(\sigma_i)\right]\\&=\sum_{i=1}^{N_T}\left[\sum_{j=1}^{N_S}\sum_{k=1}^{N_{ij}}(we_{ijk}-ws_{ijk})+CGap(\sigma_i)\right]\end{aligned} \tag{17}$$

Among them, the expression of σ_i , $CGap(\sigma_i)$ are (9), (11).

In summary, the optimization goal of this paper scheduling problem is:

$$\max J=p_{Spot}\cdot C_{Spot}+p_{\text{Resource}}\cdot C_{\text{Resource}}+p_{Time}\cdot C_{Time} \tag{18}$$

Among them, p_{Spot}, $p_{Resouce}$, p_{Time} are the penalty factors for each sub-indicator.

The following constraints must be met for the geostationary satellite scheduling problem:

1) The uniqueness constraint of the point target, that is, each point target can only be observed once:

$$\forall i=1,\cdots,N_T:\sum_{j=1}^{N_S}\sum_{k=1}^{N_{ij}} x_{ijk}\le 1 \tag{19}$$

2) Satellite storage constraints:

$$\forall j=1,\cdots,N_S:\sum_{m=1}^{N_{Bj}}(we_{jm}-ws_{jm})m_j\le M_j \tag{20}$$

Which, N_{Bj} is the number of tasks performed by the satellite sj.

3) Satellite energy constraints:

$$\forall j=1,\cdots,N_S:\sum_{m=1}^{N_{Bj}}(we_{jm}-ws_{jm})p_j+\sum_{l=1}^{N_{Bj}-1}\left|g_{j(l+1)}-g_{jl}\right|\rho_j\le P_j \tag{21}$$

The satellite consumes a certain amount of electrical energy for imaging and side-swinging activities. Therefore, the satellite's energy consumption is expressed as a function of the satellite's observation time and the side view angle of the remote sensor. The energy consumed by the satellite cannot exceed its maximum energy limit.

3. Solving scheduling problem based on genetic algorithm

The genetic algorithm is a global optimization algorithm, which has the advantages of fast global optimization and is suitable for real-time engineering applications[17]. Since the satellite's observation activities have a strict timing relationship, it is necessary to extrapolate the existing satellite resources according to the user's scheduling requirements before scheduling. Next, based on the orbital extrapolation results and the target point information, information such as the satellite resources and the observation time window that can perform the reconnaissance task are counted. The pre-processed information storage format is as follows:

$$Data\_in_{n\times 6} = \begin{bmatrix} Sat_1 & Target_1 & WS_1 & WE_1 & Continue_1 \\ Sat_2 & Target_2 & WS_2 & WE_2 & Continue_2 \\ \vdots & \vdots & \vdots & \vdots & \vdots \\ Sat_n & Target_n & WS_n & WE_n & Continue_n \end{bmatrix} \tag{22}$$

Which *Sat* represents the satellite number, *Target* represents the target point number, *WS* represents the observation window start time, *WE* represents the observation window end time, and *Continue* c represents the observation duration. Each row of the matrix represents an executable task in the satellite resource, and the subscript n only indicates the number of tasks that can be performed in the existing satellite resources, and does not distinguish between the satellite number and the target number. In summary, the main problem solved in this paper is to select the optimal scheduling scheme with the least resource consumption and satisfying the constraint conditions in the input information (22).

Based on the coding rules of the genetic algorithm, each scheduling scheme is an individual, each of which contains n chromosomes, combined with the formula (22). The chromosome coding designed in this paper is as follows:

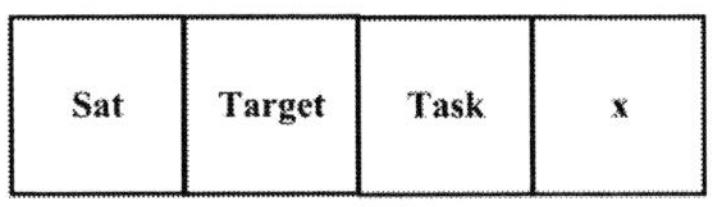

Figure 1. Schematic diagram of the coding structure

As shown in Figure 1, Sat represents the satellite number; Target represents the target number; Task represents the task number; x represents the decision variable, taking 0 or 1, which 0 indicates that the task is not executed and 1 indicates execution. It can be seen that each row of the formula (22) represents information of one chromosome. This kind of chromosome coding is relatively intuitive, and each chromosome represents whether the corresponding observation task is finally scheduled, and the encoding and decoding process is simple.

4. Satellite mission scheduling simulation

In order to verify the efficiency of the model and algorithm for point target scheduling, the following test cases are used for verification. In the example, the observation mission consists of three point targets, and the satellite resources include 10 Earth observation satellites.

The coordinate information of the target is shown in Table 1:

Table 1. Target coordinate information table.

Target number	longitude(°)	latitude(°)
1	250	8
2	300	17
3	200	20

The initial orbital information of the 10 Earth observation satellites is shown in Table 2. The initial orbital time is 22:00:00 on January 14, 2018.

Table 2. Initial orbit information of 10 satellites.

Satellite number	Semi-long axis(km)	Eccentric -ity	Inclination(°)	Argument of perigee(°)	Ascending node(°)	Mean anomaly (°)
1	7029.8	0.022	100.7	66.2	153.9	238.1
2	7843.9	0.033	104.6	126.7	4.2	28.2
3	7025.8	0.021	99.24	39.14	132.4	261.7

4	7834.6	0.034	104.5	66.1	359.9	111.4
5	7246.4	0.025	101.9	50.6	5.6	192.3
6	7837.5	0.033	104.7	107.5	0.7	51.6
7	7247.9	0.024	102.2	220.9	6.5	23.5
8	7018.7	0.019	97.7	356.1	103.2	73.3
9	6991.4	0.018	98.7	15.5	124.4	95.8
10	6995.2	0.019	98.2	51.6	115.7	46.9

For the initial orbit data of the above 10 satellites, the orbit extrapolation for 168 hours (7 days), combined with the known target point information, can count 184 track window information that meets the user's requirements, as shown in Figure 2. The red bar indicates the No. 1 target, the green bar indicates the No. 2 target, and the blue bar indicates the No. 3 target.

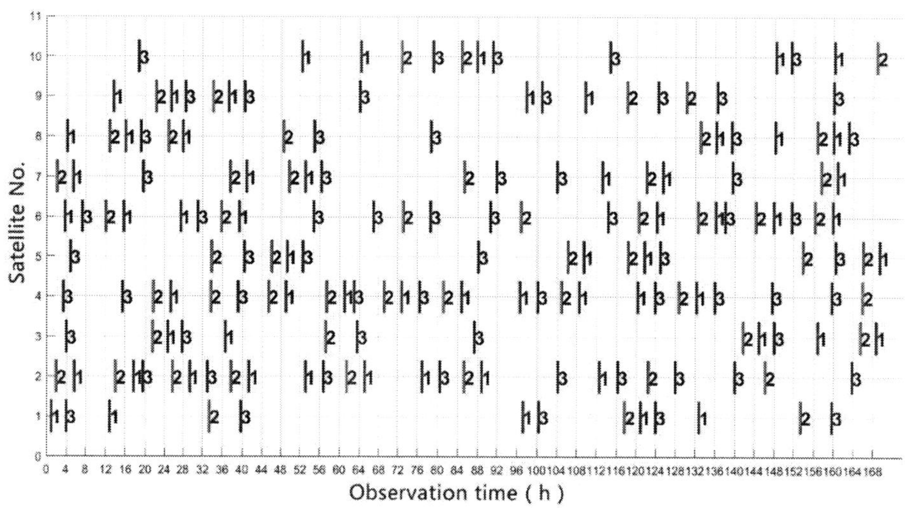

Figure 2. All orbits window information Gantt chart

In summary, the simulation calculation is carried out according to the above simulation conditions, wherein the genetic algorithm parameters are set, the population number is set to 40, the crossover probability is 0.8, the mutation probability is 0.6, and the maximum iteration number is defined as 1000. Since the results obtained by the genetic algorithm are not necessarily global optimal solutions, we have carried out multiple sets of simulation experiments under the same parameters, and compared their final fitness values respectively. The simulation results show that it is going to 1000 iterations. After that, the fitness value of the population basically reached convergence.

According to the observation task requirements, the target invisible time of No.1 and No.2 is 8 hours, and the invisible time of No.3 target is 10 hours. This paper simulates the 7 days scheduling tasks, and the simulation results are as as shown in Figure 3- Figure 4.

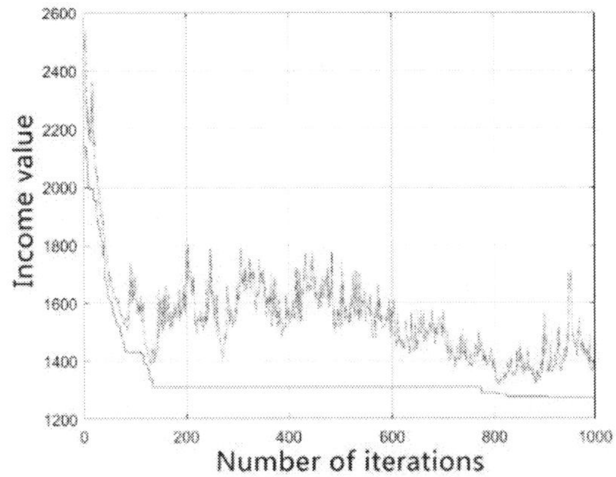

Figure 3. Scheduling 7 day – genetic Algorithm search process

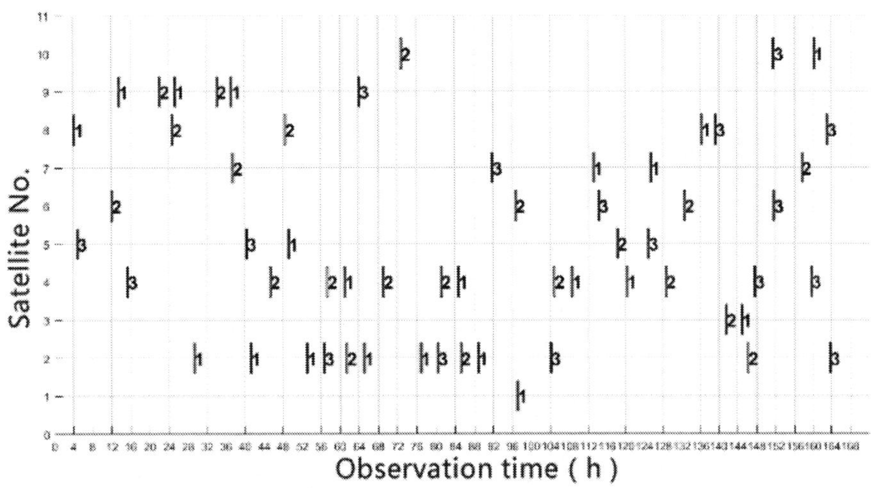

Figure 4. Scheduling 7 day – mission planning Gantt chart

Through the statistics of the last scheduling result, the forecasting schedule is 1 day, 28 window resources can be used before scheduling, and the total number of 8 window resources after scheduling is used, and 3 satellite resources are used. The forecasting schedule is 3 days. 81 window resources can be used before scheduling, and the total number of 26 window resources after scheduling is used, and 7 satellite resources are used. The forecasting schedule is 7 days. 181 window resources can be used before scheduling, and the total number of 59 window resources after scheduling is used, and 10 satellite resources are used.

Through the above simulation calculations and data statistics, the satellite resource scheduling model and intelligent optimization algorithm established in this paper can meet the user scheduling requirements and help solve the engineering application problems. From the income value curve and the scheduling task forecast Gantt chart, the final scheduled satellite reconnaissance mission can meet the maximum invisible duration requirement and use the least resources.

5. Conclusion

This paper describes the scheduling problem of Earth observation satellites. Under the constraints of satellite resource capability, observable time, conversion time between reconnaissance missions, energy, etc., a multi-satellite multi-target reconnaissance mission is established. Planning model. With the most resource-saving and high situational awareness uniformity as the optimization index, the optimized satellite reconnaissance mission planning scheme is obtained based on genetic algorithm. It is of great significance to improve the comprehensive application efficiency of spacecraft, to form uniform coverage of key areas, and to improve the uniformity of satellite resource utilization and situational awareness.

References

[1] Jinghua Tong. Research on strategy for solving satellite emergency dispatch problem [D]. 2013. Wuhan: Master of Science, China University of Geosciences.
[2] Yuhua Guo. Research on key technologies of joint mission planning for multi-type Earth observation satellites [D]. 2009. Changsha: PhD thesis of National University of Defense Technology.
[3] Vasquez M, Hao J K. A "Logic-Constrained" Knapsack Formulation and a Tabu Algorithm for the Daily Photograph Scheduling of an Earth Observation Satellite[J]. Computational Optimization & Applications, 2001, 20(2):137-157.
[4] Kavelaars A, Barnoy A M, Gregory S, et al. Centralized mission planning and scheduling system for the Landsat Data Continuity Mission[C]. America: International Conference on Space Operations, 2014.
[5] Sherwood R., Govindjee A., Yan D., Rabideau G., Chien S and Fukunaga A. Using ASPEN to Automate EO-1 Activity Planning[C]. Colorado, America: Proceedings of the 1998 IEEE Aerospace Conference, 1998.
[6] Frank J, Jonsson A, Morris R, et al. Planning and Scheduling for Fleets of Earth Observing Satellites[C]. Montreal: Proc. of the 6th International Symposium on Artificial Intelligence, Robotics, Automation and Space, 2002.
[7] Dungan, J, Frank, J, Jonsson, A, Morris. Advances in Planning and Scheduling of Remote Sensing Instruments for Fleets of Earth Orbiting Satellites[C]. Proc. of the 2nd Earth Science Technology Conference, Pasadena, CA, USA, 2002
[8] Ackermann S, Angrisano A, Del Pizzo S, et al. Digital Surface Models for GNSS Mission Planning in Critical Environments [J]. Journal of Surveying Engineering, 2013, 140(2):21-29.
[9] Florio S D, Zehetbauer T, Neff T. Optimal Operations Planning for SAR Satellite Constellations[C].In Low Earth Orbit. 6th International Symposium on Reducing the Costs of Spacecraft Ground Systems and Operations, Darmstadt, Germany, 2005.
[10] Florio S D. Performances Optimization of Remote Sensing Satellite Constellations: a Heuristic Method[C]. International Workshop on Planning and Scheduling for Space, 2006.
[11] Lee S I, Ahn J, Bang J. Dynamic selection of zero-revolution Lambert algorithms using performance comparison map [J]. Aerospace Science and Technology, 2016, 51: 96-105.
[12] Izzo D. Revisiting Lambert's problem [J]. Celestial Mechanics and Dynamic Astronomy, 2015, 121: 1-15.
[13] He Y,Xu M,Jia X,et al.High-Precision Repeat-Groundtrack Orbit Design and Maintenance for EarthObservation Missions[J]. Celestial Mechanics and Dynamical Astronomy,2017,128(2-3):275- 294.
[14] Wittig A, Colombo C, Armellin R. Long-term Density Evolution through Semi-Analytical and Differential Algebra Techniques [J]. Celestial Mechanics and Dynamical Astronomy, 2017, 158(4): 435-452.
[15] Sun Y, Kumar M. Uncertainty Forecasting in the Perturbed Two-Body Problem via Tensor Decomposition [C]. American Control Conference (ACC), IEEE, Boston, MA, 2016: 5431-5436.
[16] Koenig A W, Guffanti T, D'Amico S. New State Transition Matrices for Spacecraft Relative Motion in Perturbed Orbits [J]. Journal of Guidance, Control, and Dynamics, 2017, 40(7): 1749-1768.
[17] Mi Zhang. Research on satellite ground observation and data downlink integrated scheduling based on genetic algorithm [D]. Hefei: Hefei University of Technology, 2017: 34-69.

Orbit-Attitude Coupled Tracking and Landing Control for an Asteroid

Lingchao ZHU[1,2] and Shuquan WANG[1,*]

[1] Technology and Engineering Center for Space Utilization, Chinese Academy of Sciences, Beijing 100094, China;
[2] University of Chinese Academy of Sciences, Beijing 100049, China

[*]Corresponding author: shuquan.wang@csu.ac.cn

Abstract. Recent research on asteroid exploration has shown tremendous potential to provide humankind detailed views of unexplored worlds in the inner solar system. A potential market of asteroid mining has aroused interest from commercial companies gradually. Asteroid exploration has the potential to offer the possibility to revolutionize the supply of many resources which are vital but exorbitant for human. A typical asteroid exploration mission is composed of remote asteroid detecting, approaching, hovering, landing and sampling, take-off and return, re-entry stage. Among these stages the tracking and landing control of the asteroid is crucial for the exploration. Rendezvous, approaching, fly-around and landing problem of an asteroid is investigated in this paper. The expected motion of the tracking spacecraft determined by the asteroid' orbital and attitude status is presented firstly. According to the design of the OSIRIS-REx project, several constant thrusters are used for spacecraft to track the asteroid. Different control strategies like pulse control, saturated nonlinear control, limit cycle control and PWM based low-thrust control are developed to fulfil the asteroid's tracking mission of different stages. Numerical simulations demonstrate the effectiveness of the tracking control.

1. Introduction

Asteroid exploration and sampling have shown tremendous potential to provide humankind detailed views of the formation and evolution of planetary system and life origin and evolution in the solar system. With the continuous development of space technology and the increasing complexity of space science research mission, the asteroid exploration is developed from the early asteroid fly-by method, the impact attachment method, to the current sampling and return detection methods gradually.

The soft landing problem on the asteroid surface is a key problem for deep space exploration. According to the different instruments and equipment carried by the asteroid detector, a variety of researches can be carried out cover a large number of scientific research, such as appraising of asteroid mining and composition and characteristics analysis of asteroid soil. These research can deepen the understanding of asteroids, providing a scientific basis for the further research and utilization of asteroids, and promoting the development of space science and technology.

The close-range detection of small celestial bodies began in the 1990s. In recent years, small celestial exploration has become one of the key areas of deep space exploration, the most representative of these space exploration missions includes the Near Earth Asteroid Rendezvous (NEAR) [1] mission of NASA, the Stardust mission [2] of NASA's comet sample return research, the

Deep Impact mission [3] for studying the interior composition of the comet Tempel 1, the Hayabusa mission [4] developed by the Japan Aerospace Exploration Agency (JAXA) to return a sample of material from a small near-Earth asteroid named 25143 Itokawa and the Rosetta mission [5] built by the European Space Agency to perform a detailed study of comet 67P. Among these space missions, the most representative missions are the NEAR and Hayabusa, the former was the first space mission to achieved flying around and landing on an asteroid, the latter was the first space probe to achieve attachment and sampling return. The Origins Spectral Interpretation Resource Identification Security Regolith Explorer (OSIRIS-REx) mission was a NASA new frontiers mission launching in 2016 to rendezvous with the near-Earth asteroid (101955) 1999 RQ36 in late 2018 [6-7], On 21 September 2018, Hayabusa2 ejected the first two rovers, Rover-1A and Rover-1B, from about 55 meters altitude that fell independently to the surface of the asteroid Ryugu [8]. The necessity and key techniques of asteroid autonomous landing was elaborated by Cui [9]. Peng Z [10] designed an adaptive super-twisting controller without chartering problem for spacecraft soft landing on asteroids.

For most of the references, researchers tend to focus on the control research in the power decent phase of asteroid landing [11], in the meantime, the controller only contains a study of orbital or attitude control design. This paper will divide close proximity asteroid tracking into several stages based on the OSIRIS-REx mission planning [12], and different control methods are adopted in different stages to satisfy different task requirements.

In this paper, the close proximity control process after asteroid rendezvous and docking is divided into an approaching phase, a long distance fly-around phase, a close distance fly-around phase, a soft landing phase. At different phase, thrusters with different parameters are adopted, and the orbit-attitude coupled controllers are designed separately according to different task requirements. In the process of the spacecraft approaching the asteroids, different orbit and attitude control targets need to be met, the details are as follows:

(1). Approaching phase: The control objectives include keeping the distance between the spacecraft and the asteroid in the range of 20km, and keeping the spacecraft's certain instrument pointing to the asteroid during the process. The actuator is a 5N constant small thruster;

(2). Fly-around phase: The control objective is to achieve a full-scale observation of the asteroid, the optical camera should be able to fully cover the visible area of the asteroid. According to the requirements of optical navigation system, this phase is divided into two parts: long-distance flying and short-distance flying. The long-distance fly-around phase is to achieve relative optical navigation data acquisition. The actuator is a 5N constant small thruster, the radius of the relative circular orbit is about 20km. The close- distance fly-around phase mainly relies on relative optical navigation to check the asteroid's topography, preparing for the landing site selection and sampling; the actuator is a 0.2N constant-value micro-thruster, the radius of the relative circular orbit is about 5km;

(3). Landing phase: The control objective is to achieve soft landing within the tolerance. In the landing and sampling phase, the 0.2N constant value small thruster is used to control the spacecraft to hover over the top of the landing site about 100m, and then the 0.09N variable micro-thruster is used for soft landing, the position error of the landing horizontal direction is less than 10m, and the vertical relative speed is less than 10cm/s.

2. Lander orbital and attitude dynamics

There are three kinds of coordinate frames used to represent the spacecraft and asteroid orbit and attitude dynamics: asteroid body-fixed frame $\mathcal{B}_c$, spacecraft body-fixed frame $\mathcal{B}_d$, and inertial-reference frame $\mathcal{N}$. The inertial translational EOMs of the asteroid and spacecraft are:

$$\ddot{\boldsymbol{R}}_c = -\mu_s \boldsymbol{R}_c / |\boldsymbol{R}_c|^3, \quad \ddot{\boldsymbol{R}}_d = -\mu_s \boldsymbol{R}_d / |\boldsymbol{R}_d|^3 + \boldsymbol{U}/m_d + G \tag{1}$$

where μ_s is sun's gravitational constant, $\boldsymbol{R}_c$ and $\boldsymbol{R}_d$ are the inertial position vectors of the asteroid and spacecraft respectively with respect to (w.r.t.) the inertial-reference frame $\mathcal{N}$. m_d is the mass of the spacecraft, $\boldsymbol{U}$ is the control forces of the spacecraft, G is microgravity of the asteroid. The attitude dynamics of a rigid spacecraft can be described by the following nonlinear equation:

$$J\dot{\omega}+\omega\times J\omega+t_d=t_c \tag{2}$$

where ω is the angular velocity of the spacecraft with respect to the inertial-reference frame $\mathcal{N}$ and is expressed in the body-fixed frame $\mathcal{B}_d$, J is the inertia matrix of the spacecraft, t_c is the torque control and t_d is the bounded external disturbance. The nonlinear differential equations governing the attitude kinematics of the spacecraft in terms of unit quaternion can be expressed as:

$$\dot{q}=1/2\cdot Q(q)\omega \tag{3}$$

2.1. Asteroid gravitational field model

There are three methods for calculating the irregular gravitational field of asteroids [13]: spherical harmonic function/ellipsoid harmonic function method[14,15], particle group method[16], and polyhedral method[17-20]. Itokawa is selected as a hypothetical asteroid for landing control in this paper, the gravitational field model of asteroid is calculated based on polyhedral model. Specifically, the asteroid is decomposed into a polyhedron composed of 1846 vertices and 3688 faces, the gravitational potential on a field point due to an asteroid modelled as a polyhedron can be defined as

$$U=\frac{1}{2}G\sigma\sum_{e\in edges}r_e\cdot E_e\cdot r_e\cdot L_e-\frac{1}{2}G\sigma\sum_{f\in faces}r_f\cdot F_f\cdot r_f\cdot\omega_f \tag{4}$$

where G is the universal gravitational constant, σ is the polyhedron density, r_f is a vector from the field point to any point on the facet, r_e is a vector from the field point to any point on the edge common to two facets. F_f and E_e are the second-order tensors corresponding to the facets and their edges, L_e is a dimensionless factor that represents the potential energy of the each edge, w_f is an angle defined for each facet.

2.2. Expected orbital and attitude motion

This section is going to obtain the expected translational and rotational motion of the spacecraft, such that the position of the spacecraft tracks the landing site of the asteroid and the attitude of the spacecraft pointing towards the landing site. Therefore the landing site in asteroid body-fixed frame $\mathcal{B}_c$ is ${}^{\mathcal{B}}\rho_d=r_{land}$, due to soft landing requirements, the velocity in landing site is ${}^{\mathcal{B}}\dot{\rho}_d=0_{3\times1}$. The expected landing position, velocity and acceleration vector projected in inertial-reference frame $\mathcal{N}$ is

$$\begin{cases} R_{dd}=R_c+C_c{}^{\mathcal{B}}\rho_d \\ \dot{R}_{dd}=\dot{R}_c+C_c\left({}^{\mathcal{B}}\omega_{B/N}\times{}^{\mathcal{B}}\rho_d\right) \\ \ddot{R}_{dd}=\ddot{R}_c+C_c\left[\dot{\omega}_{B/N}\times{}^{\mathcal{B}}\rho_d+{}^{\mathcal{B}}\omega_{B/N}\times({}^{\mathcal{B}}\omega_{B/N}\times{}^{B}\rho_d)\right] \end{cases} \tag{5}$$

where C_c is the rotation matrix from asteroid body-fixed frame $\mathcal{B}_c$ to inertial-reference frame $\mathcal{N}$, $\omega_{B/N}$ is the angular velocity of the $\{\mathcal{B}_c\}$ frame w.r.t. the $\{\mathcal{N}\}$ frame.

The objective of the attitude control is to make a certain instrument point towards the asteroid. The expected vector pointing from the spacecraft to the asteroid is given by $d=R_c-R_d$. Without loss of generality, this paper defines that the z axis of the spacecraft's body frame should point to the center of mass of the asteroid. The direction of d is $\hat{e}_r$. The expected attitude maneuver can be expressed in quaternion as:

$$q_r=\left[\cos\alpha/2\quad \hat{e}_\theta^T\sin\alpha/2\right]^T \tag{6}$$

where $\hat{e}_\theta$ the rotating axis during the attitude maneuver, $\alpha=\cos^{-1}(z^T\hat{e}_r)$ is the angle between two vectors $\hat{e}_r$and z. The expected angular velocity and the expected angular acceleration of the spacecraft are:

$$\begin{cases} \omega_d=\omega-\omega_r=Q_{DB}\omega-\omega_r \\ \dot{\omega}_d=\dot{\omega}-\dot{\omega}_r=Q_{DB}\dot{\omega}-\omega_r^{\times}Q_{DB}\omega-\dot{\omega}_r \end{cases} \tag{7}$$

where $\dot{Q}_{DB} = -\boldsymbol{\omega}_r^{\times} Q_{DB}$ and Q_{DB} is the rotation matrix from the spacecraft body fixed frame $\{\mathcal{B}_d\}$ to target spacecraft body frame, $\boldsymbol{\omega}_r$ is the manoeuvre angular velocity.

3. Controller design

Given the expected trajectories and attitude of the spacecraft with respect to the asteroid, a control input U is expected to make the spacecraft land smoothly. Define the position tracking error as $\Delta \boldsymbol{r} = \boldsymbol{R}_d - \boldsymbol{R}_{dd}$. In approaching phase, limit cycle control with a 5N constant small thruster is adopted in approaching phase, the control objectives are to approach the asteroid within 20km, and keeping the spacecraft's observation axis pointing to the asteroid during the process. The analog phase plane controllers include lead correction controller and pseudo rate controller. The lead correction controller is generally composed of a lead correction network and a Schmitt trigger, therefore the limit cycle controller is designed as:

$$u_c = \begin{cases} 1 & u_i \geq u_D \\ 1 & u_i > (1-h)u_D, \dot{u}_i < 0 \\ 0 & -(1-h)u_D \leq u_i \leq u_D, \dot{u}_i > 0 \\ 0 & -u_D < u_i \leq (1-h)u_D, \dot{u}_i < 0 \\ -1 & u_i \leq -(1-h)u_D, \dot{u}_i > 0 \\ -1 & u_i \leq -u_D \end{cases} \tag{8}$$

where u_c is the output signal of controller, h is the hysteresis coefficient, u_i is input signal of the Schmitt trigger which is designed as $\boldsymbol{u}_\mathrm{i} = -K_1 \Delta \boldsymbol{r} - K_2 \Delta \dot{\boldsymbol{r}}$. In fly-around phase, the control objective is to achieve a detailed observation of the asteroid. According to the previous design, this phase is divided into long-distance fly-around and close-distance fly-around. For long-distance fly-around phase, limit cycle control with a 5N constant small thruster is adopted again, the radius of the relative circular orbit is about 20km. The controllers are same with the controllers in the approaching phase.

For close-distance fly-around phase, the actuator is a 0.2N constant-value micro-thruster, the radius of the relative circular orbit is about 5km. A nonlinear orbit control law is designed as:

$$\boldsymbol{u}_i = \ddot{\boldsymbol{R}}_{dd} + \mu_s \boldsymbol{R}_d / |\boldsymbol{R}_d|^3 - [K_1]\Delta \boldsymbol{r} - [K_2]\Delta \dot{\boldsymbol{r}} \tag{9}$$

For a large number of thrusters, it is convenient to achieve on-and-off switch without adjusting the magnitude of the thrust. Meanwhile, constant thruster reduces the requirements on the thruster and improves the reliability of the controller. Pulse width modulation (PWM) can be utilized to simulate the control signal calculated by the designed algorithm for micro thruster. Amplitude modulation ratio is an important parameter of PWM, which is defined as $K_a = V_m/V_c$, where V_m is peak of modulated wave, V_c is peak of carrier wave. Control law in landing phase has the same form with different coefficients as the previous control law in fly-around phase.

A nonlinear attitude control law can be designed as

$$\boldsymbol{T}_c = \boldsymbol{\omega} \times (\boldsymbol{J}\boldsymbol{\omega}) + \lambda \boldsymbol{J}\dot{\boldsymbol{\omega}}_d - \boldsymbol{D}(\boldsymbol{\omega} - \boldsymbol{\omega}_d) - \boldsymbol{K}\boldsymbol{q}_{ev} \tag{10}$$

where D and K are diagonal matrix with positive diagonal elements, λ is constant feedback coefficient, $\dot{\boldsymbol{\omega}}_d$ is the expected angular acceleration, Tc is the control torque.

4. Numerical simulations

The scope and effect of the previous control law will be analysed through numerical simulations. The initial orbit parameters $\boldsymbol{R}_{c0}$ and $\boldsymbol{V}_{c0}$ of asteroid are calculated by six Keplerian elements of asteroid Eros. The asteroid is 1500 kg with the moment of inertia $\boldsymbol{J}_c = [500\ 700\ 1000]\mathrm{kg} \cdot \mathrm{m}^2$. The initial orbit parameters of spacecraft $\boldsymbol{R}_{d0} = \boldsymbol{R}_{c0}$+[20;30;-30] and $\boldsymbol{V}_{d0} = \boldsymbol{V}_{c0}$, ass of landing spacecraft is 200 kg with the moment of inertia $\boldsymbol{J}_d = [3\ 3\ 3]\mathrm{kg} \cdot \mathrm{m}^2$. The initial Euler Angle and $\boldsymbol{\omega}$ of asteroid and spacecraft are zeros. The latitude and longitude of the landing site are 45 and 90 degrees. In approaching phase and long-distance fly-around phase, limit cycle control is adopted, h=0.2 is the hysteresis coefficient, $u_m = 5N$ is the constant thrust, $u_D = 4.5N$ is control parameter in limit cycle control method. PWM based constant thrust $u_{up} = 0.2N$ is used in short-distance fly-around phase and landing phase, the sawtooth amplitude is 0.1 and the carrier frequency is 0.001 Hz.

Due to paper length limitations, attitude control effect will be demonstrated by table only. Table 1 shows pointing angle mean error and maximum torques in different phase.

Table 1. Pointing angle (PA) error and torques in different phase

	approaching	fly-around	landing
PA error (Deg)	0.1	0.0012	0.01
Torque (Nm)	0.4	1.4×10^{-5}	3×10^{-4}

Tracking and landing simulation of tracking spacecraft relative to asteroid are shown as follows.

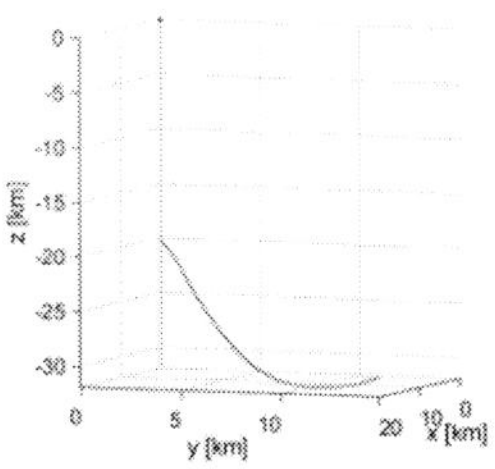

Figure 1 trace in approaching phase

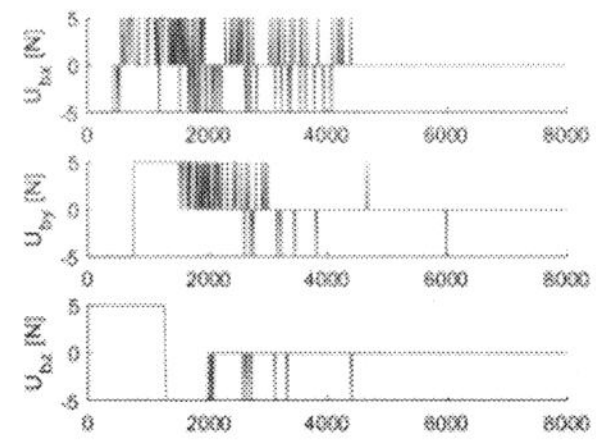

Figure 2 control force in approaching phase

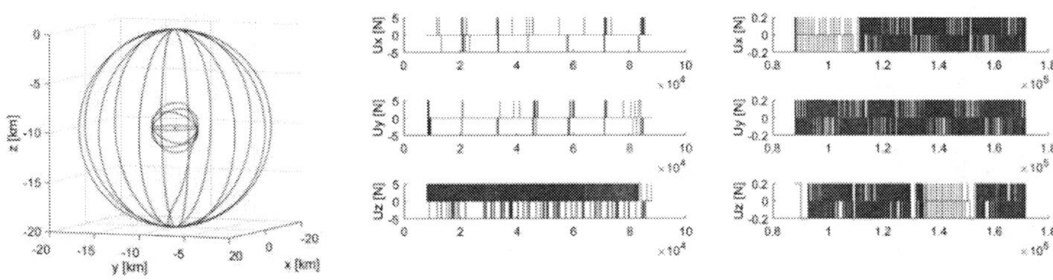

Figure 3 trace and force in long-distance (20 km) and short-distance (5 km) fly-around phase

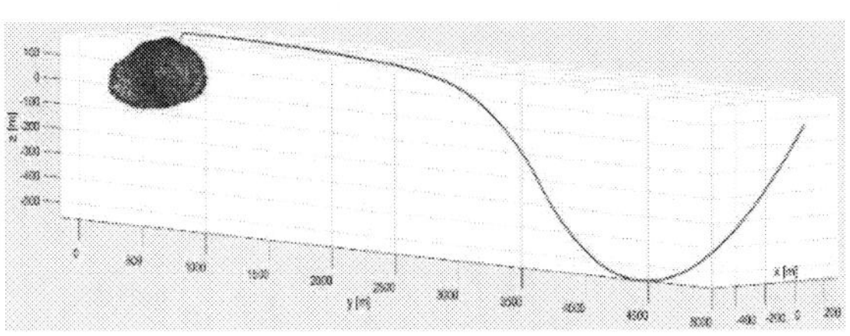

Figure 4 trace in landing phase

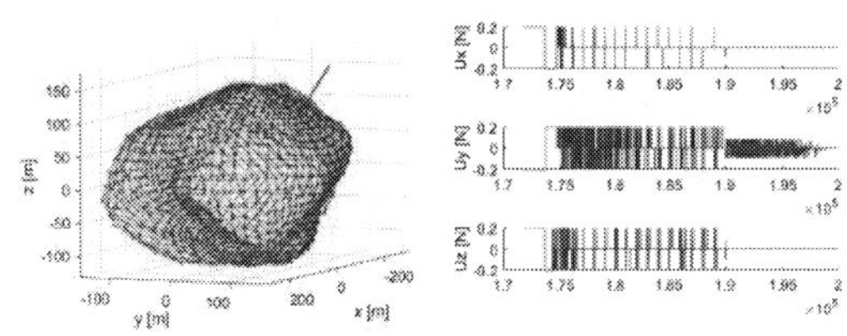

Figure 5 vertical soft landing and control force

Figure **1**-Figure **5** show the trajectories of spacecraft relative to asteroid in different phase. As we can see from these graphs, the spacecraft completed the asteroid landing mission well with pre-set thruster parameters. The pulse used in each phase are shown in the table 2.

Table 2. pulse used in each phase ($N \cdot s$)

approaching	long-fly	long-fly	hover	landing
19583.35	105584.42	18114.31	2514.81	384.96

5. Conclusion

In this study, a detailed landing detection project is proposed for asteroid exploration, we divide close proximity asteroid tracking into several stages. Different control methods are adopted in different stages to meet the limits of different thrusters. The expected motion of the tracking spacecraft determined by the asteroid' orbital and attitude status is presented firstly and several controllers is designed for the landing detection. The gravitational acceleration of asteroid is calculated by the polyhedral model. Terminal landing accuracy and attitude pointing accuracy have meet the demand by proposed control algorithm. Simulations results of asteroid detection problems are given to substantiate the effectiveness of landing control and illustrate the reliability of the multi-controllers cooperation control strategy.

Acknowledgement
This research was supported by the National Natural Science Foundation of China (under grant No. 11672294) and the Advanced Research Project China's Manned Space Program.

Reference

[1] Holdridge M E . NEAR Shoemaker spacecraft mission operations[J]. Johns Hopkins Apl Technical Digest, 2002, 23(1):58-70.
[2] Brownlee, Don. "The Stardust mission: analyzing samples from the edge of the solar system." Annual Review of Earth and Planetary Sciences 42 (2014): 179-205.
[3] A'Hearn, Michael F., et al. "Deep impact: excavating comet Tempel 1." science (2005).
[4] Kawaguchi, Jun'ichiro, Akira Fujiwara, and Tono Uesugi. "Hayabusa—its technology and science accomplishment summary and Hayabusa-2." Acta Astronautica 62.10-11 (2008): 639-647.
[5] Glassmeier, Karl-Heinz, et al. "The Rosetta mission: flying towards the origin of the solar system." Space Science Reviews 128.1-4 (2007): 1-21.
[6] Berry, Kevin, et al. "OSIRIS-REx touch-and-go (TAG) mission design and analysis." (2013).
[7] Lauretta, D. S., et al. "OSIRIS-REx: sample return from asteroid (101955) Bennu." Space Science Reviews 212.1-2 (2017): 925-984.
[8] "Japanese Probe Drops Tiny Hopping Robots toward Big Asteroid Ryugu". Space.com. September 21, 2018.
[9] Cui P Y , Yuan X , Zhu S Y , et al. Research progress of small body autonomous landing techniques[J]. Journal of Astronautics, 2016.
[10] Peng Z , Xiaosong L , Bo D , et al. Adaptive supper-twisting control for spacecraft soft landing on asteroids [J]. Journal of Jilin University (Engineering and Technology Edition), 2016, 46(5).
[11] Huang Y, Shuang L I, Jiang X. Adaptive iterative learning based fault tolerant control for asteroid orbiting[J]. Chinese Space Science & Technology, 2017.
[12] Lauretta, D. S., et al. "OSIRIS-REx: sample return from asteroid (101955) Bennu." Space Science Reviews 212.1-2 (2017): 925-984.
[13] Yu Y. 2014. Research on orbital dynamics in the gravitational field of small bodies.[PhD Thesis]. Beijing: Tsinghua University, 2014.
[14] Hobson E W. The Theory of spherical and ellipsoidal harmonics[M]. New York: Chelsea, 1965.
[15] Heiskanen W A, Moritz H. Physical geodesy[M]. Fan Francisco: W.H. Freeman and Co., 1967.
[16] Geissler P, Petit J-M, Durda D, et al. Erosion and ejecta reaccretion on 243 Ida and its moon[J]. Icarus, 1996, 120:140-157.
[17] Werner R A. The Gravitational potential of a homogeneous polyhedron or don't cut corners[J]. Celestial Mechanics and Dynamical Astronomy, 1994, 59(3):253-278.
[18] Werner R A, Scheeres D J. Exterior gravitation of a polyhedron derived and compared with harmonic and mascon gravitation representations of asteroid 4769 Castalia[J]. Celestial Mechanics and Dynamical Astronomy, 1997, 65(3):313-344.
[19] Yu Y, Baoyin H X. Orbital dynamics in the vicinity of asteroid 216 Kleopatra[J]. The Astronomical Journal, 2012, 143(3):62-71.
[20] Yu Y, Baoyin H X. Generating families of 3D periodic orbits about asteroids[J]. Monthly Notices of the Royal Astronomical Society, 2012, 427(1):872-881.

Magnetic Simulator Testbed for APSCO Student Small Satellite Project

J Uscategui[1,a] **and W Xinsheng**[2,3,4,b]

[1] School of Instrumentation Science and Opto-Electronics Engineering, Beihang University, 100191, Beijing, China.
[2] Department of Spacecraft Technology, School of Astronautics, Beihang University, 100191, Beijing, China.
[3] Deputy director of Microsatellite Technology Laboratory, Beihang University, 100191, Beijing, China.
[4] Regional Center for Space Science and Technology Education in Asia and Pacific (China), (Affiliated to the United Nations)

[a]jhouscategui@hotmail.com, [b]xswang@buaa.edu.cn

Abstract. The Student Small Satellite (SSS) is a university project of Asia-Pacific Space Cooperation Organization (APSCO), composed by one microsatellite (SSS-1) and two 3U CubeSats (SSS-2A and SSS-2B). Regarding SSS-1 and SSS-2A satellites, the active 3-axis magnetic control strategy has been applied in the Attitude Determination and Control System (ADCS). In this paper, the necessary study for the development of a Magnetic Simulator Testbed for the SSS-1 and SSS-2A satellites was performed using one pairs of Helmholtz coils for each axis of rotation (X, Y, Z). Each pair of coil can generate a sufficiently large and stable magnetic field that simulate the Earth's magnetic field but with a reliable level of accuracy, to perform the ADCS test on the satellites. Matlab software was used to generate the propagation of the orbit for the satellites, to calculate the Earth's magnetic field around that orbit and simulate a pair of Helmholtz coil in each axis of rotation to reproduce the Earth's magnetic field using the Biot-Savart Law. The results obtained for propagation of the orbit, calculation of terrestrial magnetic field and reproduction of the magnetic field through the Magnetic Simulator Testbed, were very close to the results obtained in important researches previously consulted, giving reliability to this research and allowing an analysis to determine the real physical size that Magnetic Testbed should have. The size determined, is very close to what was expected, allowing to conclude the reliability and viability of the development of the Magnetic Testbed for APSCO SSS Project.

1. Introduction

Asia-Pacific Space Cooperation Organization (APSCO) is an intergovernmental organization who promotes multilateral cooperation, enhances the space application capacity of member states through their universities and promotes the peaceful use of outer space by mankind. The Student Small Satellite (SSS) is a university project composed by one microsatellite (SSS-1) and two 3U CubeSats (SSS-2A and SSS-2B) [1]. Regarding SSS-1 and SSS-2A will have 3 axis Magnetic Attitude control with magnetorquer for stabilization and control in the three axes X, Y and Z. Each magnetorquer will generate a magnetic field that will interact with the Earth's magnetic field to modify its Attitude. The basic information can be seen in table 1. Attitude Determination and Control System (ADCS) is an important part of a nano-satellite, which usually performs attitude determination, stabilization and acquisition control [2].

In this paper, the necessary study for the development of a Magnetic Simulator Testbed for the SSS-1 and SSS-2A satellites was performed using three pairs of Helmholtz coils, one for each axis of rotation (X, Y, Z) in order to perform the tests of stabilization and attitude control to ensure the proper functioning of the ADCS after putting into orbit. Magnetic Simulator Testbed will simulate the Earth's magnetic field along the same LEO orbit that will have the SSS project, allowing to obtain a clearer vision of what the behaviour of the subsystem will be when it is in orbit, and making the relevant design corrections to ensure its proper functioning after launch.

Matlab software was used in the first stage to generate the propagation of the orbit on the studied satellites and then, it was calculated the Earth's magnetic field around that orbit. To this, the standard World Magnetic Model (WMM) of the United States' National Geospatial-Intelligence Agency (NGA) [3] was used. Later, in the second stage of the study, Matlab software was used to simulate a pair of Helmholtz coil in each axis of rotation to reproduce the Earth's magnetic field previously obtained in the first stage, using the Biot-Savart Law [4].

The Magnetic Simulation Testbed system (MST), in general, is composed of three modules; the Control Module (CM), the Interface Module (IM), and the Testbed Module (TBM). The CM, is in charge of carrying out the orbit propagation of the SSS-1 / SSS-2A satellites and the calculation of the magnetic field along the orbit, in each moment of time. The IM, is responsible for the connection and exchange of information between the CM and TBM. The TBM will be a hardware of three pairs of Helmholtz coil, one pair for each axis X, Y and Z, this will simulate the magnetic field of the Earth in any direction and orbit, according to calculations generated by the CM. shown figure 1.

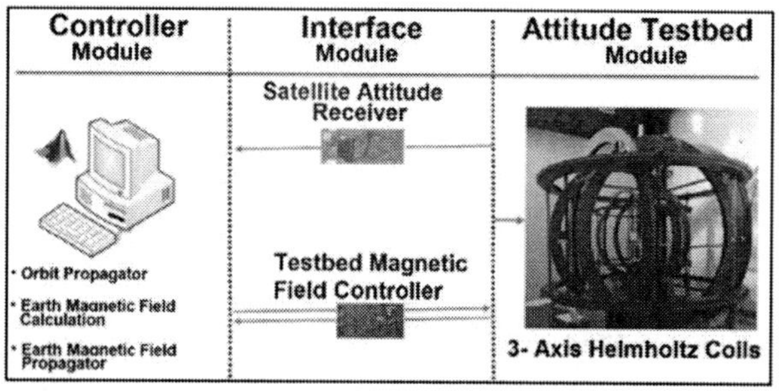

Figure 1. Magnetic Simulator Testbed structure.

For the development of the MST, the mathematical models of Helmholtz Coil and the Earth's magnetic field were studied as well as the propagation of the satellites' orbit to predict the magnetic field that the Earth will have in the orbit of the satellites at each moment of time, in this way will guarantee that the MTB will generate a magnetic field as close as possible to reality to perform the ADCS tests of the SSS APSCO project.

Table 1. SSS APSCO Project Basic information.

Altitude	500 km
Local time of descending node	10:30AM
SSS-1 Dimension	350*350*650mm
SSS-1 Mass	30kg
SSS-2A Dimension	100*100*300 mm
SSS-2A Mass	4 kg.
ADCS Control System (SSS-1 and SSS-2A)	Magnetic
Inclination	97.4065°
Eccentricity	0

2. Mathematic Theory

2.1. Satellite orbit propagation

The orbits propagation consists of predicting or estimating the future position of satellites from known ephemeris in a certain period of time. If we know all the orbital elements of a satellite, we will know its orbit precisely and we can make the propagation using the laws of orbital mechanics [5]. The orbital elements that are altered by the chosen propagator are the following [6]:

$$\Omega = \Omega_0 - 1{,}5n\frac{R_e^2}{p^2}J_2\cos(i)\Delta t \tag{1}$$

$$\omega = \omega_0 + 0.75n\frac{R_e^2}{p^2}J_2(5\cos^2(i) - 1)\Delta t \tag{2}$$

$$M = M\_0 + (n + 0.75n\frac{R_e^2}{p^2}J^2\,(2 - 3\sin^2(i)\sqrt{1-e^2})\,\Delta t \tag{3}$$

Where, Ω is right ascension of the ascending node (RAAN), ω is argument of periapsis, M is average anomaly.

From the orbital elements we can know thanks to the time laws; the true anomaly of the satellite and with it the position in the plane of the orbit in the perifocal reference system (with the XY plane contained in the orbit and the OX direction pointing perigee) [6]:

$$r^R = \frac{p}{1+e\cos(\theta)}\begin{pmatrix}\cos(\theta)\\ \sin(\theta)\\ 0\end{pmatrix} \tag{4}$$

To give the location of the satellite, the true anomaly θ is used, while the element that has propagated is the average anomaly M. The change between the two is not trivial, and is detailed in figure 2.

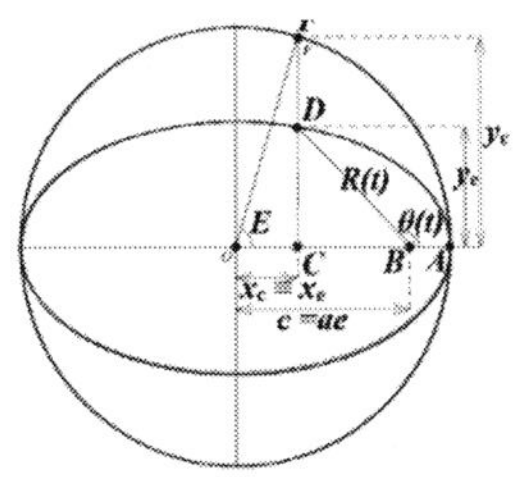

Figure 2. Definition of the Eccentric Anomaly.

The relationship between Eccentric Anomaly E and true anomaly θ, is given by the equation [5]:

$$\tan({}^{\theta}/_{2}) = \sqrt{\frac{1+e}{1-e}}\tan({}^{E}/_{2}) \tag{5}$$

Similarly, by relating E to M, using geometric relationships and Kepler's second law, we obtain that:

$$M = E - e\sin(E) \tag{6}$$

To find the trace, a change of axes is made from the perifocal reference system to the Earth's equatorial inertial reference system [3]:

$$C_E^R = \begin{pmatrix} \cos(\Omega)\cos(\omega) - \sin(\Omega)\sin(\omega)\cos(i) & \sin(\Omega)\cos(\omega) - \cos(\Omega)\sin(\omega)\cos(i) & \sin(\omega)\sin(i) \\ -\cos(\Omega)\sin(\omega) - \sin(\Omega)\cos(\omega)\cos(i) & -\sin(\Omega)\sin(\omega) + \cos(\Omega)\cos(\omega)\cos(i) & \cos(\omega)\sin(i) \\ \sin(\Omega)\sin(i) & \cos(\Omega)\sin(i) & \cos(i) \end{pmatrix} \tag{7}$$

When performing a second transformation to the geographic reference system, the transformation matrix is, in this case:

$$C_E^R = \begin{pmatrix} \cos(GST_0 + \omega_e t) & sin(GST_0 + \omega_e t) & 0 \\ -sin(GST_0 + \omega_e t) & cos(GST_0 + \omega_e t) & 0 \\ 0 & 0 & 1 \end{pmatrix} \tag{8}$$

Once you have the position in these axes, to obtain the trace you only have to change from Cartesian to spherical coordinates, where the two angles give the latitude φ and length λ [6].

2.2. Mathematical Model of Earth Magnetic Field

The Earth's magnetic field in general resembles the field around a magnetized sphere, or a dipole. As of 1999, the dipole axis was tilted approximately 11.5 degree from the axis of rotation, and grew at a rate of approximately 0.2 degrees per year [7]. A low intensity of magnetic field at approximately 25 S and 45 W which is known as the Brazilian Anomaly, and a high point of Earth's magnetic field at 10 N and 100 E. Emphasis must also be placed on the growth of global magnetic anomalies (Canadian, Brazilian, Antarctic and East-Siberian) in the Earth's magnetic reorganization. Its importance since these global anomalies make up a source's magnetic almost independent of main Earth's magnetic field, additionally these anomalies' magnetic intensity in the world substantially exceeds the entire residual non-dipolar component, what is obtained by subtracting the dipole component from the total magnetic field of the Earth [7-8].

The Earth's magnetic field is a vector magnitude and as such is characterized by its module, its direction and its sense. To module of this vector we call it total force or total intensity F, equivalent to the module of the vector resulting from the vector sum of its three Cartesian components (X,Y,Z) [9]. The geomagnetic field BT, is a vector magnitude and according to [8] the relations between components and angles are:

$$H = \sqrt{X^2 + Y^2} \tag{9}$$

$$X = H\, cos\, D \tag{10}$$

$$Y = H\, sin\, D \tag{11}$$

$$tgD = \frac{Y}{X} \tag{12}$$

$$B_T = \sqrt{H^2 + Z^2} = \sqrt{X^2 + Y^2 + Z^2} \tag{13}$$

$$H = B_T\, cos\, I \tag{14}$$

$$Z = B_T\, sin\, I \tag{15}$$

$$tg\, I = \frac{Z}{H} \tag{16}$$

The Earth's magnetic field (from the surface of the Earth and up to about 5 times the radius of the Earth) corresponds approximately to the field that generates a magnetic dipole located in the center of the Earth. In the figure 3, shows a simplified diagram of the Earth's magnetic field [8-9].

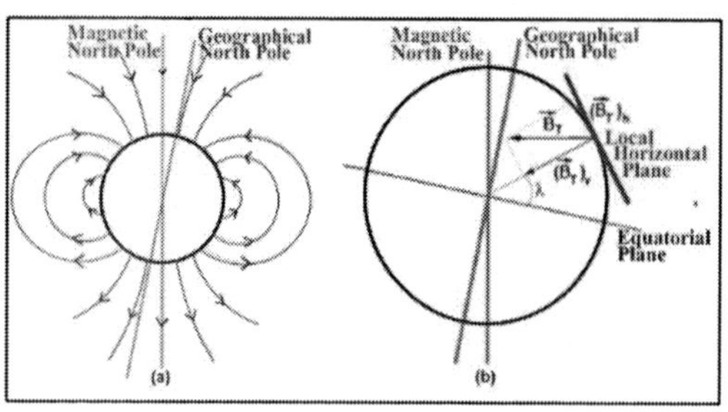

Figure 3. (a) Earth's magnetic field Simplified diagram. (b) Earth's magnetic field decomposition at a λ latitude place, in its horizontal components $(\overrightarrow{B_T})$h, and vertical components, $(\overrightarrow{B_T})$V.

The angle that forms the magnetic field vector direction with the horizontal of a place is called the angle of inclination [9]. In the same way, the magnetic field can be represented at any point by a three-dimensional vector (usually called B). The F field's intensity (also called Lorentz force) is proportional to the force exerted on the magnet of said compass. We can see this in the figure 4.

This magnetic field is measured in Tesla (T) according with the International System of Units, and in Gauss (G) according with the Cegesimal System of Units. For weak magnetic fields, it is most common use the Gauss. The relationship between these two units is as follows:

$$1\mathrm{T} = 1\mathrm{e}4\,\mathrm{G} \quad (17)$$

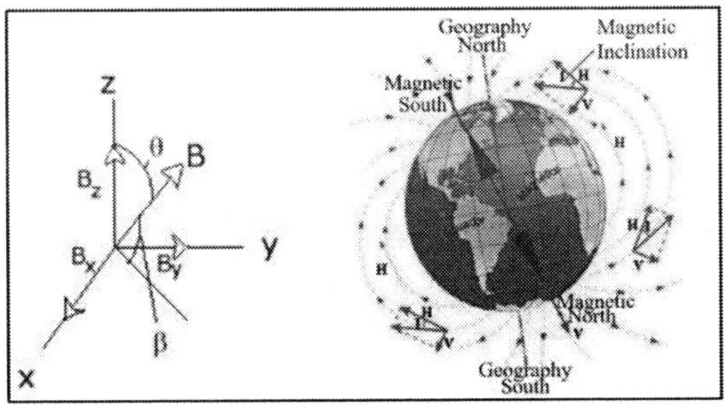

Figure 4. Earth's Magnetic Field Vectors components, H and V are the horizontal and vertical components, respectively.

The Earth's magnetic field is increasing its value as we approach the poles magnetic, while reaching its lowest values at the equator [10], observed in the table 2.

Table 2. Magnetic field on the Earth's surface in different units.

Place	μTesla	Tesla	Gaus
Poles	60μT	6 e-5 T	0.6G
Ecuador	30μT	3e-5 T	0.3G
Atlantic anomaly	23μT	2.3e-5 T	230μG

2.2.1 The International Geomagnetic Reference Field (IGRF) model. Belonging to the International Association of Geomagnetism and Aeronautics (IAGA) is a mathematical description of the main magnetic field which is widely used in interior studies of the crust, the ionosphere and the terrestrial magnetosphere [8]. The IGRF incorporates data from terrestrial, aerial, maritime and satellite exploration observatories [7]. The IGRF consists of a series of Gaus coefficients and their secular variations of degree and order from n = 1 to n = 13, since it is accepted that the terms of minor degree represent, to a large extent, the main field coming from the nucleus [7-8].

$$V(r,\theta,\emptyset,t) = R\sum_{n=1}^{k}\left(\frac{R}{r}\right)^{n+1}\sum_{m=0}^{n}[g_n^m\cos(m\emptyset) + h_n^m\mathrm{sen}(m\emptyset)]\,P_n^m(\theta) \quad (18)$$

Where, R is the radius of the Earth which approaches 6371.2 Km, r is the distance (Km) from the centre of the Earth to the point on which you want to measure the field (altitude), ϴ is the co-latitude (ϴ = 90 - latitude) and Φ is the longitude, like the figure 5.

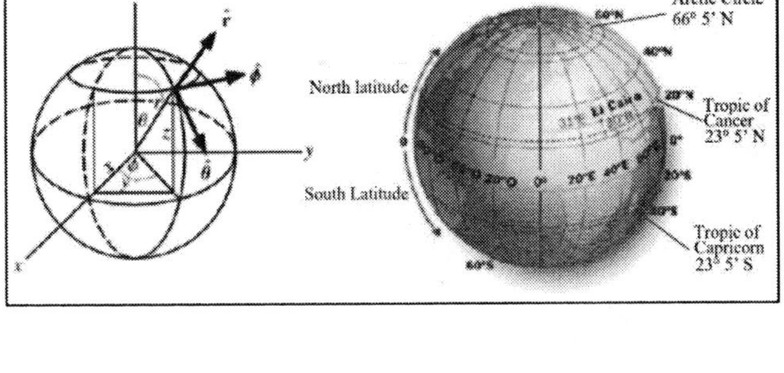

Figure 5. Location in spherical coordinates of latitude and longitude.

The Earth's magnetic field can be obtained from the negative gradient of the scalar function V as shown in the following equation [7]:

$$B = -\nabla V \tag{19}$$

If the expression given by the previous equation is used, to determine the magnetic field in each of its spherical components, according to [7, 11] we have:

$$B_r = -\frac{1\,\partial V}{\partial r} = \sum_{n=1}^{k}\left(\frac{R}{r}\right)^{n+2}(n+1)\sum_{m=0}^{n}[g_n^m \cos(m\emptyset) + h_n^m \mathrm{sen}(m\emptyset)]\,P_n^m(\theta) \tag{20}$$

$$B_\theta = -\frac{1}{r}\frac{\partial V}{\partial \theta} = -\sum_{n=1}^{k}\left(\frac{R}{r}\right)^{n+2}\sum_{m=0}^{n}[g_n^m \cos(m\emptyset) + h_n^m \mathrm{sen}(m\emptyset)]\,\frac{P_n^m(\theta)}{\partial \theta} \tag{21}$$

$$B_\phi = -\frac{1}{r\,sen\theta}\frac{\partial V}{\partial \phi} = -\frac{1}{sen\theta}\sum_{n=1}^{k}\left(\frac{R}{r}\right)^{n+2}\sum_{m=0}^{n} m[-g_n^m \mathrm{sen}(m\emptyset) + h_n^m cos(m\emptyset)]\,P_n^m(\theta) \tag{22}$$

The three previous equations are tangential components of the Earth's magnetic field in spherical coordinates for B_r, B_Θ, B_Φ, respectively. To make use of the three equations above, it is necessary to convert them to geocentric inertial components, which can be acquired through the satellite data of a GPS global positioning system, in the following way [8]:

$$B_x = (B_r cos\delta + B_\theta sen\delta)cos\alpha - B_\emptyset sen\alpha \tag{23}$$

$$B_y = (B_r cos\delta + B_\theta sen\delta)sen\alpha - B_\emptyset sen\alpha \tag{24}$$

$$B_z = (B_r cos\delta + B_\theta sen\delta) \tag{25}$$

In the equations (23), (24) and (25), B_X, B_Y and B_Z correspond to the latitude measured in the positive north of the equator and is the local sidereal time of the point of interest.

2.2.2 The World Magnetic Model (WMM). It is the standard model used by the U.S. Department of Defense, the U.K. Ministry of Defence, the North Atlantic Treaty Organization (NATO) and the International Hydrographic Organization (IHO), for navigation, attitude and heading referencing systems using the geomagnetic field [3], it is also used widely in civilian navigation and heading systems. The model, associated software, and documentation are distributed by NCEI on behalf of NGA and is produced at 5-year intervals [8, 3].

2.3. ADCS Testbed Magnetic Field Calculation

To create an ADCS Testbed Magnetic Field that generates a stable magnetic field and simulates the magnetic field of the Earth, a pair of Helmholtz Coil can be used [12]. Helmholtz coils consist of two plates circulars rolled into a conductive material by the which will circulate current in the same direction and that it will generate the almost uniform magnetic field, as can be seen in figure 6 as [11-12]. To determine the magnetic field generated in the Helmholtz pair, we establish the vector that describes the magnetic field $\vec{B}$ as a function of ρ, φ and z; in cylindrical coordinates, that is to say $\vec{B}$ (ρ, φ, z).

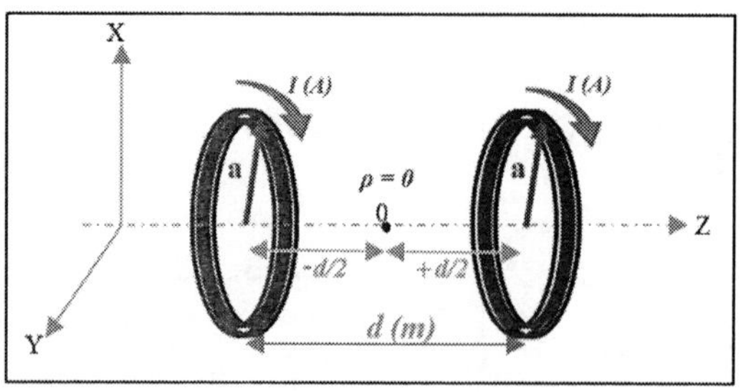

Figure 6. Helmholtz Coil configuration for SSS Project's Magnetic Testbed.

Therefore, for two turns separated from each other put a distance (d), the law of Biot-Savart is written

IOP Conf. Series: Materials Science and Engineering **685** (2019) 012004 doi:10.1088/1757-899X/685/1/012004

as follows [12-14].

$$B_z = \frac{\mu_0 I a^2 N}{2}\left[\frac{1}{\left(a^2+\left(z+\frac{d}{2}\right)^2\right)^{3/2}} + \frac{1}{\left(a^2+\left(z-\frac{d}{2}\right)^2\right)^{3/2}}\right] \tag{26}$$

Where $\mu 0=4\pi\times 10^{-7}$ NA^{-2} is vacuum's magnetic permeability, $\boldsymbol{I}$ is Current (Ampere), **a** is coil's radio (meters), $\boldsymbol{d}$ is the separation between coil (meters), N is coil's turns number total, $\boldsymbol{B_z}$ is the magnetic field (Tesla) along the Z axis. For this investigation **a** = ***b***.

To perform simulation of the magnetic field around the axis of revolution (Z), the equation (27) will be taken, according to those described [13].

$$B_z(\rho = 0; z) = \frac{8\mu_0 I}{5\sqrt{5}\,a}\left[1 - \frac{144}{125}\left(\frac{Z}{a}\right)^4\right] \tag{27}$$

Finally, to perform the magnetics' field simulation around the radial coordinate (ρ), the equation (28) can be applied, according to what is indicated in [13]. At (27) and (28) are assumed that $N = 1$.

$$B_\rho(\rho, z)_{\rho \ll a} = \left(\frac{8\mu_0 I}{5\sqrt{5}\,a}\right)\left(\frac{288}{125}\right)\left(\frac{\rho}{a}\right)\left(\frac{Z}{a}\right)^3 \tag{28}$$

3. Simulation and Results

3.1. APSCO Satellite orbit propagation and Earth's magnetic field calculation

To obtain the propagation of the orbit that will have the satellites SSS-1 / SSS2A of APSCO, simulation was carried out with a Two Line Element (TLE) file that contains the orbital information described in table 1. Later, the magnetic field of satellites around its orbits was calculated using the WMM model. The orbital simulation was carried out with Matlab from 17 May 2019 14:00:00 UTC to 17 May 2019 20:00:00 UTC. The result of the simulation can be seen in the figure 7.

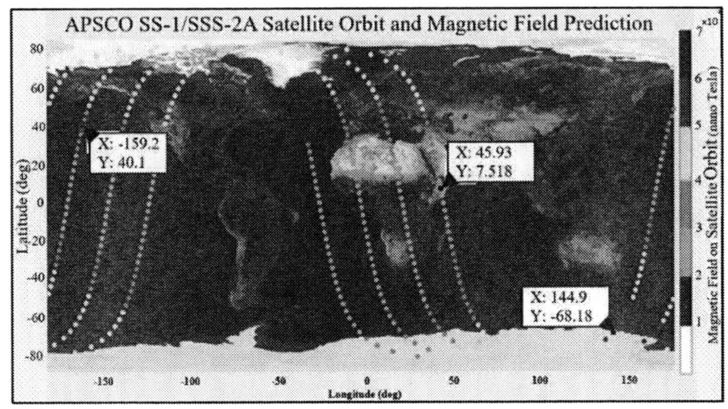

Figure 7. APSCO Satellite Orbit Propagation and Magnetic Field Results, the color at each point on the orbit represents different intensities of the magnetic field obtained.

80

Additionally, thanks to the simulation performed, it was possible to obtain the rods of the magnetic fields along the orbital line of the satellites of the SSS project.

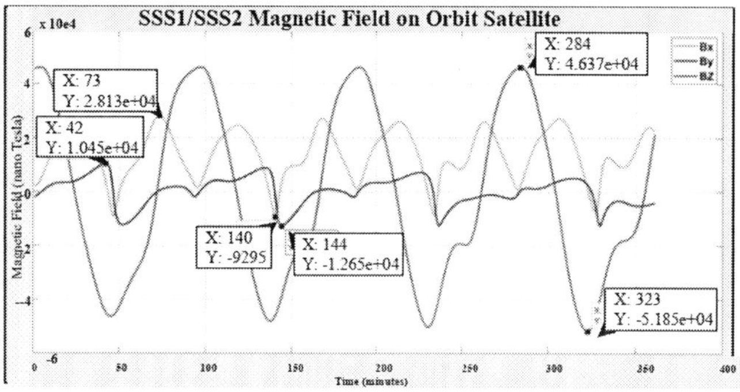

Figure 8. Components of the magnetic field X, Y and Z in the orbit of the SSS-1 / SSS-2A satellites.

The figure 8 shows the results of the magnetic fields obtained in nano-tesla on the X, Y and Z axes, the maximum and minimum values obtained can be seen in table 3.

Table 3. SSS1/SSS2A Magnetic Field on satellite orbit (nano Tesla).

Value	Bx	By	Bz
Maximum	2.8135E+04	1.0448E+04	4.6365E+04
Minimum	-9.2951E+03	-1.2646E+04	-5.1846E+04

After obtaining the results of the simulation in figure 7, three (03) random measurements were selected to form table 4 with its respective values, which were used to compare these results with those of other investigations. This will allow to verify if the results in the present research are correct.

Table 4. Satellite Orbit Propagation and Magnetic Field Results Selection.

Latitude	Longitude	Height (Km)	Magnetic Field on Satellite Orbit (nano Tesla) Results			
			X	Y	Z	F
-68.178	144.91	500	-1571.562	1676.9861	-51558.813	51610.0123
7.5180	45.921	500	27788.4276	55.6766	516.9338	27793.291
40.098	-59.187	500	18933.7793	3521.1578	29246.223	35017.5403

From the simulation we can also extract the maximum and minimum values of the magnetic field (B_T) on which the satellite will be subjected along its orbit, as shown in table 5.

Table 5. Maximum and minimum magnetic field (BT) obtained.

Value	B_T (nano Tesla)	B_T (Tesla)	Latitude	Longitude
Maximum	51926.54197	0.00005193	-64.57	148.220
Minimum	20494.51779	0.00002049	-34.09	-16.263

3.2. ADCS Testbed Magnetic Field Calculation

A simulation in matlab of a pair of Helmholtz coils was carried out in order to study the magnetic field generated in the center of the coils. Observe figure 9.

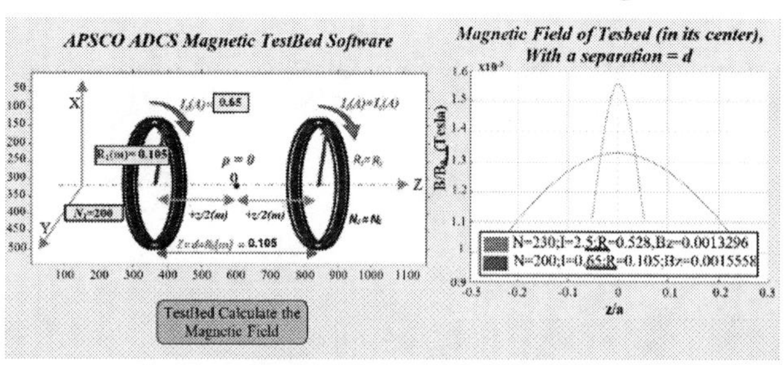

Figure 9. APSCO ADCS Testbed Magnetic Field Software.

For the simulation in figure 9, two (02) different measurements were made according to the data in table 6. The initial data for the measurements were taken from [11, 15], respectively.

Table 6. Initial data for the simulation of the APSCO ADCS Magnetic Testbed.

Parameters		Results obtained (Tesla)
Measurement 1	Radio (m)= 0.528, N= 230, I (Amp)= 2.5	0.001329
Measurement 2	Radio (m)= 0.105, N= 200, I (Amp)= 0.65	0.00155

4. Simulation Results Verification.

To ensure that the results obtained in this investigation are correct, they were compared with the results of other previous investigations. In the case of the magnetic calculation in the orbit of the satellite SSS-1 / SSS-2A, the results of this investigation observed in table 4 were compared with the results of [16]. For the case of the magnetic field generated by the Helmholtz coils in table 6, they were compared with the results of [11, 15]. The result of the comparison can be seen in table 7.

Table 7. Simulation Results Verification.

	Basic Information	My Results	External Research Results
Satellite Orbit Propagation and Magnetic Field Results on satellite Orbit	Lat: -68.178, Lon: 144.91, H: 500 Km	51,610.0123 nT	51,651.8 nT [a]
	Lat: 7.5180, Lon: 45.921, H: 500 Km	27,793.291 nT	27,719.2 nT [a]
	Lat: 40.098, Lon: -159.187 , H: 500 Km	35,017.5403 nT	35,104.9 nT [a]
APSCO ADCS Testbed Magnetic Field Simulator	R= 528 mm, N=230, I=2.5 Amp	0.001329 Tesla	= 10 [b] Gauss (0.001T)
	R= 10.5 cm, N=200, I=0.65 Amp	0.00155 Tesla	= 11.13 [c] Gauss (0.001113 T)

[a] Data obtained from [16].
[b] Data obtained from [15].
[c] Data obtained from [11].

As can be seen in table 7, the results obtained in this research are very close to the results of other investigations, so it can be concluded that the simulations are reliable.

5. APSCO SSS-1/SSS-2A ADCS Magnetic Testbed size determination.

The magnetic field generated from a Helmholtz pair is uniform in a relatively large region of space; this space is considered as spherical whose center is at the midpoint of the segment joining the centers of the coils, and the radius measurement is of the order of half the radius of the coils [13, 17]. Therefore, taking into account the dimensions of the SSS-1 / SSS-2A satellites shown in Table 1 and the maximum values of the Earth's magnetic field that will affect the satellites along its orbit, according to table 5, performed simulation in Matlab to determine the real size that the APSCO ADCS Magnetic Testbed must possess in such a way that it is capable of simulating the Earth's magnetic field and allows to perform the attitude control tests to the ADCS subsystem. As for the 3D simulation of the ADCS Testbed, it was carried out, taking as initial values a radius equal to 0.80 meters, separation between each coil equal to the radius, Current intensity of I = 2A, and number of turns 80. The two Helmholtz coils are centered on the Z axis, the first located at Z = (-r / a) and the second located at Z = (r / 2), that is, at Z = -0.45 and Z = + 0.45, respectively. The result of the simulation can be seen in figure 10.

IOP Conf. Series: Materials Science and Engineering **685** (2019) 012004 doi:10.1088/1757-899X/685/1/012004

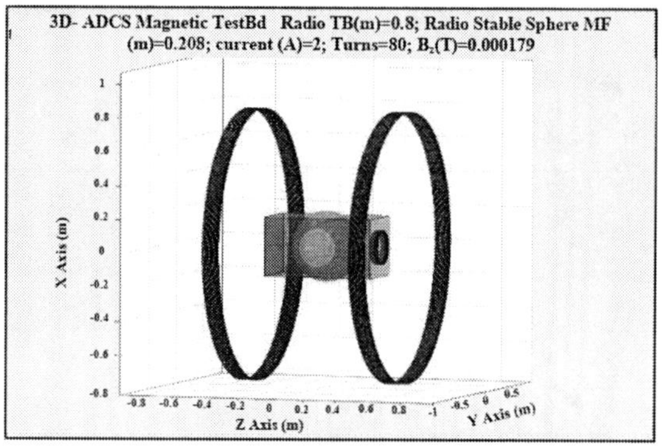

Figure 10. 3D Simulation of the APSCO ADCS Magnetic Testbed, Satellite and Stable Magnetic Field Generated.

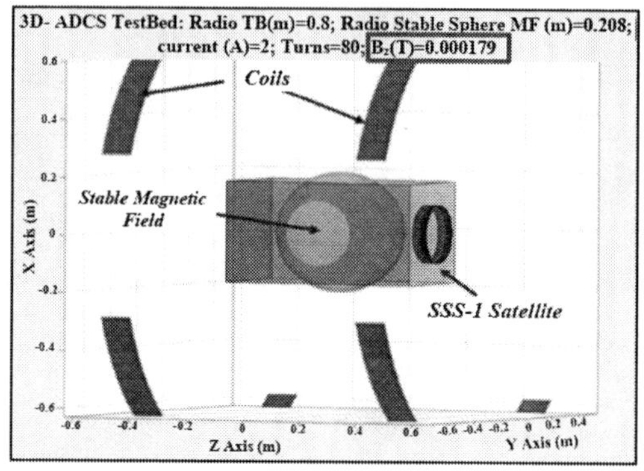

Figure 11. 3D Simulation of the APSCO ADCS Magnetic Testbed, Satellite and Stable Magnetic Field Generated.

According to figure 11, is shown each one of the parts of the APSCO ADCS Testbed, APSCO SSS-1 satellite and the stable magnetic field generated between the two Helmholtz coils in greater detail, using the parameters of Table 1, a testbed with radio coils 0.80m was created with a current of 2A and 80 turns, generating a stable magnetic field of 0.000179 Tesla. Also can be observed graphically the result of the simulation in figures 12, 13 and 14. Where the magnetic field generated by the testbed on the Z axis can be seen in more detail, simulation using nineteen contours and the magnetic field on the three coordinate axes, respectively.

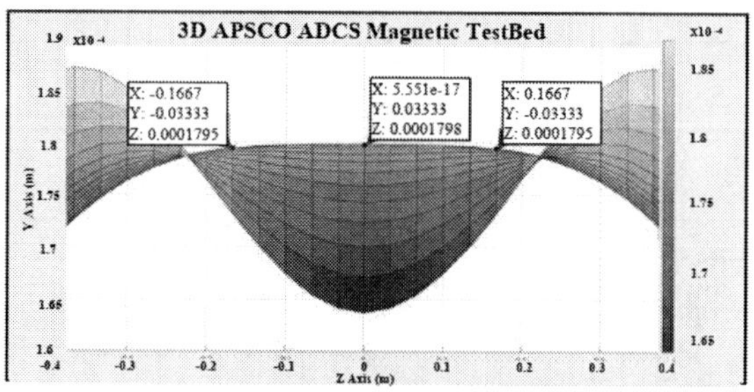

Figure 12. 3D ADCS Testbed with Length of the stable magnetic field, generated on the Z axis.

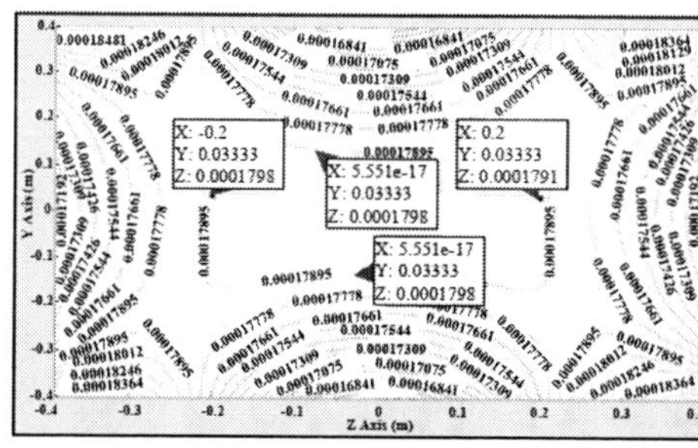

Figure 13. Graph of the simulation of the Helmholtz Coil using 19 contours.

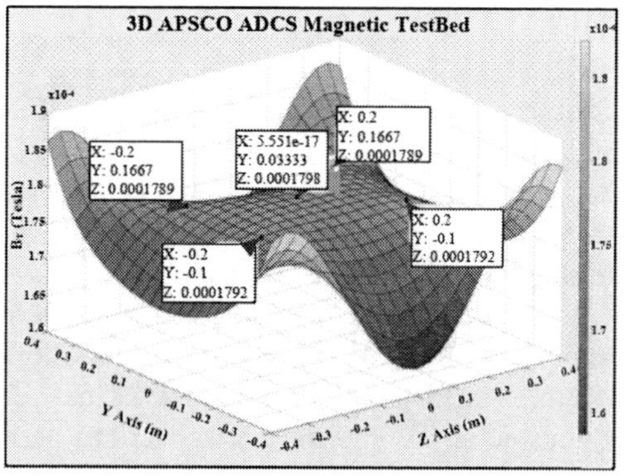

Figure 14. 3D Magnetic field generated by the ADCS Testbed with R=0.80m, I = 2A and N=80.

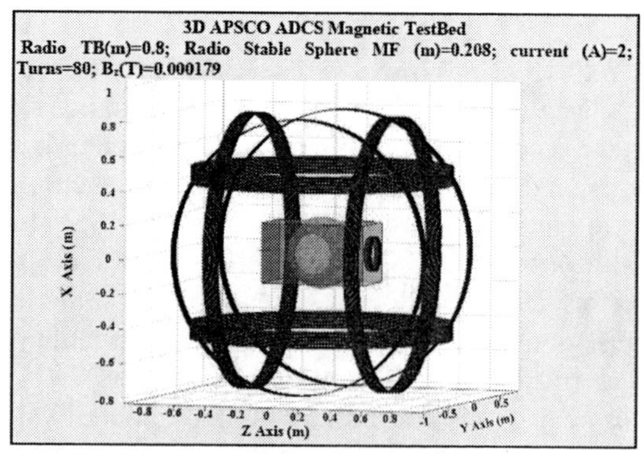

Figure 15. 3D APSCO ADCS Magnetic Testbed with X, Y and Z Axis Coils.

On the other hand, in figure 15 the result of the simulation of the APSCO ADCS Magnetic Testbed can be observed with a pair of Helmholtz coil for each axis X, Y and Z, you can also see the six coils, the SSS-1 satellite and the stable magnetic field generated near the center of the coils.

6. Conclusion

As a result of the research and simulation process, the following general conclusions can be reached:

The simulations of Earth's magnetic field as well as the APSCO SSS-1 / SSS-2A ADCS magnetic testbed, showed that the results of this investigation are very close to those obtained by other investigations, therefore, it can be considered reliable.

The development of the APSCO SSS-1 / SSS-2A ADCS Magnetic Testbed required to perform the Attitude Control tests on the satellites of the APSCO project can be designed with the following specifications: radio 0.8m, N = 80 and I = 2A. This configuration generates a maximum magnetic field of 0.000179 Tesla according to figure 11. This result is much higher than the maximum required in table 5. And the magnetic field intensity within the stable sphere shown in figure 11, can be varied simply by varying the intensity of the current flowing through the Helmholtz coils.

An identical Helmholtz array should be made for each coordinate axis (X, Y, Z), similar to those observed in figure 1 and figure 15.

References

[1] APSCO Hands-on Training - Student Small Satellite （SSS） Project Asia-Pacific Space Cooperation Organization

[2] Liu Y, Liu K-P, Li Y-L, Pan Q and Zhang J 2017 A ground testing system for magnetic-only ADCS of nano-satellites *2016 IEEE Chinese Guidance, Navigation and Control Conference (CGNCC)* 1644–7

[3] NGDC and BGS 2019 The World Magnetic Model URL https://www.ngdc.noaa.gov/geomag/WMM/DoDWMM.shtml

[4] Zatonov I, Baranov P and Kolomeycev A 2018 Magnetic field computation and simulation of the coil systems using Comsol software *MATEC Web of Conferences* **160** 01006

[5] Soler F 2015 *Herramienta para el cálculo de precisión y coberturas de un sistema de navegación por satélite* (Sevilla: Universidad de Sevilla)

[6] Rubio S 2013 *Análisis y desarrollo de algoritmos para la planificación optimizada de misiones espaciales de observación de la tierra* (Sevilla: Universidad de Sevilla)

[7] Cortés J A and Medina F A 2009 *Diseño de un magnetómetro asistido por computador* (Pereira: Universidad Tecnológica de Pereira)

[8] Contreras A C 2014 *Fundamentos teóricos y su aplicación en los métodos de potencial geomagnético y gravimétrico* (Bogotá: Editorial Universidad Distrital Francisco José de Caldas)

[9] Práctica 8. Medida del campo magnético terrestre Physics Practices for Geologists URL http://www.ugr.es/~fteorica/Docencia/Laboratorio/Web_Labo_FisicaGeneral/guiones/Geo/campo_magnetico.pdf

[10] Landínez R 2017 *Simulador de Campo Magnético Terrestre* (Granada: Universidad de Granada)

[11] Bulla, K Bobinas de Helmholtz Laboratory Report URL https://www.coursehero.com/file/11970110/laboratorio-bobinas-de-helmholtz/

[12] Fano W G, Alonso R and Quintana G 2017 El campo magnético generado por las bobinas de Helmholtz y su aplicación a la calibración de sondas *Revista elektron* **1** 91–6

[13] Martínez E C 2015 Bobinas de Helmholtz URL https://www.academia.edu/23958376/Bobinas_de_Helmholtz

[14] Restrepo A F, Franco E and Pinedo C R 2013 Study and Analysis of Magnetic Field Homogeneity of Square and Circular Helmholtz Coil Pairs: A Taylor Series Approximation *2012 VI Andean Region International Conference* 77–80

IOP Conf. Series: Materials Science and Engineering **685** (2019) 012004 doi:10.1088/1757-899X/685/1/012004

[15] Robert R 2003 Bobina de Helmholtz *Revista Brasileira de Ensino de Física* **25** 40–4
[16] GDC and BGS 2019 Magnetic Field Calculators URL https://www.ngdc.noaa.gov/geomag/calculators/magcalc.shtml#igrfwmm
[17] Strufaldi D, Granziera F, Carvalho M and de Melo L F 2018 Three-Axial Helmholtz Coil Design and Validation for Aerospace Applications *IEEE Trans. Aerosp. Electron. Syst.* **54** 392–403

IOP Conf. Series: Materials Science and Engineering **685** (2019) 012005 doi:10.1088/1757-899X/685/1/012005

Reliability Design for Data Storage in an Integrated Test System

Zhang Bing[1], Sun Xiao[1], Shi Jianwei[1] and Deng Rongjun[2]

[1] China Astronaut Research and Training Centre, Beijing 100094, China;
[2] Beijing Xuan Yu Space Technology Company, Beijing 100039, China

Email: gfkdzb@163.com

Abstract. To solve the failing errors of prompting message and storing files which are occurred in a physical-files-receiving configuration software item installed on an Integrated Test System, some specific measures have been taken. The receiving-file software item is upgraded and tested. The upgraded item is validated by many different examples, and the results show that the reliability of the item is significantly improved.

1. Introduction

The Integrated Test System is a testbed for different communication circumstances. It can receive and store files and data from spacecraft, and make an automatic classification, display, storage and inquisition for different categories of data[1-2].

The system is based on multi-platforms for Spacecraft Integrated Test and data management projects. It is an open-up and modularized structure with a stable and coincident software framework. The system can be reconfigured and rebuild with the reconfigurable management tools for the front and the rear user interfaces. The system has open-ended interfaces for the future mission with the general and standardized inner modules.

The system can be flexibly deployed in multi operating system (Linux and Windows) without running errors with other applications and software. Even under the fault or unreasonable input circumstance, it also has a higher anti-interference and anti-malfunction ability. With the robust structure, it can give many prompts for the clients without mistaken-running.

The data flow-chart and software configuration schematics are illustrated in Figure1.

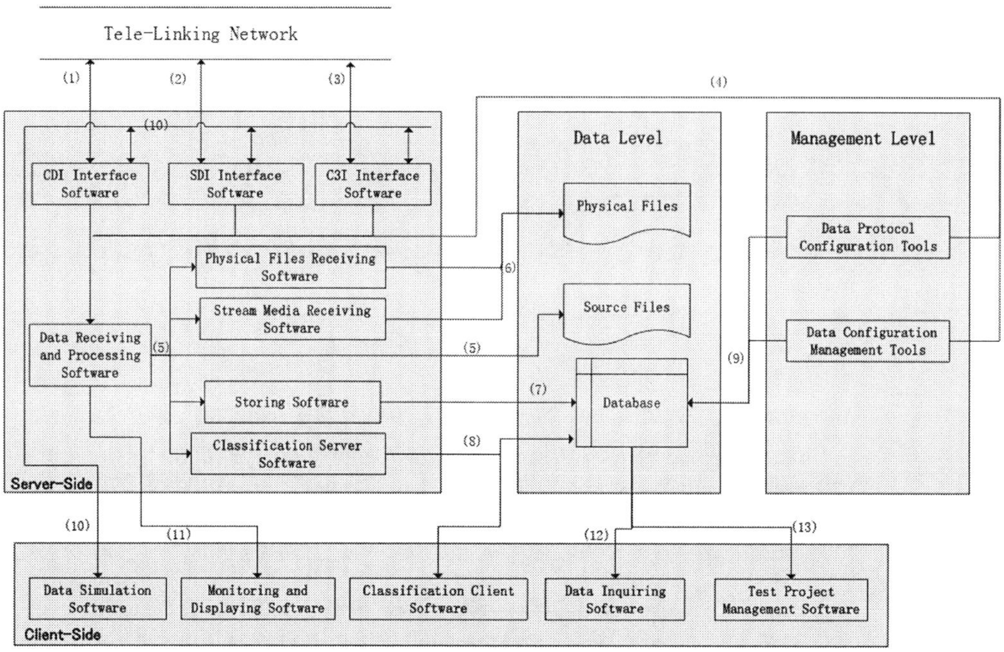

Figure1. The data flow-chart and software configuration schematics

The Interface software (CDI/SDI/C3I) receives the data from the tele-linking network, and will identify the source and categories of the data according to the fixed head-bytes. The raw data will be transferred to the Data Receiving and Processing (DRP) software.

The function of DRP software is to manage inner data transferring and decomposing. First it will store the raw data after receiving them from the Interface software. With the data protocol configuration tools and data configuration management tools, it will get the specific protocol and processing methods in the form of the configurating files. Secondly, the DRP software gets the protocol and the predefined methods in the classification, transformation, extraction and decomposition for the data and uses them to decompose and process the data. Finally, the results of the processed data will be sent to the other software, including the Classification Server software, Monitoring and Displaying software, Physical Files Receiving software, Stream Media Receiving software, Data Storing software.

The Monitoring and Displaying software start with the different managerial authentication, and will display the different kinds of data with original number, the waveforms of the physiology data. The Physical Files receiving software and Stream Media Receiving software will accept the data and make the final data files in the form of byte streams. The data files will be stored in the disk array of the data base, and the index will be established in the database.

The Classification Server software will interpret the decomposed data according the knowledge base. The interpreting results will be sent to the Classification Client software.

The function of the Data Storing Software is to store and manage all the decomposed data. The Data Inquiring software will inquire the raw data and the decomposed data under the inquiring conditions, and display the inquiring results with the displaying requirement.

The low orbit spacecraft fly around the Earth in 90 minutes. The ground system can receive data only in the observing and controlling zone. If the received data is incomplete, it will send a requirement for resending the data. Due to the limited time in the observing and controlling zone, if the data and files are received and stored successfully, but the ground system make a wrong requirement for resending the data only for the unsuccessful prompting message. The precious telecommunicating resources will be wasted, thus the reliability for receiving and storing data in the ground system is an important factor[3].

2. The problem of the traditional design

The Figure 2 is the flow chart based on the interface, function and performance requirements.

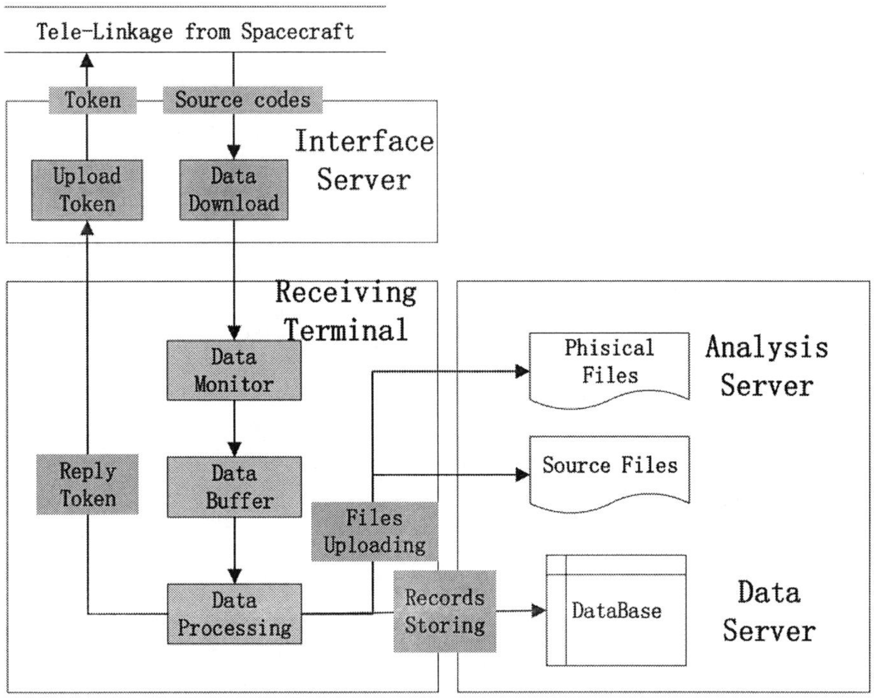

Figure2. The Receiving and Storing Data Flowchart

The process is described as below.

a. The raw data from the tele-linkage to the interface server, include: head, body and tail of the data;
b. Storing source codes after getting them, forward them to the file-receiving server.
c. Buffering the source codes in sequences after receiving them, if some packets are missing, ask for resending by a reply token.
d. Processing the buffer data until the tail of the data, send the integrity message back to the receiving server, then reload by it. If the data is incomplete, then send the failure token.
e. The receiving files are storing in the buffer memory
f. Send the files to the analysis server through FTP, and store the records in the server, mark the success message after storing files.

The procedure for recording the files in receiving software item is:

a. Verify the connection with the data base when starting up the receiving process. A connecting success message with the server will show up in case of losing any data from database.
b. The connection will be keeping through the whole process, waiting for the result.
c. Once the data is successfully stored, *Receiving Files Successfully* will prompt on the interface of the software. The connection will still be kept.

After the data is completely received, it will often find some problems:

a. The source codes are received successfully and the data file is constructed well, FTP is good, but the record is unsuccessfully written to the database, that means even after receiving files, *Receiving Files Successfully* message is still not prompted.
b. The source codes are not received successfully, the files are bad, and *Receiving Files Successfully* is still not prompted.

After checking the system's log files, the first problem is caused by the failure of writing records to the database because of the unstable of connection with the database. Once the connection is built up, it will not break with it actively and reconnect it next time. The failure for writing to database is due to the unstable or abnormal connection.

The second problem is caused by the failure of the buffer hardware in the receiving module. Since the bad sectors can be existed, and the uncompleted source codes will make the files unsuccessfully done.

Above all the two problems, the source codes are all successfully received, but no message is shown up and no follow-up methods are available. The requirements for resend the files will waste the precious tele-linkage resources between the spacecraft and the ground control system.

3. The Improvement Measures

To make sure the whole process is completed successfully from receiving source codes, uploading to FTP server and writing, three methods are proposed.

3.1. The prompting message method

The four steps for receiving, creating, uploading to FTP, and writing to the database are divided for the process of receiving and recording the physical files. The prompting message is only showed on the steps of receiving source codes and writing to the database successfully.
Now the prompting message will be showed on every step, they will help to monitor the whole process. Once the uploading to FTP is successful, it will not require to send the source codes again.

3.2. The fault-tolerant method

Using the fault-tolerant method[4], it will successfully complete the whole process even either of the creating, uploading and recording process is failed.

3.2.1The software fault-tolerant technique and recovery block. The software tolerant technique is a redundant method for the mistakes in the software. These mistakes can be found or miss-found. But the fault-tolerant technique can provide a mechanism instead of falling into a failure. It includes Recovery Blocks(RB), N-Version Programming(NVP), Consensus Recovery Block(CRB), Acceptance Voting(AV) and N-Self-Checking(NSC).

The strategies include: *Forward Recovery* and *Backword Recovery*. *Forward Recovery* is to continue the current computing process to make it consistently in a right way even the current state is incorrect. *Backward Recovery* is to return back to the previous right state. As is showed in Figure 3.

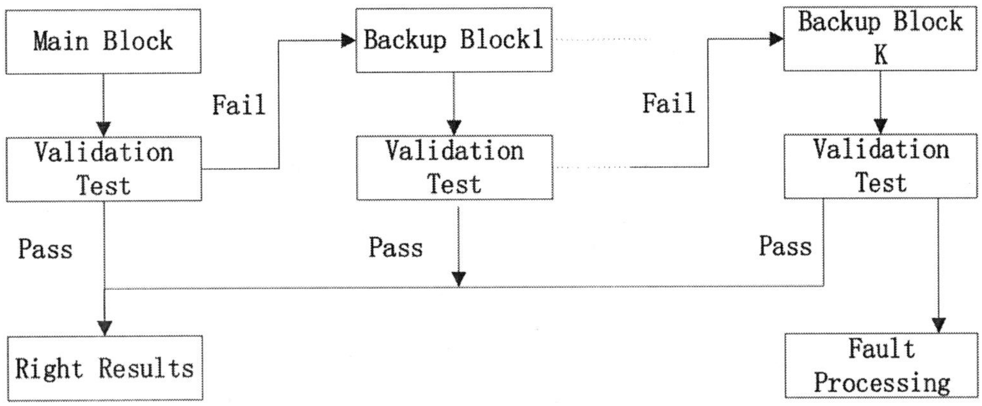

Figure 3. Flow Chart of the Software Recovery Blocks

The *Recovery Block* is to choose a group of blocks as fault-tolerant blocks, changing them from normal program blocks into *Recovery Blocks*. The blocks can be software module, sub-program, program segments or procedures. One *Recovery Block* include many blocks with the similar function and different design. Only one block is running at a specific moment. The dynamic redundancy is made by replacing a failure one by the backup *Recovery Block* while the whole system is keep running properly.

3.2.2 Increasing the recovery button for the receiving and storing procedure . Three recovery buttons are added for the receiving, creating and FTP uploading files procedures. Once the time is over 15 minutes and there is no message for the next step, the recovery buttons can be pressed to implement the process due to the file capacity is no more than 1Gigabytes. These buttons can make sure that the above three processes are successful completed.

3.3. The modified connection mode for the receiving software and the database

The failure is caused by the unstable connection between the receiving software and the database. To solve this problem, the connection should be confirmed before writing the records to the database. Add the corresponding software codes for the confirmation of the connection, and reconnect the database until it is normal. In this way the connection is constantly working well before any actions.

4. Validation for The Improvement Measures

4.1. The improved connection processes

Adding the software codes for the connection with the database, it will reconnect to the database until it is normal.

In a simulated testing circumstance, the transmitting process is normal even the connection is compulsory disconnected.

4.2. The improved message prompting process

The following optimized steps are applied for the receiving source codes, creating files, uploading to FTP, and writing records to the database:

a. Normal testing: updating the software, simulating the procedure of the tele-linkage of the spacecraft, implementing several times, the results are normal.

b. Simulated loss packets testing: transmitting a large file with the capacity over 1.5G. In order to simulate the loss packets, the cable line is pulled out. The error message is fed back to the linkage of the communication network. The data will be re-sent until they are successfully received and stored.

4.3. Adding the recovery buttons

There will be a prompt message for over 15 minutes since the capacity of the file is less than 1G. If this happened, press the recovery button.

a. Simulated test for the failure of creating and uploading files to the FTP server: The test is simulating the procedure of creating and uploading files to the FTP server with a large file (over 0.5G). While the message for the source code is normal, pull out the cable line for the receiving software, the cable line is unconnected with the FTP server. Install the cable line again, press recovery button, the files will be re-created in the receiving software item, re-uploading to the FTP server. The procedure will insure the whole process is completed normally.

b. Simulated test for the failure of writing to the database[5-6]: After the source codes are completely received, the prompt message is shown up. Stop the service of Oracle in database using the command *Shutdown Immediate*. Then start the database with SQL command *Start up*, press recovery button, the files will be processing again by the receiving software item, the results will be sent to the database successfully.

The above two tests can show that the files can be processed properly even the database is abnormal.

5. Conclusion

The Integrated Test System is optimized with three proposed measures including the prompting message method, the fault-tolerant method, the modified connection mode for the receiving software and the database. The three proposed methods are validated with many examples. It has shown that the reliability of the system is improved, while the precious communication resources are preserved.

Reference

[1] Hua Mao Wang, Jindong Yan, 2017. *Electrical Test Technology of Spacecraft.* (China: National Defense Industry press, Beijing)

[2] Hua Mao Wang, 2018. *Spacecraft electrical test technology.* (China: Beijing Institute of Technology press, Beijing)

[3] Gary Musgrave Ph. D, Axel Larsen, Tommaso Sgobba, 2011. *Safety Design for Space Systems.* (UK: Elsevier Limited of the Boulevard, Langford lane, Kidlington, Oxford)

[4] Lorenzo Strigini, 2004. *Fault Tolerance Against Design Faults.* Available online.

[5] Saikat Basak, 2010. *Oracle DBA Concise Handbook Covers 9i to 11g.* Available online.

[6] http://docs.oracle.com/cd/E11882_01/appdev.112/e13995/oracle/jdbc/OracleConnection.html.

Mechanical System Performance Analysis of Commuter Aircraft's Rear Landing Gear Design for Unpaved Runway

N A Rammon[1] and W Nirbito[1,*]

[1]Department of Mechanical Engineering, Universitas Indonesia, Depok, West Java, 16424, Indonesia

[*]Corresponding author: wahyu.nirbito@ui.ac.id

Abstract. Landing gear design of this commuter aircraft was developed from the previous model, to carry more passengers. The total load of the aircraft will increase as passenger capacity increases. With a variety of runways in Indonesia, it is necessary to have a good landing gear performance for landing on a paved and unpaved runway. Performance of the landing gear design is analyzed through kinematics and dynamics methods. Main concern of this analysis is to find the capability of the landing gear shock absorber to damp the vibration response due to the impact of the landing gear system with the wavy surface of the unpaved runway, as well as the actuator ability as the extension and retraction mechanism driver. The analysis concluded that the landing gear design capable to perform during landing on a paved and unpaved runway, with a modification of the shock absorber and actuator design.

1. Introduction

Public and goods transportation is increasing in Indonesia nowadays. In Indonesia, there are still many airfields that limited in length and pavement [1]. Existing Airplanes need to be improved in the capacity to support air transportation to remote areas. This modification affects the airplane's center of gravity [2], the total load of the airplane, the structure, as well as the landing gear design especially for the unpaved runway on the remote region airport. The landing gear is one of the most important components in the airplane, which will absorb large energy to reduce the impact effects in the landing event. Landing gear design needs to be right on target and accordance with regulation set by Federal Aviation Regulation. Good landing gear must have good damping characteristics, also a good extension and retraction mechanism. The performance was analyzed through the general Kinematics and Dynamic method.

2. Methodology

The analysis begins with the landing gear design of a previous model airplane. Landing Gear Performance is analyzed from the actuator design aspect as well as suspension systems to calculate its suitability for the unpaved runway.

2.1. Landing Gear's Design General Information

The airplane particular specification, as shown in Figure 1 is needed to determine the boundaries of the performance are a follows maximum take-off weight of 18,700 kg, maximum payload of 5,500 kg with maximum 58 passengers, and landing velocity of 55.5 m/s.

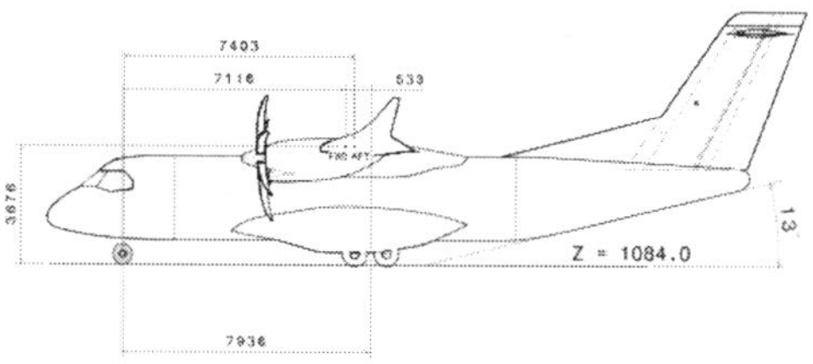

Figure 1. Aircraft General Dimension is needed to calculate the suspension's maximum load. [3]

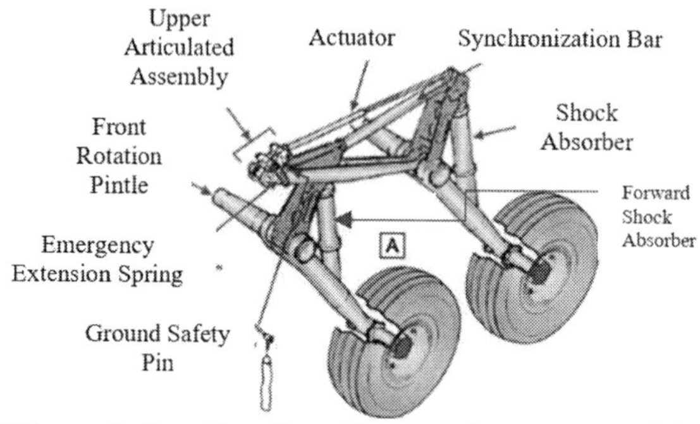

Figure 2. Landing Gear General Component of the Commuter Aircraft with 2 Engine and 35 Passengers Engineering Data. [3]

2.2. Kinematics Analysis Landing Gear System

Extension and Retraction have a time target to find pressure needed of the actuator's specification, including actuator's hydraulic pressure. Therefore, the knowledge of landing gear's main component is needed as shown in Figure 2. These aircraft need 9 seconds to perform extension, meanwhile, 14 seconds were needed to perform full retraction on landing gear system.

Kinematics and Dynamics use to determine the performance of the landing gear extension and retraction system by analyzing the velocity, acceleration, angular velocity, and angular acceleration.

Few forces are affecting the performance of extension and retraction mechanism, such as Aerodynamic Drag Force, Weight Force, and Inertia Force. These forces determine the size of the actuator, length of the rod's, and actuator's pressure

2.3. Suspension's Performance

The suspension is needed to cushion the landing of the aircraft. Generally, commuter aircraft use an oleo-pneumatic type of suspension because of high efficiency. Suspension in landing gear is divided into two sections, the tire, and the shock absorber. Both components have their damping and spring force constant, affecting the performance of unpaved runways' landing.

With special aviation shock absorber, hence the force distribution for each strut is needed. The tire also contributes to suspension's performance, the right type of tire will result in better suspension.

3. Results and Discussion

3.1. Kinematics Analysis of Extension and Retraction Mechanism

Figure 3, and 4 are illustration when retraction and extension occur. The hydraulic fluid pushes the actuator piston and rod, therefore the connection strut rotates as goes by pushing synchronize bar until the wheel moves 90 degrees from its initial position.

Figure 5, and 6 illustrate free body diagram of retraction and extension mechanism, to perform kinematic analysis.

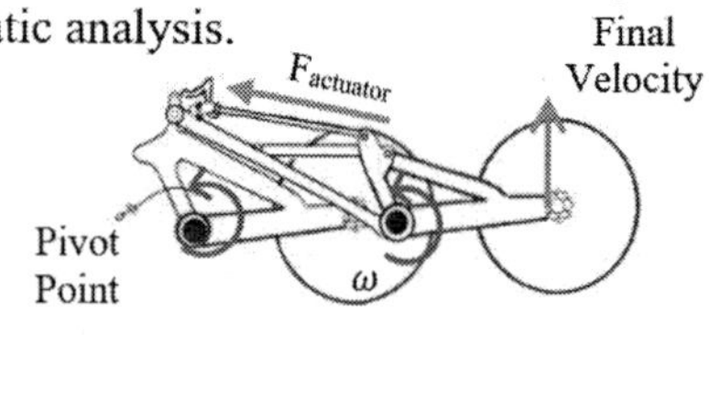

Figure 3. Kinematics Analysis of Retraction Mechanism.

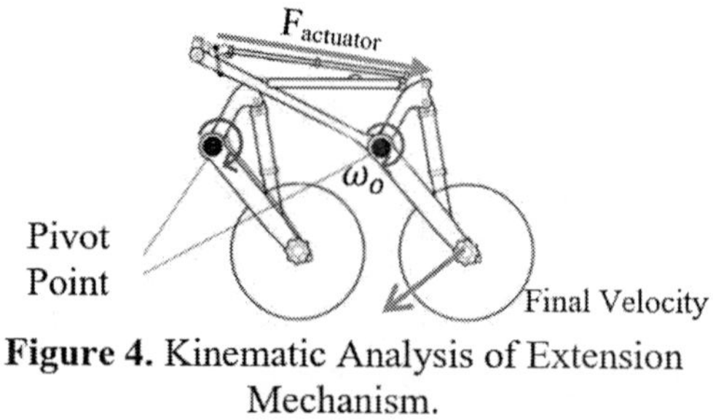

Figure 4. Kinematic Analysis of Extension Mechanism.

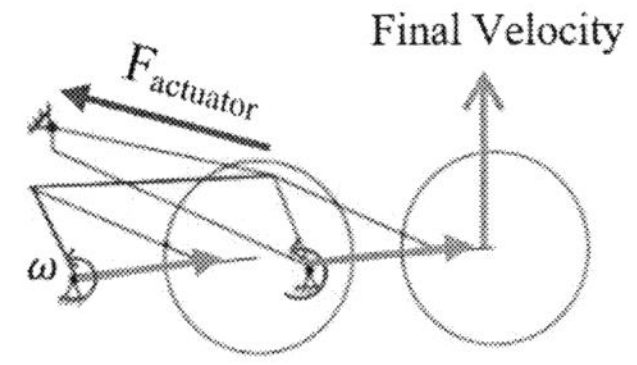

Figure 5. Retraction Mechanism Free Body Diagram

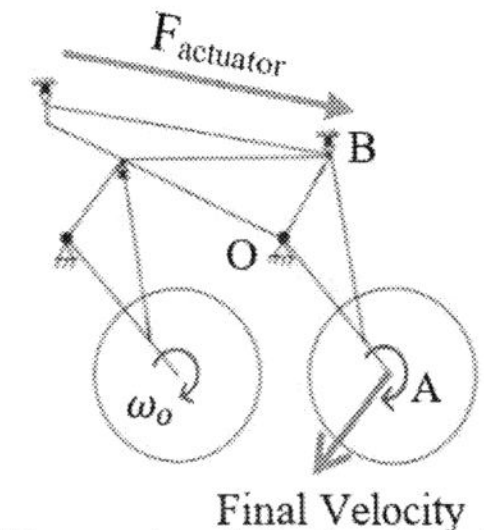

Figure 6. Extension Mechanism Free Body Diagram.

Rods with the same shaft will produce equal angular velocity, with length different will produce different linear velocity. Angular velocity of point O is 3.375 rad/s, then angular acceleration and velocity of point A are can be determined as below. With radius of movement from the static point to extension point of 800 mm and the travel time to do an extension of 9 seconds, then;

$$F.r = I\alpha \tag{1}$$

$$\alpha = 0.375\frac{rad}{s^2}$$

$$V_A = Vo + V_{AO} \tag{2}$$

$$V_A = Vo + \omega_{AO}\, r_{AO}$$
$$V_A = 2.7\frac{m}{s}$$

$$a_A = a_o + a_{AO-t} + a_{AO-n} \tag{3}$$

[4]

$$a_A = a_o + \alpha_{AO}.r_{AO} + \omega^2{}_{AO}.r_{AO}$$
$$a_A = 9.4125\ m/s^2$$

From the above calculation, final linear velocity of the extension component is 2.7 m/s in order to achieve 9 seconds targeted time, and the final acceleration need to perform extension mechanism of 9.4 m/s^2.

3.2. Dynamics Analysis of Extension and Retraction Mechanism

Landing Gear's Actuator needs to overcome external forces and internal forces to perform an extension and retraction mechanism. Aerodynamics Drag Force, Weight Force, and Inertia Force affect the performance and determine the actuator size needed.

3.2.1. Aerodynamic Drag Force

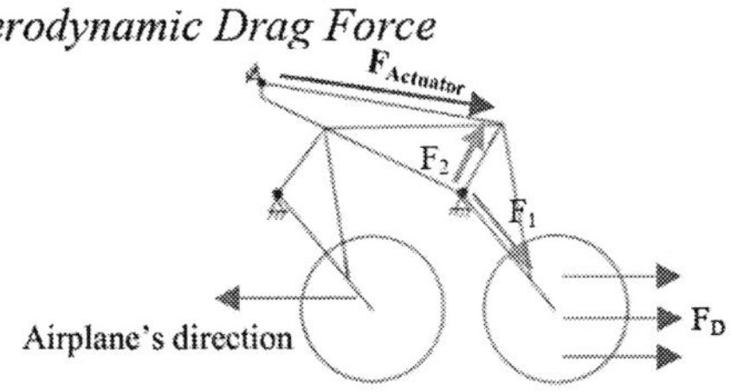

Figure 7. Dynamics Analysis of Aerodynamic Drag Force

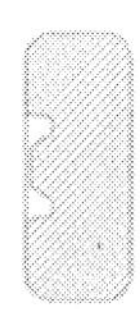

Figure 8. Tire Frontal cross-sectional area [5].

As shown in Figure 7, the aerodynamic drag force on the landing gears in full extended position occurs from the force opposite to the airplane's direction. This drag force inhibits a solid body movement

IOP Conf. Series: Materials Science and Engineering **685** (2019) 012006 doi:10.1088/1757-899X/685/1/012006

through fluids, i.e. air, works due to the air dynamics pressures against the frontal cross-section area of the body, as shown as Fig 8, that will be formed as the so-called bluff body in the context of aerodynamics. Figure 8 shows the tire frontal cross-sectional area based on Tandem Landing Gear type that will be the area of the air dynamics pressures work on. According to Spagnolo in his journal about Unsteady Force and Flow Features of Single and Tandem Wheels, the coefficient drag (C_D) for tandem main gear as a bluff body is 0.336 [5]. The drag force of a bluff body due to it's cross-sectional (A), flow velocity (v), and air density (ρ), as is using the following formula **[6]**,

$$F_D = C_D \frac{1}{2} \rho v^2 A \tag{4}$$

So, therefore the drag force of landing gear tires from the calculation above formula is 287 N.

3.2.2. Structure Weight Force. As shown in Figure 9, to obtain the structure force, the Engineering Dynamics general formula can be used

- $\Sigma Fx = 0$
- $\Sigma Fy = 0$

$$Ray + Rby = W \cos 45$$

- $\Sigma M = 0$

$$Wcos\ 45(0.05)cm = Rby(0.8)cm$$
$$Rby = 91.8\ N$$
$$Ray = 1379.7\ N = F_3$$

Therefore,

$$F_{Structure} = F_3 . cos45$$
$$F_{Structure} = 975.6\ N$$

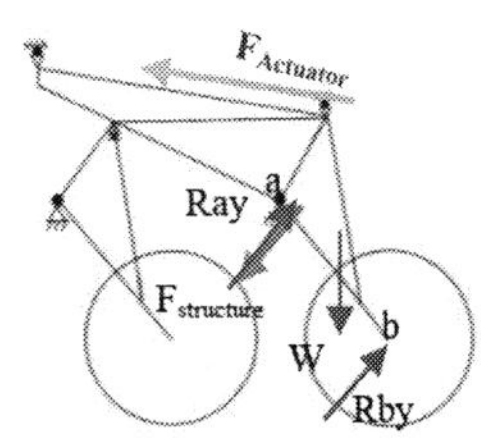

Figure 9. Dynamics Analysis of Structure Weight Force. The force distribution of the Structure Weight Force occurs because of gravity. Dynamics analysis begins with finding a reaction force of the weight on cantilever force in point a (Ray) and point b (Rby), the force produces in Ray equal to the force that actuator ($F_{actuator}$) should compensate.

3.2.3. Inertia Force

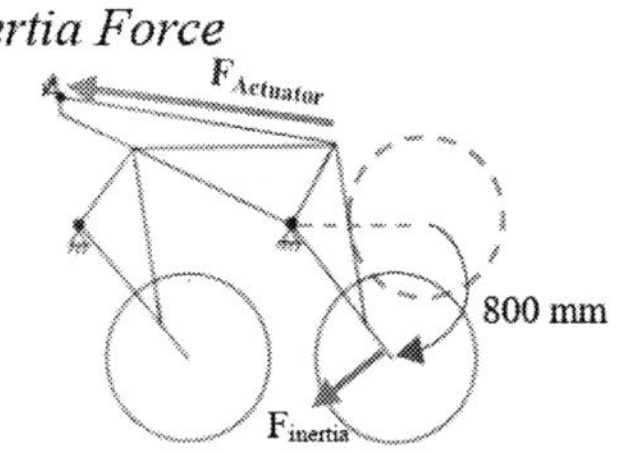

Figure 10. Dynamics Analysis of Inertia Force. Inertia force is a force in the opposite direction to an accelerating force acting on a body. To find inertia force, displacement from retraction position to extension position (r) and inertia of body (I) is needed. With r = 800 mm and time needed to extend (t) is 9 seconds. [3]

As shown in Figure 10, to obtain the Inertia Force, the following Dynamics general formula is used

$$\omega = \alpha . t \quad \omega = 3.375 \frac{rad}{s} \tag{5}$$

$$W = \frac{1}{2} I \omega^2 \quad W = 252.87 J \tag{6}$$

IOP Conf. Series: Materials Science and Engineering **685** (2019) 012006 doi:10.1088/1757-899X/685/1/012006

$$W = F.s \tag{7}$$

$$F = 5057.4\ N$$

The calculation above represents one structure of strut, whereas in every landing there are two struts, so this following calculation is needed

$$F_{inertia} = 2\ x\ 5057.4\ (cos45) = 7026.32\ N$$

To find total force (F_{total}) that needed to do extension and retraction mechanism, the following formula can be used

$$F_{total\ actuator} = F_{drag} + F_{structure} + F_{inertia} \tag{8}$$

$$F_{total\ actuator} = 8207\ N$$

3.3 Suspension System

Unpaved runways are still many in Indonesia mostly in the remote region. Therefore, the analysis of a landing gear response performance on the unpaved runway is important. Almost all unpaved runways have wavy surface due to the soft structure of soil material composition. As shown in Figure 11, the wavy unpaved runway is illustrated as the sinusoidal waves model to simplify the calculation.

The suspension system has a damping coefficient (C) of the damper and a spring constant (k) of the rigidity.

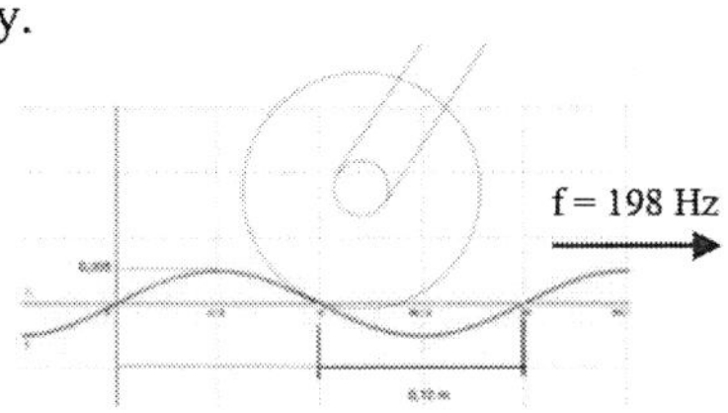

Figure 11. Landing Illustration on Unpaved Runway. Deflection of the soil for unpaved runway mostly has a maximum depth of 0.005 m (Y) [7] and a maximum length of 0.04 – 0.06 m (λ) [8].

This following Dynamics general Impulse-Momentum formula can be used to find the impulse force that works against the suspension system.

$$mv_l = F\Delta t \tag{9}$$

Mass (m), and speed of rolling (v_l) of landing gear is the basic parameters in the application of the above formula. With the airplane approaching speed of 55.5 m/s, and time for deceleration (t) is 15 seconds, therefore, the impulse force (F) of the landing gear is 41,466 N.

According to Sivakumar S (2014) in The Journal of Analysis of Active Landing Gears Chapter 3, the damping ratio for main landing gear is 0.35 [9], and then the damping coefficient and spring constant can be found with this general Mechanical Vibration following formula.

$$F = k.\Delta x \tag{10}$$

$$\omega_n = \sqrt{\frac{k}{m}} \tag{11}$$

$$\zeta = \frac{c}{2m\omega_n} \tag{12}$$

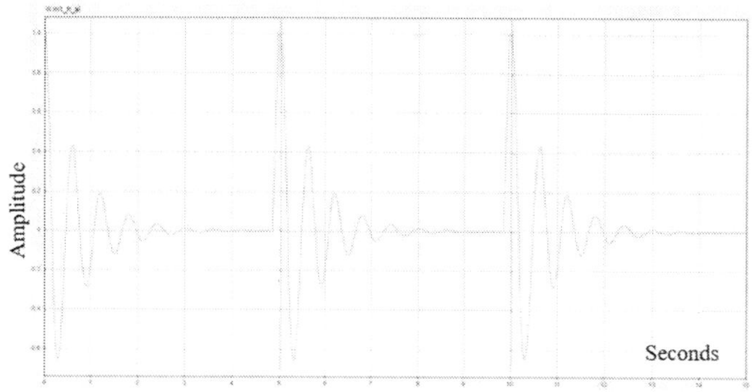

Figure 12. Vibration Response on Unpaved Runway. The vibration response of the landing gear is showed in the wavy unpaved runway condition as a repetition of 3 transient sinusoidal waves for 15 seconds, which l representing the stroke length of 0.065 m. With spring constant k of 197,457, 2 N/m and damping coefficient C of 89,400Ns/m, therefore the damping occurs in 4 seconds.

The response shown in Figure 12 happens repeatedly because of repeated excitation due to the wavy runway. The suspension consists of two-component, tire and shock absorber, which the shock absorber consists of spring and damper. Type of tire that suitable for the unpaved runway is the Radial type with diamond shape tread as shown as Figure 13, that the pressure distribution towards the side of the tire and the tread will increase the traction.

Figure 13. The diamond shape tread is suitable for the unpaved runway since the tread shape is increasing adhesion in the wet area, acting as a hydroplaning compensator, as well as preventing excessive heat occur due to friction. [10]

According to Aircraft Tire Engineering Data, the pressure also affect the performance of the suspension. Under pressured or deflated tire may flex beyond design limits, causing excessive build-up that weakens the tire construction, whereas over pressured or over-inflated tire may cause burst out due to high expansion [10].

4. Conclusion

To perform the landing gear mechanism movement as designed for this aircraft, the actuator needs to compensate 8207 N of the total force. The actuator's wall has to be strong enough to compensate pressure from the hydraulic pump. So, the actuator for this application is needed to have a working pressure of 3000 psi, cross-section area 4 cm^2, 2 mm wall thickness, as well as 10 cm rod length, and 0.5 cm rod diameter with linear velocity and constant acceleration. This specification has a wall strength that 2.5 times of the normal pressure, as a safety factor, that comply with the best practice standard according to Roskan and Jan (1986) [11].

To perform safe landing on the unpaved runway, the improved suspension system with shock absorber of a spring and a damper, as well as the specially designed tire as discussed above are used in the landing gear system. This landing gear suspension system gives a relatively comfortable transient response in every 5 seconds for each hump of the wavy runway surface.

Acknowledgment.
The authors would like to thank PITTA grant program of the University of Indonesia for funding this study with contract number NKB-0796/UN2.R3.1-HKP.05.00/2019, and also to PT. Dirgantara Indonesia (Indonesian Aerospace) high appreciation for the facility, opportunity, and guidance for the development of this study.

References

[1] Sunara, L.L, and Nirbito W. 2018. *An Estimation of the Maximum Take-off Weight Commuter Category Aircraft Based on Airstrip Condition on Remote High Terrain Area*, ATCES 2018 Shanghai ROC, IOP Conference Series.

IOP Conf. Series: Materials Science and Engineering **685** (2019) 012006 doi:10.1088/1757-899X/685/1/012006

[2] Aditya, N, and Nirbito W. 2018. *Determining the Horizontal Tail Optimum Dimension of Civil Transport Class Aircraft Based on the Previous Model for Upgrading the Passengers Number.* ATCES 2018 Shanghai ROC, IOP Conference Series.

[3] Commuter Aircraft with 2 Engine and 35 Passengers Engineering Data. Indonesian Aerospace.

[4] Kraige, L.G., and J.L. Meriam. *Engineering Mechanics: Dynamics.*

[5] Spagnolo, Stefano & Zhang, Xin & Hu, Zhiwei & Stalnov, Oksana & Angland, David. (2015). *Unsteady Force and Flow Features of Single and Tandem Wheels.* 53rd AIAA Aerospace Sciences Meeting.

[6] Currey, Norman S. January 1982 *Landing Gear Design Handbook.*

[7] Bentsen, Ross A., Bush, III, Albert J., Harrison, J A. January 1989. Evaluation of *Non-Destructive Test Equipment for Airfield Pavement.* Phase 1. Calibration Test Results and Field Data Collection. Army Engineer Waterways Experiment Station Vicksburg MS Geotechnical Laboratory Report.

[8] *Proposed Construction and Operation of a gravel runway in the area of Mario Zuccheli Station, Terra Nova Bay, Victoria Land, Antartica.* Asian Technology Conference in Mathematics XXXVIII 2016. Santiago.

[9] Sivakumar S. 2014. Thesis. *Analysis of Active Landing Gears Chapter 3* (India: Anna University, Faculty of Mechanical Engineering). Page XXV, 179.

[10] Good Year Aircraft Tire Engineering Data. 2018.

[11] Roskan, Jan. 1986. *Airplane Design Part IV: Layout Design of Landing Gear and System.*

IOP Conf. Series: Materials Science and Engineering **685** (2019) 012007 doi:10.1088/1757-899X/685/1/012007

Derated Thrust: Method Analysis for Optimizing Turbofan Engine Takeoff Performances (SFC, EGT) Due to Lower Maximum Takeoff Weight (MTOW) Requirement

R F Irsyadi[1] and W Nirbito[1,*]

[1]Department of Mechanical Engineering, Universitas Indonesia, Depok, West Java, 16424, Indonesia

[*]Corresponding author: wahyu.nirbito@ui.ac.id

Abstract. Reduced thrust or derate has been practiced commonly nowadays as a standard procedure of takeoff. Since power is being used at maximum during takeoff phase, the practice has many benefits for the operation of aircraft. Not only it reduced fuel consumption during the takeoff phase, but also reduced engine's wear and tear that caused by stress and a very high temperature. Exhaust Gas Temperature (EGT) is one of the most important parameters that used to monitor the engine deterioration. There is a definite correlation between reducing takeoff thrust and EGT reduction, where both values are contradicting one another. However, there must be some calculation by the flight crew on how much percentage of derate is eligible prior to takeoff. The goal of this paper is to find the effectiveness of derate to reduce EGT in order to improve performance, prolong engine life and reduce operating costs of the operators.

1. Introduction

Every airplane has its thrust setting based on the design of what the airplane meant for. For a long-range wide-body twin-engine aircraft, it was built purposely to carry a heavy load and fly in a very long-distance flight. To do that, it requires an engine that could generate enormous thrust to lift the aircraft. In the modern era, most of the commercial aircraft are equipped with an efficient high bypass ratio turbofan.

The high temperature in the hot section of the engine could deteriorate and shorten the engine's life. Exhaust Gas Temperature (EGT) is used as a reference to monitor the engine condition at the High-Pressure Turbine (HPT) outlet [1]. EGT level is proportional to how much thrust inputted, meaning if an airplane uses almost its maximum thrust, then the EGT will be high and approaching its near maximum allowable level.

In the takeoff phase, a full rated thrust is applied only if aircraft carries a load in its maximum takeoff weight (MTOW). It is the ultimate design limit of the aircraft where the aircraft carries a maximum payload, applied its full thrust and the temperature inside of the engine peak almost to the limit. However, in some cases such as the runway conditions is limited, which is still many in Indonesia mostly in the remote regions [2], the airplanes are not allowed to takeoff with its specified MTOW. It means that the airplane should also reduce its thrust to perform efficiently. Otherwise, the high temperature from the excessive thrust could wear and tear the engine.

That is where reduced takeoff thrust or derate technique comes. This technique reduces the thrust of the airplane during takeoff sufficiently, thus reduce the stress inside of the engine during takeoff. This could prolong the engine's life, therefore saving cost for maintaining the engine [3]. There is a definite correlation between weight, thrust, and EGT level in terms of aircraft performance during takeoff.

By applying reduced thrust, not only the decreasing EGT could be calculated thermodynamically, but also every temperature at each section using The Brayton cycle calculation. The assumed temperature will be inputted to the flight management computer (FMC) to 'trick' engine to perform a reduced-thrust takeoff can also be predicted.

2. Methodology

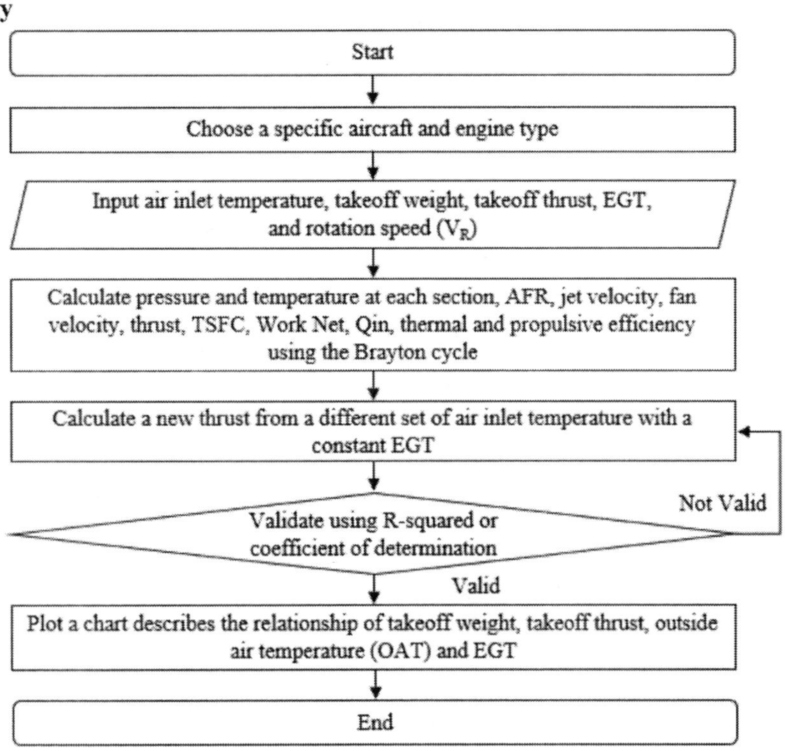

Figure 1. Flowchart of the methodology

The whole process of this research is illustrated in figure 1. A long-range wide-body twin-engine aircraft uses a high bypass ratio turbofan engine as shown in Figure 2. A turbofan engine consists of two flow; primary flow that goes through the hot core, and secondary flow that bypass through the fan. The airflow that goes through the engine will experience a change of temperature and pressure due to how the engine was designed and constructed. In this paper, the analysis is particularly based on the typical high bypass ratio turbofan engine of a long-range wide-body twin-engine aircraft.

To conduct the thermodynamic calculation using The Brayton cycle, a variety of data with different circumstances must be set first. Five samples are used in this calculation. The first sample is when the aircraft carries a maximum takeoff weight (MTOW), that is 351,533 kg, with a maximum takeoff thrust of 513.9 kN. The other three samples were from real flight data that was obtained from one of Indonesian airline's fleet with three different flight. The last sample is when the aircraft performs a 25% of derated takeoff thrust with an assumed takeoff weight based on the regression method from the other four data. 25% is the maximum reduction thrust allowed for an aircraft based on the FAA's Advisory Circular 25-13 [4].

The calculation took place at the rotation speed (V_R), where aircraft have gained enough momentum and speed to lift off. Table 1 displays the data of the aircraft in V_R during takeoff phase with the information of takeoff weight (TOW), derated takeoff thrust, EGT, and the outside air temperature (OAT) during that time. The ambient pressure is set constant for this calculation since the three data was taken from the same departing runway and the same flight destination that is Jakarta (CGK) to Jeddah (JED).

Table 1. Flight data of the aircraft during takeoff phase (V_R)

TOW	TO Thrust	Derate Thrust	TOW	TO Thrust	V_R			EGT		OAT
kg	kN	%	%	%	KTAS	M/S	M	^{0}C	K	^{0}C
351533	513.9	0	100	100	191	98.25	0.286	1090	1363.15	-
318167.8	464.63	9.5876	90.5	90.4	177.22	91.16	0.266	995	1268.15	31.5
294798.8	423.94	17.5061	83.86	82.49	167.49	86.16	0.251	954	1227.15	34.2
294508.5	416.27	18.9977	83.78	81	165.66	85.23	0.248	947	1220.15	32.95
282148.3	385.43	25	80.26	75	154.46	79.45	0.232	937	1210.96	-

The Brayton cycle calculation is based on the design of a typical high bypass ratio turbofan engine [5]. The changes of temperature and pressure of the airflow are divided into 10 sections as shown in Figure 2 followed by the information of the airflow in Table 2. Various values can be calculated from the pressure and temperature changes, and variables from Table 1. Those values are: Air-to-Fuel ratio (AFR), Jet Velocity, Fan Velocity, Thrust, Thrust-Specific Fuel Consumption (TSFC), Work Net, Heat Addition (Qin), Thermal and Propulsive Efficiency [6].

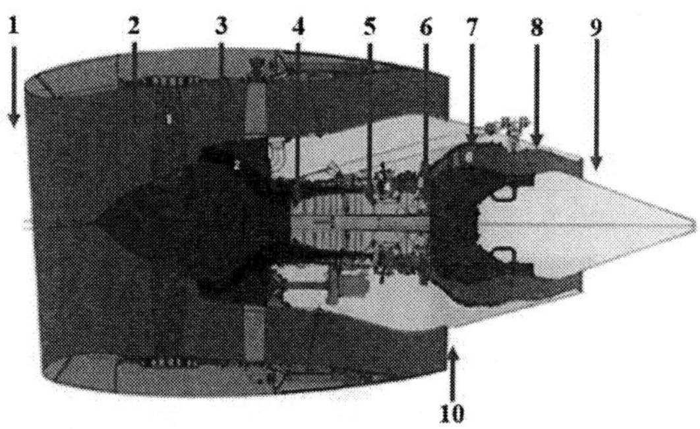

Figure 2. Cross-Sectional View of High Bypass Ratio Turbofan Engine [5]

Table 2. Section of Turbofan Engine

Stage	Flow
1	Ambient
2	Diffuser
3	Fan Out
4	LPC Out
5	HPC Out
6	CC Out
7	HPT Out
8	LPT Out
9	Nozzle Hot Core
10	Nozzle Fan

The calculation was done on a spreadsheet and the Brayton cycle calculation is specific for the typical High Bypass Ratio Turbofan Engine model, referencing from A.F. El-Sayed's formula for Forward Fan Unmixed Double-Spool Configuration [6].

After the calculation is done and the necessary values are calculated, the next step is to calculate a new set of takeoff thrust from a different set of air inlet temperature that will act as the 'Assumed Temperature' but with a constant EGT from four takeoff weight variations. The goal of this method is to find a linear line of EGT that could explain the relationship between temperature and thrust.

3. Results and Discussion

The results from the Brayton cycle calculation are presented in table 3. The results are determined based on a collective set of values that have been previously mentioned. A full rated thrust of the particular aircraft is 513.9 kN, so there is no way that the thrust will be exceeding that number. However, in this case, since the goal is to find a linear EGT line, a hypothetical result is necessary.

Table 3. Thrust, TSFC, Thermal and Propulsive Efficiency from a different set of Air Inlet Temperature

MTOW; EGT = 1090 °C						
Air Inlet Temperature (°C)	15	20	25	30	35	40
Thrust (kN)	521.5275	518.9875	516.4379	513.9	511.3092	508.73
TSFC (Kg/kN Hr)	200.382	200.324	200.268	200.430	200.167	200.122
Thermal Efficiency (%)	26.538	27.227	27.928	28.711	29.364	30.101
Propulsive Efficiency (%)	30.219	30.319	30.421	30.523	30.627	30.732
TOW = 318168 kg; EGT = 995 °C						
Air Inlet Temperature (°C)	20	25	31.5	35	40	45
Thrust (kN)	470.5214	467.9671	464.63	462.8253	460.2377	457.6387
TSFC (Kg/kN Hr)	191.725	191.680	191.627	191.602	191.569	191.540
Thermal Efficiency (%)	30.503	31.314	32.391	32.981	33.838	34.711
Propulsive Efficiency (%)	29.775	29.886	30.033	30.112	30.228	30.345
TOW = 294799 kg; EGT = 954 °C						
Air Inlet Temperature (°C)	20	25	30	34.2	40	45
Thrust (kN)	430.8954	428.4565	426.0066	423.94	421.0731	418.5892
TSFC (Kg/kN Hr)	187.018	186.971	186.929	186.897	186.859	186.830
Thermal Efficiency (%)	32.094	32.961	33.845	34.601	35.665	36.602
Propulsive Efficiency (%)	28.982	29.097	29.212	29.311	29.448	29.569
TOW = 294508 kg; EGT = 947 °C						
Air Inlet Temperature (°C)	20	25	32.95	35	40	45
Thrust (kN)	422.5312	420.1225	416.27	415.272	412.83	410.3764
TSFC (Kg/kN Hr)	186.172	186.125	186.060	186.045	186.012	185.983
Thermal Efficiency (%)	32.381	33.258	34.689	35.065	35.995	36.944
Propulsive Efficiency (%)	28.818	28.933	29.118	29.167	29.286	29.407

The increasing temperature of the air inlet will result in a smaller thrust, smaller TSFC, higher thermal, and propulsive efficiency. When the air inlet temperature is increased, the air-to-fuel ratio is decreased due to the small temperature difference in the inlet and the outlet of the combustion chamber. Therefore, the fuel that is injected to the combustion chamber is smaller than the ratio of air, creating a smaller thrust and more efficient fuel consumption. Less fuel injected to the combustion chamber means less heat energy is added to the system. When the air inlet temperature is increased, it increases all temperature at each section inside of the engine. The compressor work will increase, and the turbine will also receive higher work, resulting in a higher total work (Wnet). Consequently, when the work done is getting higher and higher while the heat supplied to the system is getting smaller and smaller, the thermal efficiency increases. For the propulsive efficiency, a higher air inlet temperature will result in a slower jet velocity. The slower the jet velocity, the higher the propulsive efficiency.

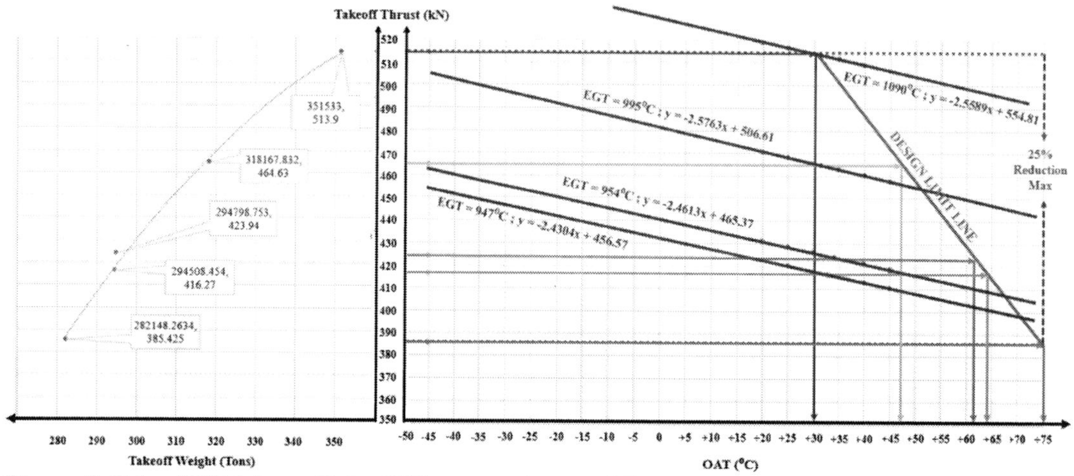

Figure 3. Takeoff Weight vs. Takeoff Thrust vs. Assumed Temperature allowable to perform reduced thrust or derate [7]

Figure 3 is a combined visual representation of Table 1 and Table 3. The left side of the chart presents the relationship between the Takeoff Weight (x-axis) and the Takeoff Thrust (y-axis), while the right side of the chart presents the relationship between the Takeoff Thrust (y-axis) and the OAT (x-axis) that can be used as a reference to choose an "Assumed Temperature" that will be inputted to the Flight Management Computer (FMC). When performing a full rated thrust, the OAT must be 30°C or below to guarantee its maximum thrust.

The way this chart is read is by starting it from the Takeoff Weight axis, decide how much payload the aircraft will be carrying, then pull a straight line until it touches the curved line, then pull a perpendicular line to the Takeoff Thrust. That is how much thrust required for the aircraft to fly. Most of the time, airplanes do not carry a full payload, therefore, it is not necessary to deliver a full 100% of thrust. Then the next step is to choose the assumed temperature that will be inputted to the FMC by simply pull a straight line until it touches the design limit line, then pull a perpendicular line downward, and that is the assumed temperature of the air inlet temperature.

These assumed temperature values are the upper limit to be inputted to the FMC. External factor such as wind speed and wet surface runway can be a consideration for the pilot to decide a lower assumed temperature than the values that have been mentioned above since it is a matter of safety issue.

An additional line is added and that is the EGT line. EGT line is obtained from the Brayton cycle calculation on the engine during takeoff, however, the value of EGT is set constant while the air inlet temperature that acted as the assumed temperature is varied. This linear line can be used similar to the design limit line, simply pull a straight line from the Takeoff Thrust axis, until it reaches the line then pull a straight line down to the value of the outside air temperature.

Traditionally, the selection of the assumed temperature is higher than the actual air temperature on the runway. Therefore, choosing an assumed temperature lower than the actual air temperature is very unlikely in the day-to-day operation and has no benefit whatsoever to improve the performance of the aircraft during takeoff.

IOP Conf. Series: Materials Science and Engineering **685** (2019) 012007 doi:10.1088/1757-899X/685/1/012007

4. Conclusion

The plotted graphic of the analysis results shows how the effect of reduced takeoff thrust lowers temperature level at each section, especially EGT. By performing reduced takeoff thrust, it also decreasing TSFC, making it more efficient. Performing reduced takeoff thrust does not necessarily mean lowering the performance of the engine, instead, it improves it by making it more effective. Therefore, inputted power is delivered base on how much power required. Finally, it is undeniably beneficial for long-term usage as engine deterioration is retarded. The plotted graphic can be used as a guide for the pilot in determining the assumed engine inlet temperature for inputting the FMC.

Acknowledgement

The authors would like to thank PITTA grant programme of the Universitas Indonesia for funding this research with grant number NKB-0796/UN2.R3.1/HKP.05.00/2019 also PT. GMF AeroAsia for the facility, chance, and guidance for the development of this research.

Reference

[1] Qiang, F., Ding, F. (2014). Takeoff Performance Analysis Based on Aeroengine EGT Trend Monitoring and Control. ISSN: 1662-7482, Vol. 680, pp 307-310 2014 Trans Tech Publications, Switzerland.

[2] Sunara, L. L, Nirbito, W. (2018). An Estimation of the Maximum Take-Off Weight Commuter Category Aircraft Based on Airstrip Condition on Remote High Terrain Area, ATCES 2018 Shanghai ROC, IOP Conference Series: Materials Science and Engineering.

[3] Chenghong, Y. (2002). Reduced Trust Takeoff. ICAS 2002 CONGRESS.

[4] Administration, F. A. (1988). *Reduced and Derated Takeoff Thrust (Power) Procedures AC No. 25-13.* Retrieved from Federal Aviation Administration, US Department of Transportation.

[5] Benarwa, S. & Wang, J. (2016). Dynamic Modeling and Simulation on G90 Engine. The International Journal of Engineering and Science (IJES), Volume 5, Issue 12, 111-119.

[6] El-Sayed. A.F. (2016). Fundamentals of Aircraft and Rocket Propulsion. London: Springer.

[7] Daidzic, N. E. (2012). Jet Engine Thrust Ratings. Retrieved from AAR Aerospace Consulting, LLC: *https://www.researchgate.net/publication/316515941*

Design Analysis of Ducted Propeller for Bicopter Drone Propulsion

M R Priatmoko[1]and W Nirbito[1,*]

[1]Department of Mechanical Engineering, Universitas Indonesia, Depok, West Java, 16424, Indonesia

*Corresponding author: wahyu.nirbito@ui.ac.id

Abstract. Drones nowadays have become an important part of people's lives. Drones for transporting people or goods are considered to develop into reality. To be effective and efficient, the drone should be light but have big propulsion for big lift and thrust forces. This paper presents a way to improve the performances by designing usage of only 2 propulsion motors with ducted and special fan blades. The special designed fan blades theoretically will produce more thrust with smaller propeller size and lower RPM since combine the lift blade and momentum thrust effects. The required propulsion force to fly the drone with its payload is analyzed and detemined. Then the torque, rpm, airfoil type, Thrust Coefficient, and blade twist angles can be determined based on that propulsion force. Afterward, an evaluation is performed with fluid dynamics simulation tools to predict the forces acting on the propeller.

1. Introduction

Unmanned Aerial Vehicle (UAV), commonly called a drone, is an aircraft system that is being flown from afar. Ranging from simple hand-operated short-range systems to durable and high-altitude systems. If previously the drone was only used for a variety of needs that were military or reconnaissance, currently drones were found for other activities such as photography, regional mapping and even for entertainment media [1].

Indonesia currently needs to improve their air transportation sector. Especially on the logistics distribution. The biggest problem that Indonesia facing now in terms of air transportation is limited space. It is difficult to find a suitable place for aircraft runways, especially in remote areas [2]. To overcome this, UAVs or drones can be used because of their flying mechanism that can do VTOL (Vertical Take-Off and Landing).

The development of UAV leads to the solution of two main problems, namely increasing battery life and increasing payload [3]. To overcome this problem, it can be done by making propellers that have huge thrust and high efficiency. To get a large thrust force, a duct can be used which is believed to increase thrust by 94% [4] as well as using special designed blades. Better battery capability and larger payload are increased by reducing the number of propeller drones to 2 motors or commonly called as bicopter, but with the consequences cost of uneasy stability control. The current bicopter aircrafts, i.e. Chinook and V22-Osprey, all are operated nicely wit their heavy weights but still with some technology secrets.

The propulsion thrust is produces from a combination of momentum theory and lift force blade such as rotary wing. The so called Blade Element Momentum Theory is implemented in the design analysis that is determining the designed ducted propeller blades, which produce big thrust and suitable for flying the bicopter drone.

2. Methodology

The model is based on the approximation calculated thrust force that is required by the drone. The dimension of the drone is designed to be of 45.8cm x 30 cm x 15 cm. With the weight of 1.75 kg, the thrust to weight ratio is 2: 1, and the total drag force is predicted based on calculation of 0.0715 N. So the total thrust required is 34.33 N, or 17.16 N for each propulsion motor. The dimensions and shape of the design of the drone can be seen in the illustration shown in figure 1.

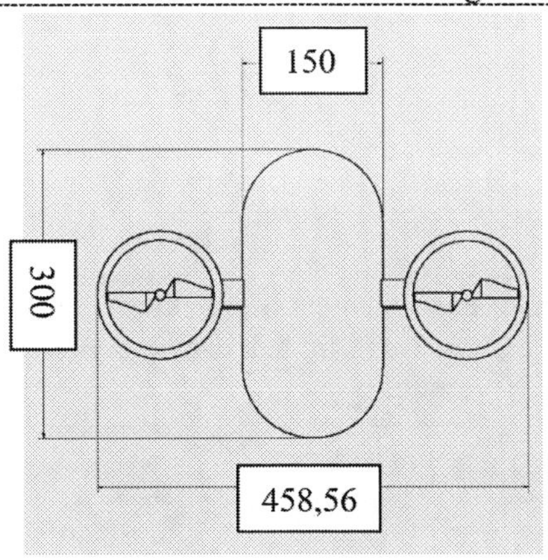

Figure 1. Drone Dimensions.

The total drag is devided into 2 section which is body drag and propeller drag. We took a note on the body with the exposed surface area A, drag coefficient Cd, take off velocity V and air density ρ. Then, we can define the body drag [4]:

$$D = \frac{1}{2}\rho V^2 ACd \tag{1}$$

To calculate the propeller drag, first we need to determine the resultant velocity W with the velocity triangle theory. To do so, we used the Blade Element Momentum Theory to calculate the local forces on a propeller blade. It uses the Blade Element Theory. We focused our attention on a propeller with radius R, rotating at an angular speed Ω, With air velocity at the inlet V_{in}, and tangential velocity U [5]:

$$U = \Omega R \tag{2}$$

$$W = \sqrt{U^2 + {V_{in}}^2} \tag{3}$$

Then, we can get the propeller drag with [6]:

$$D = \frac{1}{2}\rho W^2 Rc(r)Cd \tag{4}$$

Where c(r) is the chord of the airfoil.

The elementary thrust and torque can be obtained alternatively from the lift and drag acting on the blade element as below [7]:

$$dT = Bc\,\frac{1}{2}\rho W^2 R(Cl\,cos\phi\; - \;Cd\,sin\phi)dr \tag{5}$$

$$dQ = Bcr\frac{1}{2}\rho W^2 R(Cl\,sin\phi\; + \;Cd\,cos\phi)dr \tag{6}$$

IOP Conf. Series: Materials Science and Engineering **685** (2019) 012008 doi:10.1088/1757-899X/685/1/012008

where B is the number of blades, W is the resultant velocity with reference to the blade and ϕ is the angle of the vector W with the plane of rotation of the propeller, as proposed by Houghton et al.

Half of the blade thrust is created by momentum theory because blower blade type is also used in this design to uniform the lift distribution. Which is [8]:

$$T = \dot{m}(W - V_{in}) \tag{7}$$

Where $\dot{m}$ is the mass flow rate through the cross sectional area.

Comparisons can be made between the characteristics of open and ducted propellers, as a function of the ducted propeller's expansion ratio which is the comparison between duct exit area and rotor area. By holding any two variables fixed between the two configurations, the behavior of the other variables can be analyzed [9].

$$\frac{T_{op}}{T_{dp}} = (2\sigma_d)^{\frac{1}{3}} \tag{8}$$

Where T_{op} is open propeller's thrust, T_{dp} is ducted propeller's thrust and σ_d is the expansion ratio.

2.1. Calculation

The calculation of the blade using Blade Element Momentum Theory, are divided into 3 parts, namely: root, MAC (Mean Aerodynamic Chord), and tip that can be seen in the illustration (Figure 2a, 2b, 2c), along with the speed triangle that it produces.

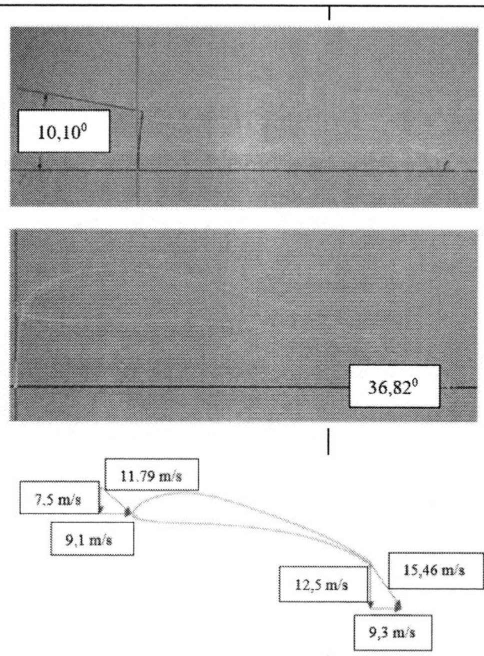

Figure 2a. Angle and Velocity Triangle at Root

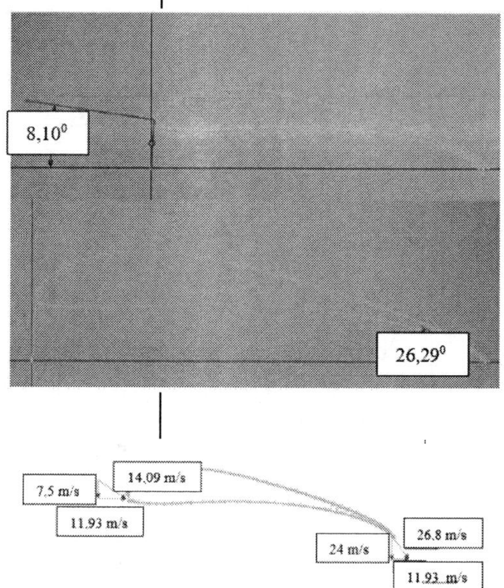

Figure 2b. Angle and Velocity Triangle at MAC

IOP Conf. Series: Materials Science and Engineering **685** (2019) 012008 doi:10.1088/1757-899X/685/1/012008

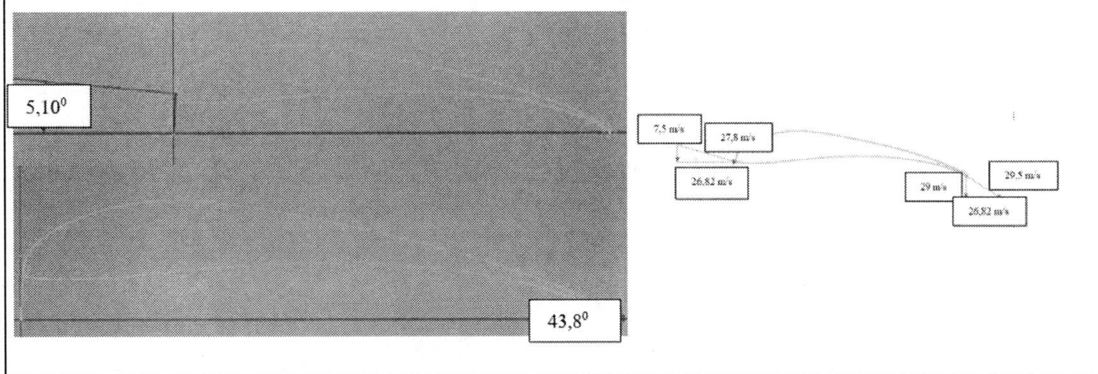

Figure 2c. Angle and Velocity Triangle at Tip

The airfoil type Selig S1223 was being modified a little bit to accomplish the desired angle. As stated at the introduction, the thrust from the propeller is a combination between lift that is produced by the rotary wing which is from the leading edge to a certain point around the trailing edge, and momentum theory that is produced by the blower blade shape that is around the trailing edge. That is why the angle at leading edge and trailing edge is different.

2.2. 3D Modelling & Simulation.

The 3D modelling is done by using Autodesk Inventor as the software. Selig S1223 with the characteristic shown in figure 3, is being used in designing this propeller. With the consideration of Reynolds number, alpha, and Lift coefficient needed by the propeller. and the airfoil coordinate to design the propeller can be obtained from the UIUC Airfoil Coordinate Database. The solid model then become an input on ANSYS Fluid Dynamics (CFX) Simulation for predicting the thrust and the torque that being produced by the propellers.

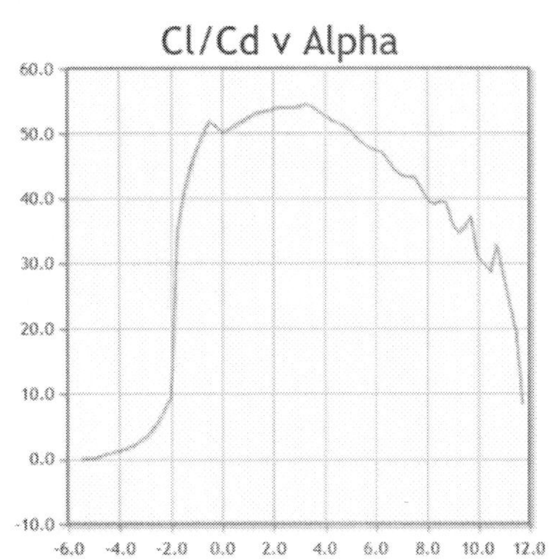

Figure 3. Airfoil characteristic of Selig S1223

The boundary conditions (Figure 4) that being used was set according to daily air condition in Jakarta.

Domain Interface - Domain Interface 1	
Boundary List1	Domain Interface 1 Side 1
Boundary List2	Domain Interface 1 Side 2 1
Interface Type	Fluid Fluid
Settings	
Interface Models	General Connection
Frame Change	Frozen Rotor
Mass And Momentum	Conservative Interface Flux
Pitch Change	Specified Pitch Angles
Pitch Angle Side1	3.6000e+02 [degree]
Pitch Angle Side2	3.6000e+02 [degree]
Mesh Connection	GGI
Domain Interface - Domain Interface 2	
Boundary List1	Domain Interface 2 Side 1
Boundary List2	Domain Interface 2 Side 2 1
Interface Type	Fluid Fluid
Settings	
Interface Models	General Connection
Frame Change	Frozen Rotor
Mass And Momentum	Conservative Interface Flux
Pitch Change	Specified Pitch Angles
Pitch Angle Side1	3.6000e+02 [degree]
Pitch Angle Side2	3.6000e+02 [degree]
Mesh Connection	GGI
Domain Interface - Domain Interface 3	
Boundary List1	Domain Interface 3 Side 1
Boundary List2	Domain Interface 3 Side 2 1
Interface Type	Fluid Fluid
Settings	
Interface Models	General Connection
Frame Change	Frozen Rotor
Mass And Momentum	Conservative Interface Flux
Pitch Change	Specified Pitch Angles
Pitch Angle Side1	3.6000e+02 [degree]
Pitch Angle Side2	3.6000e+02 [degree]
Mesh Connection	GGI

Domain - Default Domain	
Type	Fluid
Location	B42
Materials	
Air at 25 C	
Fluid Definition	Material Library
Morphology	Continuous Fluid
Settings	
Buoyancy Model	Non Buoyant
Domain Motion	Stationary
Reference Pressure	1.0000e+00 [atm]
Turbulence Model	k epsilon
Turbulent Wall Functions	Scalable
Domain - Domain 1	
Type	Fluid
Location	B101
Materials	
Air at 25 C	
Fluid Definition	Material Library
Morphology	Continuous Fluid
Settings	
Buoyancy Model	Non Buoyant
Domain Motion	Rotating
Angular Velocity	-5.7000e+03 [rev min^-1]
Axis Definition	Coordinate Axis
Rotation Axis	Coord 0.2
Reference Pressure	1.0000e+00 [atm]
Turbulence Model	k epsilon
Turbulent Wall Functions	Scalable
Domain Interface - Domain Interface 1	
Boundary List1	Domain Interface 1 Side 1
Boundary List2	Domain Interface 1 Side 2 1
Interface Type	Fluid Fluid

Figure 4. Boundary Condition

3. Results and Discussion

The calculation has been done and so the result of the blade and duct design can be seen in figure 5a & 5b.

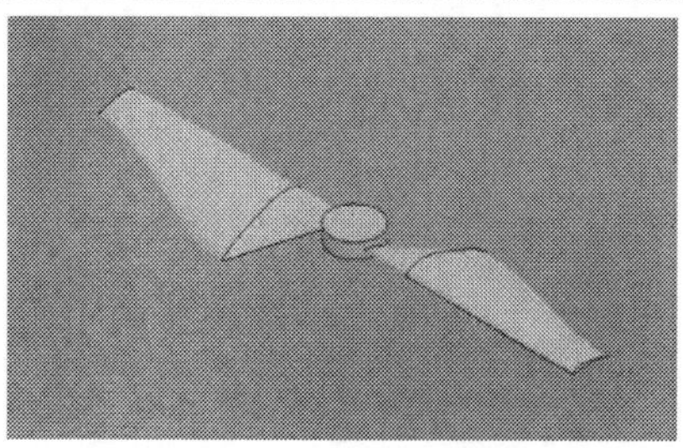

Figure 5a. Blade Design

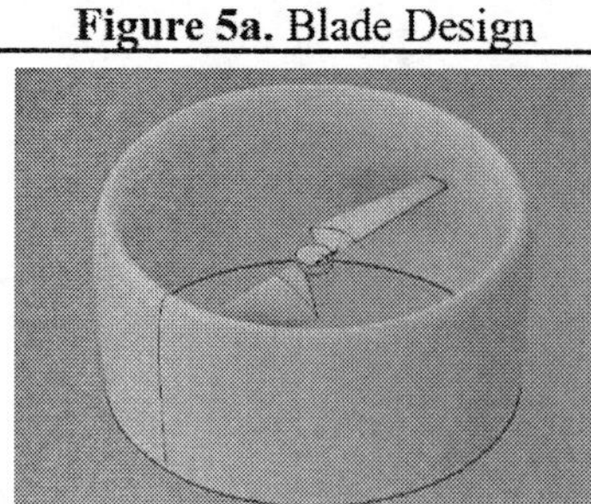

Figure 5b. Duct Design

Based on the initial calculation, one ducted propeller should be able to generate 18,858 N of thrust at 5700 RPM which is the minimum required thrust for the drone to hover. The gyroscopic effect working on the drone should be neutralized by the propeller one another.

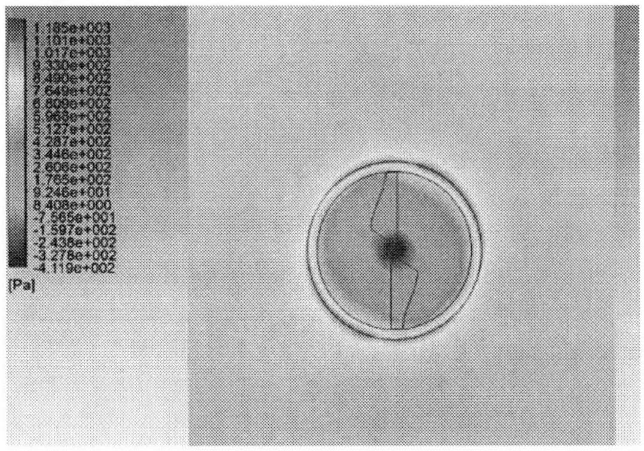

Figure 6a. Pressure Contour

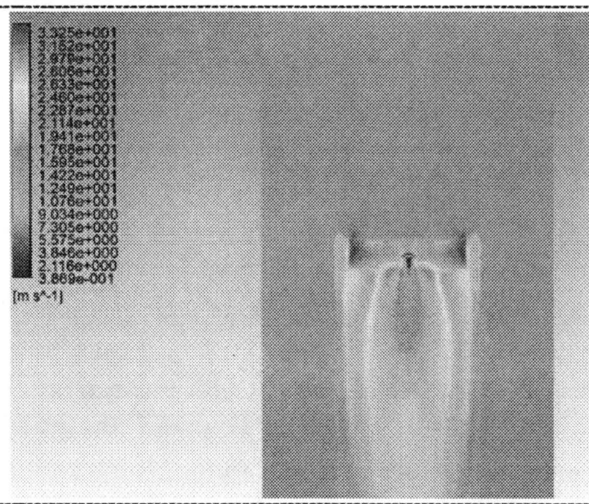

Figure 6b. Velocity Contour

It can be seen from figure 6a and 6b, that the biggest pressure occurs at the tip. This shows that it is true duct reduces tip losses so that it can produce maximum thrust, as should be the case where the biggest thrust is generated in the tip section. So that it will produce a thrust of 15,657 N. This result differs quite far from the expectations of manual calculation which is equal to 18,858 N. So that it gets an error of 15%. However, this is still in accordance with the Blade Element Momentum Theory error, which is an error that is still included in the range of 29%.

Can be seen in the illustration (Figure 2a, 2b, 2c), the angle is made into 10.10^0, 8.10^0, and 5.10^0 at the trailing edge (root, MAC, tip respectively) and $36{,}82^0$, 38^0, and $43{,}8^0$, at the leading edge. So that it produces a speed of 12.5 m/s, 24 m/s, and 29 m/s. The results of this manual calculation are still included in the speed value range in the simulation results as can be seen in the Velocity Contour illustration (Figure 6b).

4. Conclusion

a) Bicopter can also be used as a alternative drone type for everyday use just like the current model on the market, i.e. quadcopter.

b) The design verification of the resultant speed between theoretical calculation and computer simulation is not converged yet. 15% difference is merely due to inacuracy of the number of calculation points on blade, i.e. root-MAC-tip, and less optimum number of the meshing nodes as well. However this is relatively adequate for developing the capacity and performance improvements for scalng up a prototype.

c) Bicopter drones are most suitable for point to point air transportation of passengers or goods, since with only 2 propulsion motors the maneuverability of bicopter is more difficult.

Acknowledgment

The authors would like to thank PITTA grant programme of the University of Indonesia for funding this study with contract number NKB-0796/UN2.R3.1-HKP.05.00/2019.

References

[1] Fall, A. (2018). *Assistive drone technology: Using drones to enhance building access for the ph ysically disabled* (Order No. 11005335). Available from ProQuest Dissertations & Theses Globa l.(2129648180). Retrieved from https://remote lib.ui.ac.id:2155/docview/2129648180?accountid =17242

[2] Sunara, L .L, Nirbito, W. (2018). *An Estimation of the Maximum Take-Off Weight Commuter Category Aircraft Based on Airstrip Condition on Remote High Terrain Area,* ATCES 2018 Shanghai ROC, *IOP Conference Series: Materials Science and Engineering*
[3] Andria, G., Nisio, A. D., Lanzolla, A., Spadevecchia, M., Pascazio, G., Antonacci, F., & Sorrentino, G. M. (2018). Design and Performance Evaluation of Drone Propellers. *2018 5th IEEE International Workshop on Metrology for AeroSpace (MetroAeroSpace).*
[4] Pereira, and Jason Louie. "Hover and Wind-Tunnel Testing of Shrouded Rotors for Improved Micro Air Vehicle Design." Digital Repository at the University of Maryland, 29 Aug. 2008, drum.lib.umd.edu/handle/1903/8752.
[5] Andria, G., Nisio, A. D., Lanzolla, A., Spadevecchia, M., Pascazio, G., Antonacci, F., & Sorrentino, G. M. (2018). Design and Performance Evaluation of Drone Propellers.
[6] Andria, G., Nisio, A. D., Lanzolla, A., Spadevecchia, M., Pascazio, G., Antonacci, F., & Sorrentino, G. M. (2018). Design and Performance Evaluation of Drone Propellers.
[7] Andria, G., Nisio, A. D., Lanzolla, A., Spadevecchia, M., Pascazio, G., Antonacci, F., & Sorrentino, G. M. (2018). Design and Performance Evaluation of Drone Propellers.
[8] Nagpurwala, Q. H. "Ducted Fans and Propellers." PEMP RMD 2501
[9] Pereira, and Jason Louie. "Hover and Wind-Tunnel Testing of Shrouded Rotors for Improved Micro Air Vehicle Design." Digital Repository at the University of Maryland, 29 Aug. 2008, drum.lib.umd.edu/handle/1903/8752.

IOP Conf. Series: Materials Science and Engineering **685** (2019) 012009 doi:10.1088/1757-899X/685/1/012009

A Design Method for Twist Distribution along Wing Span

Xiaoyan Liu[1,a], Pan Cheng[1,b] and Yang Zhao[1,c]

[1]Flight Physics Department, Shanghai Aircraft Design And Research Institute, COMAC, No.5188 Jinke Road, Pudong District, Shanghai, China

[a]liuxiaoyan@comac.cc, [b]chengpan@comac.cc, [c]zhaoyang2@comac.cc

Abstract. A design method for wing jig shape twist is proposed. Target cruising aerodynamic loads are applied to structure model, and the deformation converge gradually using predictor-correction displacement iterative method. When ensuring the relative twist distribution in the streamwise section, a wing jig shape with straight leading ledge is created, then an aeroelastic coupling iteration is started to achieve the deformed elastic cruising shape. Taking a large aspect ratio swept wing as an example, analysis results show that the deformed elastic cruising wing had an identical streamwise twist distribution compared with rigid designed cruising shape. Sectional lift distribution along with pressure distribution matches well with that of the designed cruising shape, and the drag gap among them is not significant. There might be some change in shock wave chord position between elastic cruising wing and designed cruising wing for outboard wing section. By slight adjusting twist with unchanged airfoil profile, the difference above could be improved.

1. Introduction

Large aspect ratio wings are widely used in modern civil aircraft, which can cruise at high transonic Mach number. Due to the influence of the static aeroelasticity of the structure, the wing deformation includes deflection and torsion along the span, and the geometric twists change accordingly, which will cause the redistribution of aerodynamic load [1, 2]. The aerodynamic characteristics of aircraft will change obviously due to the effect of aeroelastic deformation. The conventional design flow of civil aircraft wing is to obtain the optimal aerodynamic shape designed at certain cruising point through aerodynamic optimization [3]. Based on the designed cruise shape, the design of the jig shape used for manufacture can be carried out, so that the aerodynamic performance of the cruise design state can be restored. The static aeroelastic effects on elastic deformation and load redistribution are considered in jig shape design [4].

In this paper, a predictor-corrector displacement iterative method is proposed. The aerodynamic load of rigid cruising design state is obtained by N-S equations. The load is applied to the structural model and the calculated displacements are reversed to put into the original structure model. After the twist solution is obtained by iterative solution until the displacement converges, the jig shape is created. The twist distribution, load distribution, and pressure distribution are compared between the elastic cruising shape deformed from the jig shape and the rigid designed cruising shape. There may be differences in the position of the shock wave in sections along the wing span near the wingtip, and the twists are slightly changed to improve the difference of the shock wave in the chord without changing the airfoil.

2. Technical methods

2.1 Predictor-Corrector displacement iterative method

The design cruising shape has a straight leading edge, but is twisted in each section to achieve the desired pressure distribution. The jig shape is obtained by changing the twists of the designed cruising wing.

The aerodynamic distribution of the designed cruising shape is taken as the target design load, which would be calculated based on N-S equations. The aerodynamic load is extracted and interpolated to the finite element model using an integral method. The structural deformation is obtained by solving the static equilibrium equation (1).

$$Kq = F \tag{1}$$

Where, K is a structure stiffness matrix; q denotes finite element node displacement, F is a load vector. The main procedures are as follows: a) Prediction process: the cruising design load is applied to the finite element model (abbreviated as FEM), thus the wing is accordingly deformed, and then the displacements are reversed to deform the original model so as to produce a new shape; b) Correction process: When applying the target load to the new shape from the previous step, the model is deformed to a new position again, and compared it with the initial FEM model. If the location difference is larger than the tolerance, this iterative process will continue until convergence.

The initial jig shape obtained by the iterative process above is bent down, and the wing would not be designed to be bent because of the requirements from manufacturing and assembly process. The wing leading edge of a jig shape is needed to maintain the same straight line as that of the designed cruising shape. The streamwise twists of the swept wing are affected by both the bending and torsion of the wing. Therefore, it is necessary to measure the geometric twist angle distribution of the initial jig shape and the designed cruising shape in the streamwise direction. The difference between them represents the streamwise twist increment caused by the target aerodynamic load. When the relative twist distributions along wing span are applied to the designed cruising shape, the modified jig shape can be achieved by ensuring the same straight leading edge [5]. Predictor-Corrector design flow for entire jig design is shown in figure 1.

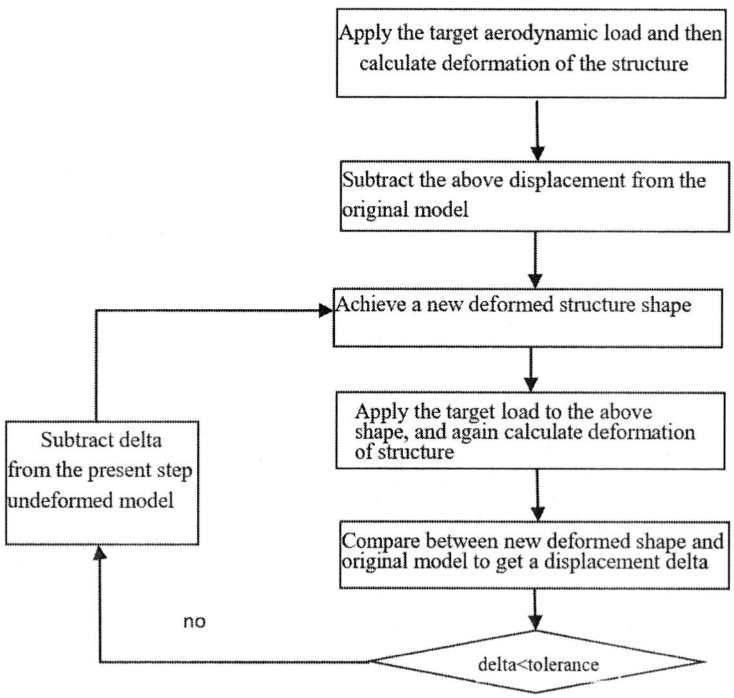

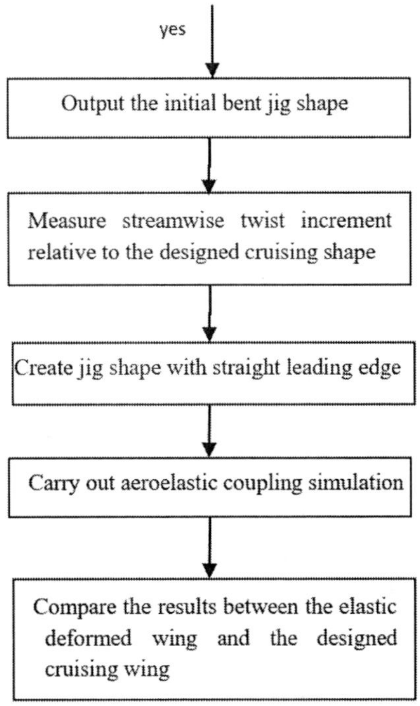

Figure 1. Predictor-Corrector design flow.

2.2 shock position adjustment method

The objective of this section is to study whether additional design iterations can produce better shock position matching. Firstly, the function between the shock wave position and the wing twist is created. In order to simplify the analysis, the wing twist increment is replaced by the angle of attack increment. The angle of attack sweep for the rigid designed cruising shape is calculated by CFD, and the change of shock position is measured. These examples are used to create the ratio of the chord position of the shock wave to the angle of attack, as shown in the Equation (2).

$$k = \frac{\Delta x}{C}\left(\Delta alpha\right)^{-1} \tag{2}$$

Where , k is a shock wave position ratio, Δx is a shock chord position difference, C is a local chord length, $\Delta alpha$ is an angle of attack increment.

Then, the chord position difference between the elastic cruising wing and the designed cruising wing is measured, and the twist increment is calculated along the wing span. Finally, the twist angle increment is applied to the original jig shape. By CFD/CSD aeroelastic coupling simulation, a new elastic cruising shape is obtained, which is compared with that of the rigid designed cruising shape. The shock position adjustment process is as shown in the figure 2.

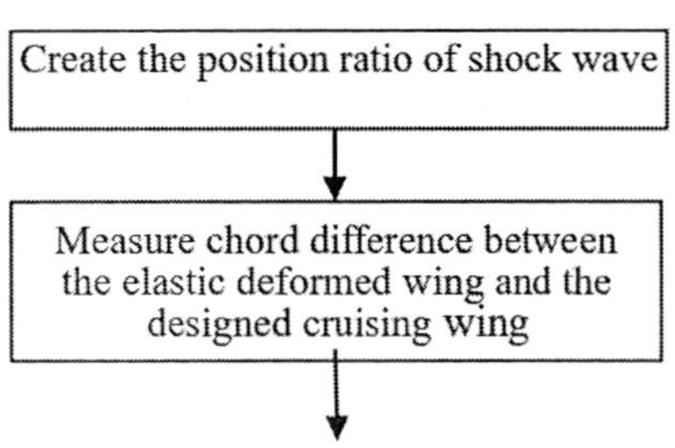

IOP Conf. Series: Materials Science and Engineering **685** (2019) 012009 doi:10.1088/1757-899X/685/1/012009

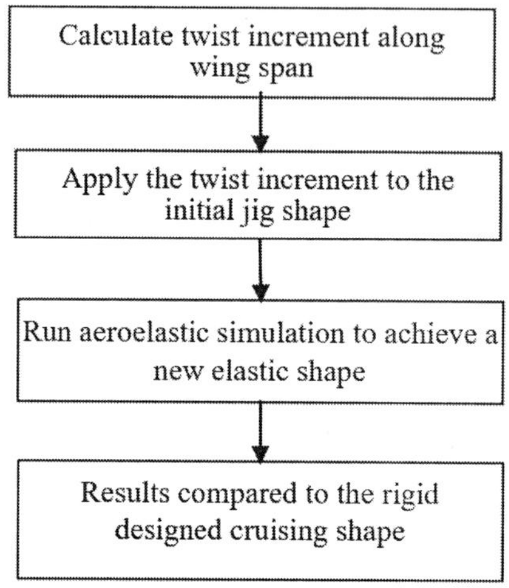

Figure 2. Shock position adjustment process.

3. An example of application

Taking a large aspect ratio swept wing as an example, the jig shape is designed. The finite element model adopts a shell-rod model, and the aerodynamic model is dispersed into structured grids. The cruise Mach number is 0.84 and the lift coefficient is 0.49. The constant lift calculation is used to ensure that the aircraft is in the same flight condition, and the angle of attack of the elastic wing is changed so that the total wing lift can meet the design requirements in static aeroelastic balance.

The aeroelastic effect of the swept wing in flight results in a negative additional twist along wing span, which makes the effective angle of attack decrease and the actual lift become smaller. In order to meet the design requirements of flight characteristics, it is necessary to design a positive pre-twist angle for jig shape, so that the flight quality of an elastic wing is not affected by aeroelastic deformation.

After the wing jig shape is created, the aeroelastic coupling iteration is started until the result converges, and the deformation of the wing reaches a stable state, called as the elastic cruise shape. The comparison of different shape is shown in figure 3, as seen the elastic cruising shape is above the designed cruising shape. The streamwise twist distribution of the elastic cruising wing is compared with that of the designed cruising shape. As shown in figure 4, they have almost the same twist distributions, the difference is less than 0.03degs.

Figure 3. Comparison of the elastic cruising shape and the designed cruising shape.

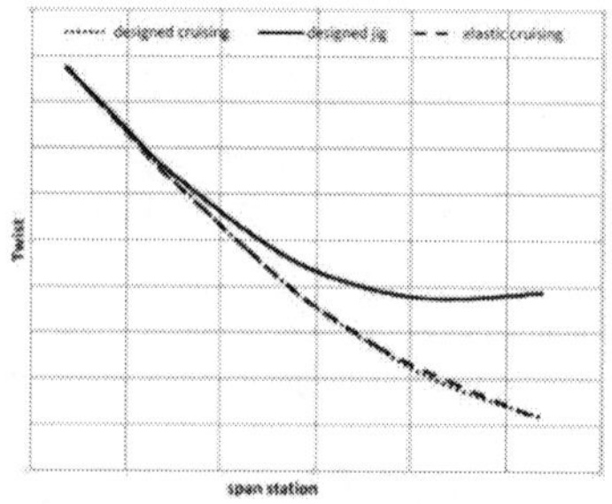

Figure 4. Comparison on twist under different wing shapes.

The load distribution along the wing span has a significant effect on the vortex-induced drag. Therefore, the elastic cruising shape and the designed cruising shape should have a similar span load. The sectional normal lift of both shapes are extracted along the wing span. As shown in figure 5, overall, the lift span distribution of the elastic cruising shape matches well with that of the designed cruising shape. The drag difference between them is less than 1.5 counts.

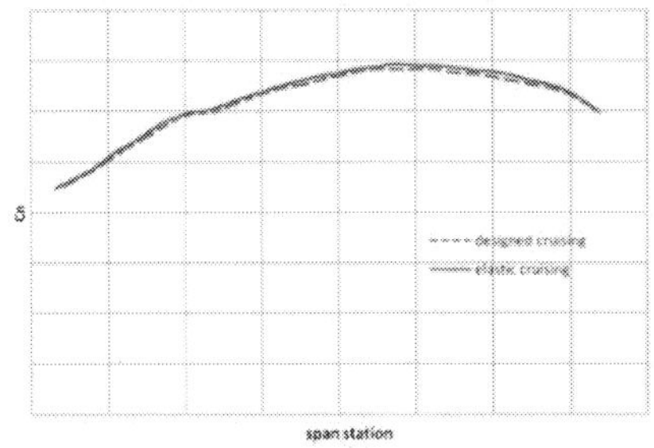

Figure 5. Sectional normal force distribution under different wing shapes.

From figure 6 to figure 7, the pressure distributions are compared. The overall pressure distribution matches closely. There is a slight difference in the position of shock waves in the outboard wing which may be due to the wing deformation resulting in interference between the spanned lift vectors.

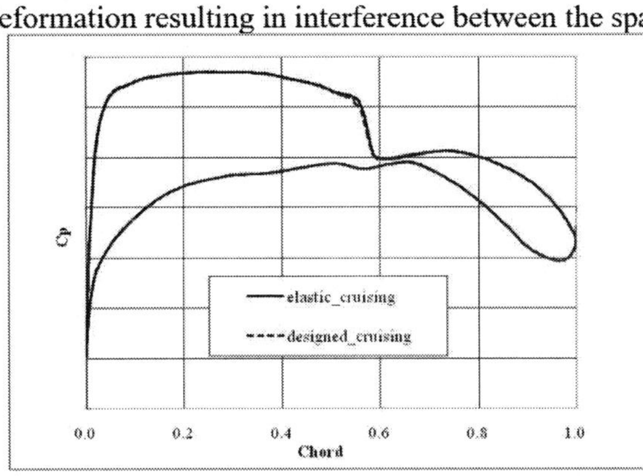

Figure 6. Comparison of the pressure distribution in wing root section.

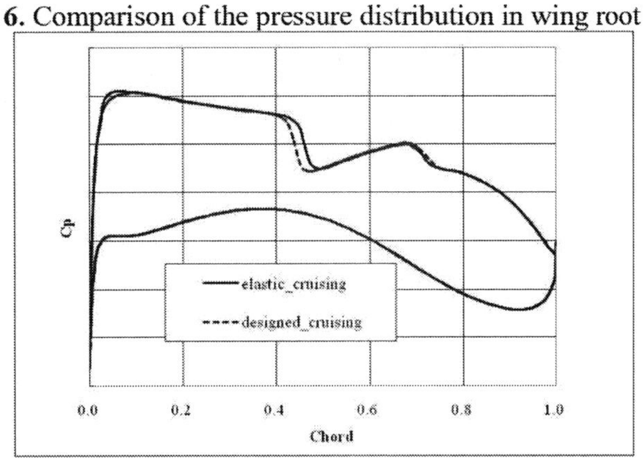

Figure 7. Comparison of the pressure distribution in the outboard wing section.

As shown in figure 7, the shock position is adjusted from the initial offset from the original designed wing chord to about 2.5%, reducing to about 1.5%. The twist of the jig shape is adjusted to achieve a slight improvement of the shock position, but the influence of the wing span load is negative, and the twist of the outboard wing is slightly reduced. Figure 8 shows a slight increase in the sectional lift of the inboard wing and a significant decrease in the outboard wing. The difference in shock position does not mean that the jig shape is improperly designed.

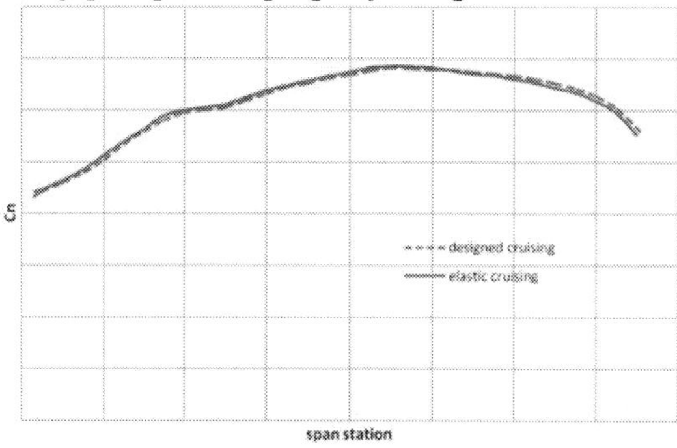

Figure 8. Cn distribution after adjustment of the shock wave chord position.

4. Conclusion

In this paper, the predictor-corrector displacement iterative method for jig shape design is developed, which can quickly obtain the jig shape that meets the requirements of engineering design. The correctness of the method is verified by an example. The streamwise twist is consistent with the designed cruise shape, and the aerodynamic characteristics of the design cruising shape are effectively restored including the lift distribution and drag characteristics. Based on the results of this paper, the slight adjustment of twist for the jig shape can improve matching of the shock wave position of the outboard wing, but the shock position and the span load can not match at the same time. This may be caused by the 3D effect of the deformable wing.

Reference

[1] Xiong Juntao, Qiao Zhide, Yang Xudong, et al. An aerodynamic shape optimization of transonic wing design method for aeroelastic sysem[J]. *Acat Aerodynamica Sinica*, 2009, 27(2): 154-159. (in Chinese)

[2] Chen Guibin, Zou Congqing, Yang Chao. Aeroelastic design basis [M], Beijing: *Beijing University of Aeronautics and Astronauics Press*, 2004. (in Chinese)

[3] Xie Meng Cheng Pan Xue Fei. Design and Analysis of Wing Jig-Shape by Different CFD Methods[J]. *Civil Aircraft Design and Research*, 2011, 4:16-17. (in Chinese)

[4] LiangQiang, YangYongnian,YeZhengyin. Analysis of Jig-Shape Design for Elastic Wing[J]. *Journal of Northwestern Polytechnical University*, 2002, 20(2):262:264 (in Chinese)

[5] Yang Guowei, Zheng Guannan. Aircraft Jig Shape Correction Method Based on Static Aeroelastic Analyses [J]. *Advances In Aeronautical Science And Engineering,* 2011, 2(2):143-150. (in Chinese)

IOP Conf. Series: Materials Science and Engineering **685** (2019) 012010 doi:10.1088/1757-899X/685/1/012010

Optimization of Heated Structures for Several Load Cases

Anatolii Kretov*

Nanjing University of Astronautics and Aeronautics, College of Aerospace Engineering, 29 Yudao St., Nanjing 210016, P.R.China

*kretov-ac@nuaa.edu.cn

Abstract. Optimization methods providing maximum strength and rigidity of the thin-walled structure are considered. A critical overview of the works related to the optimization including the heated design is done. To optimize for several cases of loading, a minimization of functional is applied, which is the sum of the total potential strain energies for each case of loading. The optimization process begins with the initial basic structure, which includes all possible variants of the location of the power elements. In the optimization process, inefficient elements are eliminated by minimizing their rigidity. Numerical optimization studies on the example of truss construction are carried out.

1. Introduction

The task of designing and obtaining the load-carrying structure of the minimum mass is paramount at all stages of the creation of the flight vehicle (FV). Typically, this process is called optimization, and it means the structure must be optimal. Under the term "optimal" in this paper we will understand the design of maximum strength and stiffness (minimum compliance) within a given mass or Vice versa – the minimization of the mass at a given strength and stiffness. Obviously, the design can be optimal only for one case of loading. In the process of functioning of the structure, it can be affected by various loads both in magnitude and in the nature of application. Moreover, these loads can be both mechanical and thermal. In this regard, the obtaining a rational design in the optimization process is very important problem.

The optimal solution of the problem is achieved primarily by choosing the optimal topology of the structure (that is, the location in a given design area of the load-bearing elements and their relationship) – this process is usually called topological optimization. Further refinement and selection of structural parameters of load-bearing elements is carried out within the selected structure (parametric optimization). Historically, these two stages of load-bearing structure design have always been carried out separately by different departments of design organizations and different specialists: designers and strength specialists. The emergence of computational and numerical methods, in particular FEM, allowed to bring these two processes closer together. However, the leading role in the process of topological optimization is left to the person with experience and intuition, which is quite difficult to formalize.

The presence of temperature effects due to significant heating or cooling can significantly complicate the process of finding the optimal structure. In the conditions of rapid development of super-and hypersonic FV, this direction requires careful research.

IOP Conf. Series: Materials Science and Engineering **685** (2019) 012010 doi:10.1088/1757-899X/685/1/012010

2. Overview of optimization metods

The methods of nonlinear mathematical programming which called the direct optimization methods are the most common to solve the problem of parametric optimization [1]. For the design variables within the specified types of FE most often accept parameters that describe the dimensions of the cross sections of finite elements (plate thickness or cross-sectional area of the rod). In this case, the optimization problem is usually described by a large number of design variables and functional constraints, estimated for real structures in hundreds of thousands, which complicates a detailed study of the search area and, consequently, obtaining accurate results. This circumstance is complicated by the fact that the methods of nonlinear mathematical programming use a very time-consuming procedure for analyzing the sensitivity of stress constraints for the set of design points for all design variables. In this regard, the use of indirect methods of optimization of elastic systems, which can have higher efficiency due to the purposeful use of the characteristic properties of the design object, is of practical interest.

As part of the indirect approach, the necessary conditions to be met by the desired design are recorded, and the procedure for finding this design is constructed. Optimality conditions are either derived from the mathematical formulation of the problem, or based on the behaviour characteristics of some classes of elastic systems [2, 3, 4]. For solving the above problem are most often used two criteria: of full-stress structure or of the greatest rigidity.

In most cases, at the initial stages of power design of thin-walled structures such as the wing, tail, body of the aircraft is most often used FEM with two types of finite elements. One-dimensional elements describe the work of shelves spars, stringers, belts ribs and frames. Two-dimensional membrane are used to describe the work of the skin, walls of longerons and ribs. The construction is called full-stress (equal-strength), if in all its elements, in which the cross section is greater than the minimum allowable value, the limiting stress state is realized in at least one of the cases of loading [4]. To search for full-stress project uses the classic formula of the stresses relationship:

$$x_i^{j+1} = x_i^j \max_c(\sigma_{\mathrm{eq}\,i\,c}) / \underline{\sigma}_i, \ c = 1, 2, \ldots, s \tag{1}$$

where x_i^{j+1} is design parameter (area for one-dimensional element A, thickness for two-dimensional element t – figure 1), $\sigma_{\mathrm{eq}\,ic}$ is the equivalent stress in the ith element in the cth case of loading, $\underline{\sigma}_i$ is the allowable equivalent stress for the ith element, s is the number of loading cases, j is the iteration number.

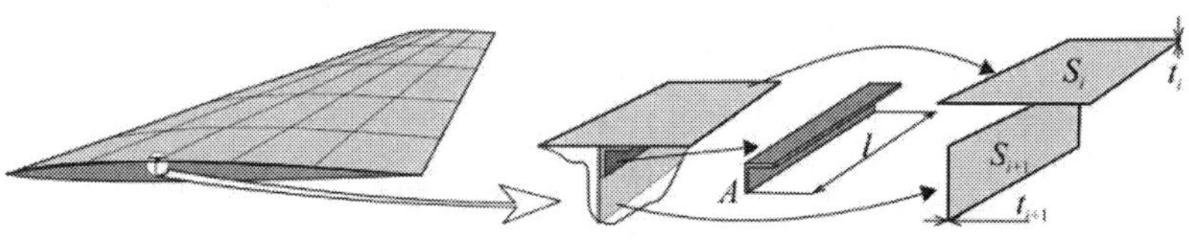

Figure 1. Structure, its fragment, and two types load-carrying elements with main parameters (one dimensional element: area A and length S=l, two dimensional element: thickness t and plane area S)

Since the 1990s, the scientific direction of topological design began to dominate approaches that use the distribution of material in a fixed reference area in the format of the so-called "homogenization method" [5]. This representation of the structure is similar to that of a gray scale in discrete form, corresponding to the raster representation of the geometry of the supporting scheme. Although the concept of homogenization is very popular, it has a number of disadvantages. One is the question of the existence of a solution. Another drawback is related to the method of solution. In many applications, the optimal structure topology must consist solely of macroscopic variation of a single material and void. This means that the density of the structure is given by the integer parameterization "0-1" (often called black-and-white construction).

Currently, the most widely used methods of structural topology optimization based on "variable density". Like the homogenization methods, these methods work on a fixed finite element domain

(FE). However, instead of a set of properties, microstructure, each FE contains only one of the design variables. This variable is often understood as the density of the element material.

When tracing back the idea of "body of variable density", the vast majority if authors refer to the paper by [5] published in 1988. However, this idea was suggested a little earlier in a paper by Komarov [6] published in Russian in 1984.

The critical aspect of the methods based on the structures of variable density is the choice of an appropriate interpolation and a method for connection of physical characteristics with the continuously calculated values of density. Most often, the distributed function for topological design is interpreted as the density ρ_i of each FE in the form

$$E(\rho_i) = \rho_i^p \overline{E} \tag{2}$$

Here, $E(\rho_i)$ is the scaled modulus of elasticity of the material, $\overline{E}$ is the modulus of the solid material, p is the penalty parameter. For $0 \le \rho_{i\min} \le \rho_i \le 1$, the modulus of elasticity E falls between zero at zero density (the absence of the body) and its "solid" value $\overline{E}$ at ρ_i=1. A minimum density value is required to prevent difficulties associated with a possible singularity in FE matrices and with the inability of the material to appear in a region with zero density. With the choice of such parameterization, there is a need to direct the problem to a solid/empty solution. This is usually achieved by optimizing the topology based on implicit penalization methods, the most common of which is a solid isotropic material with the penalization method "Solid Isotropic Material with Penalization" (SIMP) [7, 8].

3. Optimization of head structure

3.1. Full-stress structures

As is known, the main influence of heating is associated with changes in mechanical characteristics, redistribution of stresses and increase in deformations. Formula (1) allows you to easily take into account the first factor. And as shown by the computational studies in the case of the presence of low level heating recurrent formula (3) is quite workable for the receiving of full-stress structures. However, for a high level temperature field is more efficient recursive formula proposed by Adelman [9]. For multiply load cases it can be written as:

$$x_i^{j+1} = x_i^j \max{}_c(\sigma_{\mathrm{eq}\,i\,c\,M})/(\underline{\sigma}_i - \sigma_{\mathrm{eq}\,i\,c\,T}),\ c = 1, 2, ..., s \tag{3}$$

where $\sigma_{\mathrm{eq}\,i\,c\,M}$, $\sigma_{\mathrm{eq}\,i\,c\,T}$ are the equivalent stresses only from mechanical and from thermal loads on the jth iteration.

The formula (3) is based on the assumption that when the structural elements parameters change, the temperature stresses change slightly or do not change at all.

For highly heated aircraft it is very important to provide the necessary rigidity due to special requirements for aerodynamics. In some cases, rigid deformation requirements can be crucial for spacecraft. In this regard, we will focus more on the method of optimization of the structure, aimed at obtaining the structure of the maximum rigidity.

The works [10] give examples showing that a temperature change of only a few degrees radically changes the optimal structure. This once again confirms the idea that to ensure maximum rigidity of the structure at the same time from mechanical and thermal loads is quite difficult.

For the structural optimization of load-carrying elements with heating made of dissimilar materials for one load case, the following formulation is proposed basing on the optimality-criteria method:

$$\min_{\mathbf{x}}: \quad F(\mathbf{x}) = \int_V \boldsymbol{\varepsilon}^{\mathrm{T}} \boldsymbol{\sigma} dV = \sum_{i=1}^{n} \frac{\sigma_{\mathrm{eq}i}^2 V_i}{e_i} = \sum_{i=1}^{n} \frac{N_{\mathrm{eq}i}^2 S_i}{e_i x_i} \tag{4}$$

under

$$\frac{M(\mathbf{x})}{M_0} = \frac{\sum_{i=1}^{n} \rho_i x_i S_i}{M_0} = 1 \tag{5}$$

$$\mathbf{K}(\mathbf{x})\mathbf{U} = \mathbf{F}_{\mathrm{M}} + \mathbf{F}_{\mathrm{T}}(\mathbf{x}) \tag{6}$$

$$0 \le \mathbf{x}_{\min} \le \mathbf{x} \le \mathbf{x}_{\max}\,. \tag{7}$$

where F is the target function (in this case, it is the compliance),$\mathbf{x}$ is the vector of the design variables, $\mathbf{x}_{\min}$ is the minimum of $\mathbf{x}$ introduced to avoid singularity when solving a system of algebraic equations, $\mathbf{x}_{\max}$ is the maximum of $\mathbf{x}$, $\mathbf{F}_{\mathrm{M}}$ is the load vector from mechanical load, $\mathbf{F}_{\mathrm{T}}$ is the load vector from thermal load, $\mathbf{U}$ is the global displacement vector, $\mathbf{K}$ is the global stiffness matrix, M is the current mass, M_0 is the initial mass of the computational domain, ρ_i is the density of each FE, V_i is the volume of the ith FE, which is for every type of FE $V_i = x_i S_i$, e_i is the the pseudo-elastic module (generalized module), introduced in [11].

For the example of one-dimensional element, the physical meaning of e_i is explained in figure 2. The value of e_i does not depend on the sequence of application of the mechanical and thermal loads. The value $\varepsilon_i\sigma_i$ also does not depend on this sequence. The value of $\sum\varepsilon_i\sigma_i$ in equation (4) is the sum of the specific potential strain energy and specific additional potential strain energy. It is obvious that the expression of F will be equal to the sum of the potential strain energy (PSE) and the additional PSE.

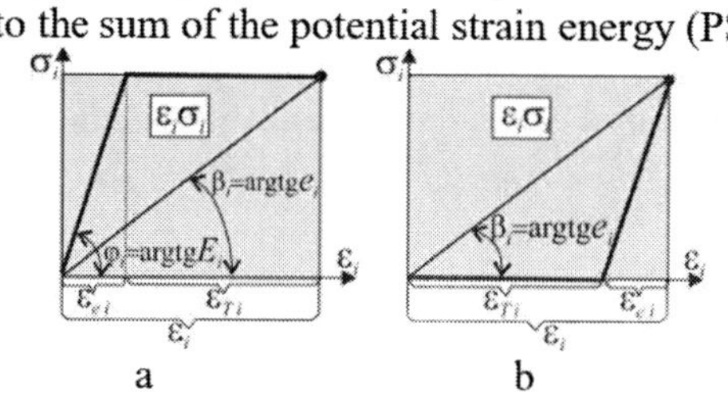

Figure 2. Geometric interpretation of a pseudo-elastic module ei on the stress-strain diagram depending on the action sequence of loads: (a) mechanical load is followed by heating, (b) mechanical load follows heating

The Lagrange multipliers method is used in order to find the optimal values of x_i satisfying equations (4) – (7). The iterative procedure of this method assumes that the equivalent loads of $N_{\mathrm{eq}\,i}$ and the pseudo-elasticity modules e_i do not change in the FE in the transition from one iteration to another. All changes occur only due to the design parameters x_i. Thus, the Lagrangian can be given in the form

$$L(\mathbf{x},\lambda) = \sum_{i=1}^{n} \frac{N_{eqi}^2 S_i}{e_i x_i} + \lambda\left(\sum_{i=1}^{n} \rho_i x_i S_i - M_0\right), \tag{8}$$

which yields the following system of equations

$$\frac{\partial L}{\partial x_i} = -\frac{N_{eqi}^2 S_i}{e_i x_i^2} + \lambda\rho_i S_i = 0, \quad (i = 1, 2, \ldots, n), \tag{9}$$

$$\frac{\partial L}{\partial \lambda} = \sum_{i=1}^{n} \rho_i x_i S_i - M_0 = 0 \tag{10}$$

where λ is the Lagrange multiplier.

From (9) follows that

$$x_i = N_{\mathrm{eq}\,i}(\lambda\rho_i e_i)^{-1/2}, \quad (i = 1, 2, \ldots, n)\,. \tag{11}$$

It is important that in some heated FE, the generalized module e_i can be (or become) negative at a current iteration. For example, a clamped heated rod has $e_i = -\infty$. Obviously, as the design parameters, such items are to be as small as possible $x_i \Rightarrow x_{\min i}$ $(i=p+1, p+2, n)$, and the elements themselves should be excluded from the list of active elements whose parameters are calculated by the final recurrent formula. But this question calls for additional studies, which will be discussed later.

The Lagrange multiplier is calculated from (10) and (11), which is then substituted into (11). The final expression for finding the structural parameters, taking into account several cases of loading, can be written as

$$x_i^{j+1} = [\sum_{c=1}^{s} N^j_{\mathrm{eq}\,i\,c}(\rho_i e^j_{i\,c})^{-1/2}](M_0 - \sum_{c=1}^{s} M^-_{i\,c})\{\sum_{i=1}^{p}\sum_{c=1}^{s} N^j_{\mathrm{eq}\,i\,c}(\rho_i / e^j_{i\,c})^{1/2} S_i\}^{-1}, \quad (i = 1, 2, ..., p) \tag{12}$$

where $M^-_{i\,c}$ is the mass of the FE with design parameters of negative modulus e_i in the cth load case.

4. Testing of method to structural optimization

By the one load case, the criteria of rigidity and strength of the structure lead to the same design parameters, and the structure degenerates into a statically definable one. Studies show that in several load cases of the parameters the obtained by the criteria of stiffness and strength do not always coincide. We can check this fact and the reliability of the proposed optimization algorithm is tested on the well-known example of Razani – a three-rod truss which is loaded with either from force F_1 or F_2 without heating (figure 3, a). The analytical solution of this is given in [12]. For the numerical solution of this problem, we assume that a=1 m, b=2 m, $F_1=F_2=2\cdot10^5$ N, the mass of the structure is M_0=23.7 kg, modulus of elasticity is $E=2\cdot10^{11}$ Pa, and material density is ρ=7800 kg/m^3. The result of optimization by the criterion of full stress structure is the statically determinate truss, in which the first rod is fully functional only at load F_2, and the third rod – at load F_1 (figure 3, b). The second rod is completely turned off from work.

In the optimal design with the criterion of maximum stiffness, the second rod is preserved with an area approximately two times smaller than in rods 1 and 3, while its ultimate stress state is not realized in any of the cases of loading (figure 3, c). But this option allows you to get a more rigid design. It should be noted that a similar distribution of design parameters is obtained if the forces act along the x and y axes (figure 3, d).

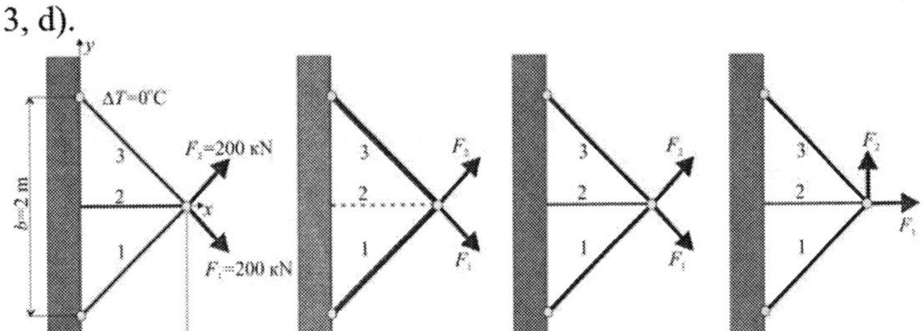

a b c d

Figure 3. Razani's trust (a), optimal decision according to criterion of full stress structure (b), optimal decision according to criterion of maximum stiffness (c, d)

It is obvious the presence of an underloaded element in the previous example is due to the fact that the original structure was not optimal. To obtain better results, we need to perform a topological optimization. As an initial scheme, we take a variant of a symmetric form with a uniform arrangement of 25 elements (figure 4, a). As the result of the optimization for two load cases we receive the statically definable two-rod truss with angle $\beta\approx35°$ (figure 4, b).

Next, we consider the loading of the same 25 rod truss first by force, then by the action of a uniform temperature field with $\Delta T=50°$ (figure 5, a). The design of the maximum stiffness will be the same as in the previous example - the two-rod truss with an angle β≈35° (figure 5, b). It is noted that with the increase in the temperature level, the convergence deteriorates markedly.

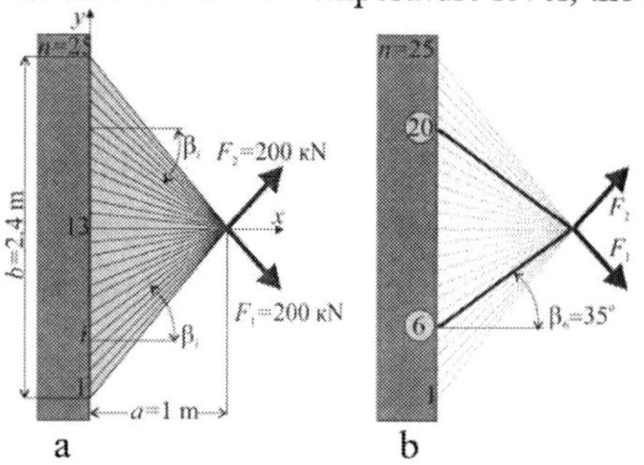

Figure 4. Truss optimization for two cases of mechanical loading: a – initial scheme, b – maximum stiffness design

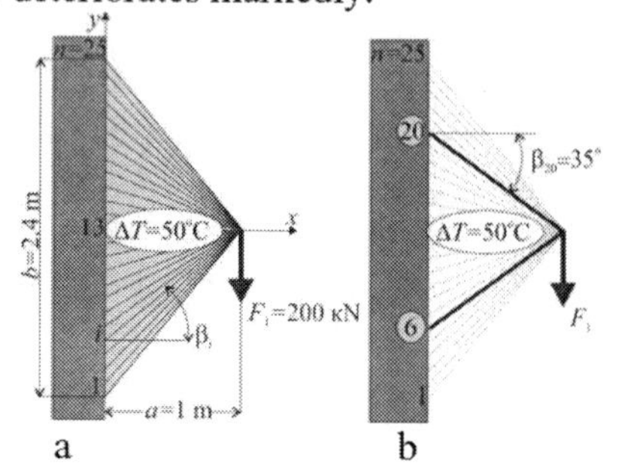

Figure 5. Truss optimization for two load cases of force F_1 and uniform temperature field: loading: a – initial scheme, b – maximum stiffness design

5. Conclusion

A criterion method of topological optimization of heated structures with elements from different materials for several cases of loading based on parametric synthesis by eliminating inefficient elements from the most common initial structure is proposed.

References

[1] Haug E J and Arora J A 1979 *Applied Optimal Design* (John Wiley & Sons Ltd, Chichester) p 520

[2] Wasiutynski Z and Brandt A 1963 *The Present State of Knowledge in the Field of Optimum Design of Structures* (Appl. Mech. Rev.) vol 16(5) pp 341–350

[3] Komarov A A 1965 *Foundations of load-bearing unit design* (Kuibyshev publishing house) p 66 [In Russian]

[4] Komarov V A 1968 *Rational load-hearing units of small aspect ratio wings* (Kuibyshev: KuAII transactions) vol 32 pp. 8-12 [In Russian]

[5] Bendsoe M P and Kikuchi N 1988 *Generating Optimal Topologies in Structural Design Using a Homogenization Method* (Computer Methods in Applied Mechanics and Engineering) vol 71 pp 197-224

[6] Komarov V A 1984 *Design of aircraft load-bearing structures*. (Actual problems of aviation science and technology. Mechanical Engineering) pp 114-129 [In Russian]

[7] Bendsøe M P 1989 *Optimal shape design as a material distribution problem* (Structural Optimization) 1 pp 193–202

[8] Rozvany G Zhou M and Birker T 1992 *Generalized shape optimization without homogenization*. (Structural Optimization) vol 4(3-4) pp 250-254

[9] Adelman H M Sawyer P L Shore C P 1979 *Optimum Design of Structures at Elevated Temperatures* (AIAA Journal) , vol 17(6), pp 622-629

[10] Rodrigues H and Fernandes P 1995 *A material based model for topology optimization of thermoelastic structures*. (International Journal for Numerical Methods in Engineering) vol 38(12) pp 1951–1965.

[11] Wang Z Kretov A S Shataev P A 2018 *Pseudoelasticity and its use for the design of heated thin-walled structures*. (Russian Aeronautics) vol 4 pp 7-14. [In Russian]

[12] Razani R 1965 *Behavior of Fully Stressed Design of Structures and Its Relationship to Minimum-Weight Design* (AIAA Journal) vol 3(12) pp 35-39

IOP Conf. Series: Materials Science and Engineering **685** (2019) 012011 doi:10.1088/1757-899X/685/1/012011

Adaptive Refinement based Integral Method for Load Transfer in Fluid/Structure Interaction Problems

Pan Cheng[1], Zekai Ding[2] and Wu Zhang[2,*]

[1]School of Computer Science and Engineer, Shanghai University, 99 Shangda Road, Shanghai, 200444, P.R. China

[2] Shanghai Institute of Applied Mathematics and Mechanics, Shanghai University, 99 Shangda Road, Shanghai, 200444, P.R. China

[*]wzhang@shu.edu.cn

Abstract. In fluid-structure interaction simulations using a partitioned approach, especially aeroelastic analysis, each physics component solves on its own mesh, and the interface between these meshes are in general non-matching. Force and displacement must be transferred from one mesh to another. Generally speaking, two classes of methods have been developed and applied in engineering: point wise interpolation and mapping method. For practical problems, point wise interpolation is not good enough in conversation and mapping method requires detailed meshes to describe surfaces of both fluid side and structure side. This paper presents a flexible, conservative and robust method which only requires surface mesh on fluid side. The merits of new method in terms of conversation properties were discussed. Also, the numerical results of practical problems are demonstrated in this paper.

1. Introduction

For simulations of complex system that require integration of multiple physical disciplines, such as fluid dynamics and structural mechanics, two approaches were adopted to solve this kind of problems. First is the monolithic approach which couples the governing equation of both fluid dynamics and structural mechanics, discretizes both parts as a single computational domain and solves on a whole mesh. The monolithic approach allows exact time synchronization but requires a new solver and mesh[1]. Moreover, the associated matrices of the whole strong-coupling system is likely to be ill-conditioned. However, these disadvantages are addressed by partitioned approach which allows to use the existing solver on both sides. Although partitioned approach cleverly circumvented the development of a new solver, it raises a difficult question in terms of data transfer between fluid mesh and structure mesh.

In general, fluid side is solved by computational fluid dynamics (CFD) which requires relatively fine spatial discretization, while structure is normally modelled as finite element model or even simpler as beams and then solved by stiffness equation. Since fluid and structure are solved on different meshes due to difference in characteristics, it is necessary to pass the aerodynamic loads from the aerodynamic surface to the structural grids consistently.

Data transfer is an indispensable subroutine in fluid structure interaction (FSI) problems. For a typical FSI problem like static aeroelastic analysis, one should first obtain the force distribution on the surface mesh that represents wing model through computational fluid dynamics (CFD). Then, the force distributions must be transferred consistently to the structure model so that the deflection may be computed through computational structure dynamics (CSD). Similarly, the deflections then need to be transferred to the surface mesh even to the volume mesh. For existing methods that dealing with data transfer in FSI problems, it can be generally divided into two main categories: point wise method and mapping method. Point wise methods only require arbitrary point clouds of both aerodynamic mesh and structural mesh[2] so that point wise method is easy to be realized. Although the total loads can be transferred conservatively through point wise method, it is short for transferring the distribution of quantities from the source mesh to the target mesh[3]. Mapping methods not only require data on points, but also connectivity information on the interface. The topology information on both sides help to improve conservation and minimize error during data transfer. But in practical problems, structure part is likely to lack description of surface so that mapping methods are not able to deal with certain kind of problems.

The purpose of this paper is to present an innovative load transfer method that is capable of transferring data from CFD surface mesh to incomplete finite element(FE) model without surface description. The basic idea of this method is to rebuild surface mesh for scattered FE grids by adaptive refinement of existed CFD surface mesh. With a rebuild mesh, the concentrated loads on scattered grids are calculated by integral. The newly created surface mesh is an approximation of Voronoi diagram. This load transfer method is easy to be implemented and parallelized, and it is also able to keep balance between accuracy and efficiency.

2. Problem Statement and Summarization

The FSI problem is solved by CFD and CSD alternatively and exchange data on the boundary Γ between each other. On structure side, the elastic equation is formulated in Lagrangian coordinates in which grids is moving with structure and boundary is denoted as Γ_s. On fluid side, the Navier-Stokes/Euler equation is solved in Eulerian coordinates in which meshes are fixed and fluxes are calculated in those volumes cells. The boundary of fluid side is denoted as Γ_f. The pressure and viscous stress are calculated by CFD and stored on the surface mesh of fluid side. The loads on the grids on structure side need to be transferred from the known pressure on the boundary mesh. The total force and moment should be conservative and more than that, the distributive force and moment should be matched before and after transferring. Usually, the fluid side is called source and structure is called target.

Mapping method includes many typical algorithms such as mortar element methods[4] and common refinement method[5][6]. This class of methods have strong mathematical foundation and conservation requirement are proved to be strictly satisfied. The quantities on source and target side are treated as function f and g separately. The two functions are denoted by the linear combination of the shape functions and quantities on n nodes on target and m nodes on source, which are $f = \sum_{i=1}^{m} f_i \phi_i$ and$g = \sum_{i=1}^{n} g_i \psi_i$. The error of load should be minimized during transfer so that the L_2 norm of $g - f$ should be minimized. The requirement leads to a linear system $\boldsymbol{Mx} = \boldsymbol{b}$, in which $\boldsymbol{M}_{ij} = \int_\Omega \psi_i\psi_j \, dx$ and $\boldsymbol{b}_i = \int_\Omega \psi_i f \, dx$. The discretization of L_2 minimization is the key to calculate integral on right and left side of the equation above[3].

Different discretization are available, including source-based, target-based and common refinement. The first two classes calculate$\int_\Omega \psi_i f \, dx$ by integrate over source mesh or target mesh. The common refinement is different from above two, it utilizes the overlap of meshes on both side and integrate over the smallest subdomains created by intersection. The accuracy, stability and efficiency are proved to be best among most discretization schemes since the common refinement fully use the geometry information on the surface mesh. However, this feature also limits the application of common refinement because the surface mesh on both side are necessary and should be based on the exactly same geometry.

Point wise method, also called interpolation or mesh free methods, is based on radial basis function (RBF)[7], multivariate spline[8] or any type of multivariate interpolation method[9][10]. These methods are used to interpolate displacement of the points on the fluid structure boundary and the concentrated force are transferred by virtual work conservation over the boundary. Discretization of displacement on Γ is formulated as $\delta\mathbf{u}_f = H\delta\mathbf{u}_s$, where the H is the interpolation matrix. The virtual work performed by the fluid loads and the structure stress should be equal

$$\delta W = \delta u_f^T \cdot F_f = \delta u_s^T \cdot F_s \tag{1}$$

where F_f is the discretized concentrated force on fluid points and F_s are the same one on structure grids. Then the force relation can be deduced as

$$F_f = H^T \cdot F_s \tag{2}$$

In different interpolation methods, the H matrix can be constructed based on variant basis functions. The earliest interpolation in aeroelastic is proposed in[11], in which thin plate surface(TPS) function are used as the basis function to transfer loads. TPS function is the fundamental solution of infinite plate equations and a special case of RBF, which minimizing the bending energy of thin plate and has the form $\phi(\|x\|) = \|x\|^2 log(\|x\|)$. Then many interpolation based on different RBFs are proposed, such as multiquadric, Gaussian or Wendland C^n functions. Some of them are compact supported, which means the affect range (circle in 2D and sphere in 3D) of these basis function is adjustable to balance between accuracy and efficiency[12]. Besides RBFs, some other methods use kriging interpolation or BEM to construct H matrix.

The interpolation method is easy and flexible to use since the equation is simple and only points are used. However, interpolation also have some drawbacks in practical problems. The first is that the distribution of quantities is not strictly conservative because it uses a different basis function instead of the source or target basis function in mapping methods. The basis functions used in mapping implies the boundary geometry information while the basis functions used in point wise method have no relevant to the geometry. It is not big deal when the points are dense and smooth enough but some time it is hard to achieve and big error will come out. Second, a $m \times m$ linear equations need to be solved to gain H matrix where m is the grids number on structure side. When m is very large, it cost too much time to solve a large linear equation which is not worthwhile in FSI coupling process. Although reduce method are introduced[13], the cost is to lower the accuracy. Third, in some interpolation such as RBFs based, three or more noncolinear points must be specified while it is not satisfied when the structure is a beam like model and auxiliary point should be added.

3. Adaptive refinement based method

In engineering environment, the mapping and point wise method are not satisfied in some special cases mentioned above. A better method should be designed to avoid the drawbacks listed above and a list of requirements are introduced as follow.

1. The loads received on structure side should be point wised. This means that the surface mesh on structure side will not be necessary, only points are needed to transfer loads.
2. The conservation should be satisfied, say $\int_\Omega |f-g|^2 \, d\Omega$ and $\int_\Omega |f-g| \, d\Omega$ should be minimized or closed to minimization, which means the loads should be conservative before and after transfer.
3. The computation should be robust and efficiency enough to be adaptive to most situation encountered in engineering.

Consider discretized Γ_s is represented by discrete point set P_s, while Γ_f is also discretized as points set P_f and cells set C_f. On structure side, only points are given while on fluid side, both points and cells are given. The boundary is shown as figure 1.

The basic ideal of the new method is similar to common refinement method, but only surface mesh on fluid side are utilized. There should be a surface mesh on structure side so that to perform intersection, however it is not existed on structure side and it should be reconstructed by existed surface mesh on fluid side. A simple way to reconstruct the surface mesh on structure side is to apply Voronoi diagram[14]. In Voronoi diagram, the space is decomposed to polygon cells and every cell match to one of given points. Arbitrary point on the edges of these cells has the same distance to two nearest points in given set. The Voronoi diagram of Γ_s in figure 1 is shown as (a) in figure 2. This tessellation treats every structure points as equality based on Euclidian distance.

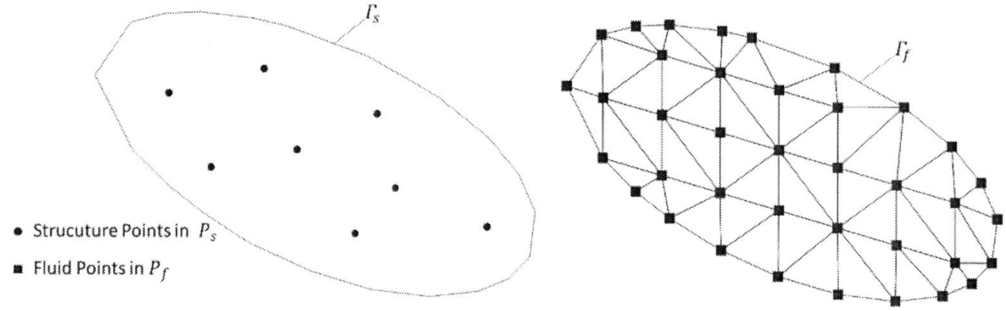

Figure 1. The boundary of structure and fluid side and their discretization

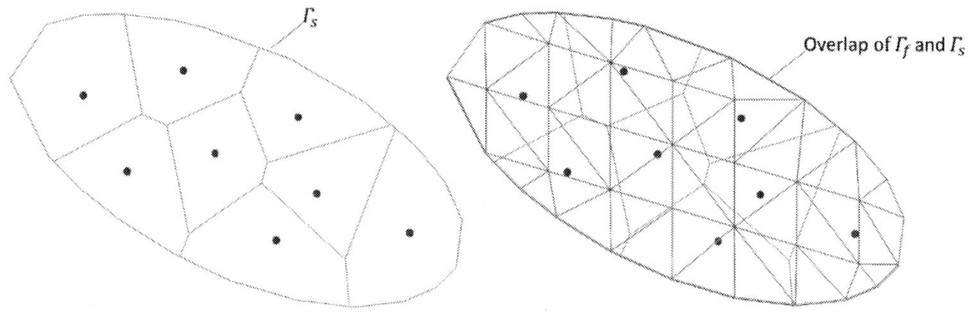

(a) Voronoi diagram of Γ_s (b) Overlap of Γ_f and Γ_s

Figure 2. The Vonoroi diagram of structure points and overlap of two meshes.

After the surface mesh reconstructed by Voronoi diagram, the common refinement scheme could be applied by intersecting two meshes theoretically. However, generation of Voronoi diagram on three dimensional surface could be very difficult and integration operation need to be performed after surface mesh generation and intersection. A natural thought is to combine these two steps together, so the adaptive refinement on surface mesh on fluid side is introduced. The edge of Voronoi diagram can be approximated by splitting and the edge will be displayed as zigzag curve along the small cells generated by adaptive refinement. The implementation of the procedure will be described as following.

Let's consider a case of only one fluid cell. For an arbitrary cell $c_i \in C_f$ on fluid mesh, a geometry center of c_i is defined as $\hat{c}_i$ and its two nearest neighbour points $p_{\hat{c}_i}^k|k = 1,2$ can be found in P_s if the distance

$$\mathrm{d}(p_{\hat{c}_i}^k, \hat{c}_i) = \min_k(d(p|p \in P_s, \hat{c}_i)) \tag{3}$$

where distance is defined as Euclidean distance. Let's denote $\mathrm{d}(p_{\hat{c}_i}^k, \hat{c}_i)$ as $\mathrm{d}_{\hat{c}_i}^k$. The edge of Voronoi diagram of this cell should be the intersection of the middle plane of $p_{\hat{c}_i}^k|k = 1,2$ and the cell. The intersection line on the cell could be approximated by recursively splitting the cell. First, we define a measurement of the cell size as s_i for cell, for example the longest edge of triangular cell or longest diagonal of quadrilateral cell. If $s_i > |\mathrm{d}_{\hat{c}_i}^1 - \mathrm{d}_{\hat{c}_i}^2|$, the cell should be split into smaller sub cells and we classify the pieces by distance. Then, for an arbitrary structure point $p_j \in P_s$, the nearest cells(including split and non-split cells) will be found through the process above and we put them into a subset denoted as C_{sub}^j. Recursively perform the process above until the measurement of the cell size

s_i smaller than a predefined threshold ε. The Voronoi cell of p_j is approximately covered by the sub cells in C^j_{sub}.

A sample case with one square cell with square side l and three structure points in 2D space is shown in figure 3. The three structure points p_1, p_2 and p_3 are marked by color blue, green and red. Respectively, their sub cell sets are marked by the same colors. At the beginning, the fluid cell split into 4 pieces and their color shows which structure point they belong to. But the distance threshold didn't satisfied, so keep splitting until the threshold reached. After 12 time splitting, the size of smallest cell is less than

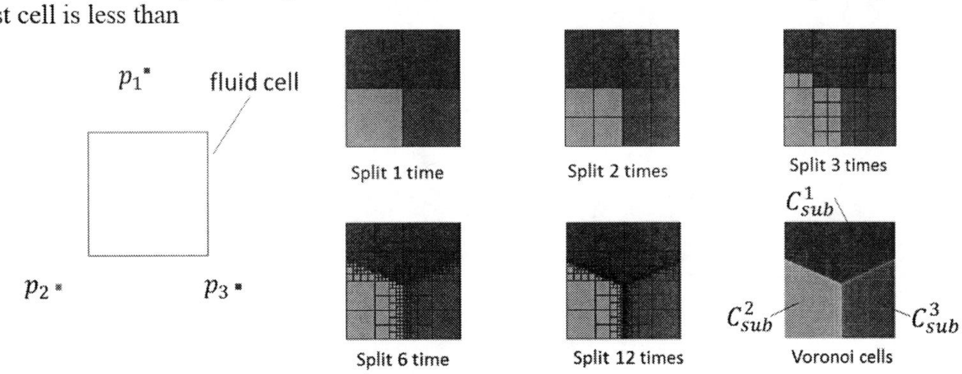

Figure 3. A square cell with 3 structure points sample case

threshold $\varepsilon = l/4096$ and the splitting is stopped. The square cell are cut into 3 blocks and every block containing split cells can be regarded as approximation of Voronoi cells. The edge of the Voronoi diagram is not straight lines, but zigzags curve with small steps. Because the splitting termination condition is $\varepsilon > s_i > |d^1_{\hat{c}_i} - d^2_{\hat{c}_i}|$ and $|d^1_{\hat{c}_i} - d^2_{\hat{c}_i}|$ always greater than 0, $\lim_{\varepsilon \to 0} |d^1_{\hat{c}_i} - d^2_{\hat{c}_i}| = 0$. This guarantee the geometry center of the splitting cells on edge will be on the Voronoi edges when ε approaches 0.

In a full fluid mesh, splitting should be performed on each cells until threshold reached, then the approximated Voronoi diagram for the whole surface will be generated. Because the splitting operation on one cell is irrelevant to other cells, it is easy to parallelize the procedure. The splitting procedure is also available on other type of mesh or even hybrid mesh. For triangular cells, the splitting cells can be generated by connect the middle point of each edge. For polygon cells, triangular cells will be generated first by connect each vertex and the geometry center then split these triangular.

After splitting procedure, the loads on structure point will be naturally calculated by integration. Suppose we have m fluid cells in C_f and n structure points in P_s. Each cell in C_f with pressure vector $\bar{\sigma}_i$ and viscous stress vector $\bar{\tau}_i$ will be split into m_i sub cells. The jth sub cell in these m_i sub cells have area of a_{ij} and the vector from reference point to its geometry center defined as r_{ij}. Normally fluid side is solved by finite volume method, the pressure and viscous stress is averaged, so the values on sub cells are the same to their parent cell. The total loads from the fluid side on boundary, including force and moment is discretized as

$$\boldsymbol{F}_f = \int_\Omega f\,dx = \sum_{i=1}^{m} \sum_{j=1}^{m_i} (\bar{\sigma}_{ij} + \bar{\tau}_{ij})\, a_{ij} \tag{4}$$

$$\boldsymbol{M}_f = \int_\Omega f \times r dx = \sum_{i=1}^{m} \sum_{j=1}^{m_i} (\bar{\sigma}_{ij} + \bar{\tau}_{ij})\, a_{ij} \times r_{ij} \tag{5}$$

While on the structure side, for each structure point p_i, an approximated Voronoi cell combined by split fluid cell is created and contain n_i sub cells.

$$\boldsymbol{F}_s = \int_\Omega f\,dx = \sum_{i=1}^{n} \sum_{j=1}^{n_i} (\bar{\sigma}_{ij} + \bar{\tau}_{ij})\, a_{ij} \tag{6}$$

$$\boldsymbol{M}_s = \int_\Omega f \times r dx = \sum_{i=1}^{n} \sum_{j=1}^{n_i} (\bar{\sigma}_{ij} + \bar{\tau}_{ij})\, a_{ij} \times r_{ij} \tag{7}$$

Since $\sum_{i=1}^{n} n_i = \sum_{i=1}^{m} m_i$ and every sub cell are covered in integration, the integral of force and moment on both side are the same. Thus, the conservation of total force and moment $\boldsymbol{F}_f = \boldsymbol{F}_s$ and $\boldsymbol{M}_f = \boldsymbol{M}_s$ are satisfied. The conservation of distribution is not strictly satisfied due to the grain size of the minimum split cells. We will show more detail in next section.

4. Numerical results of Common Research Model

We have implemented the adaptive refinement based method into a code called AECS. In this section, we will apply load transfer method to NASA Common Research Model(CRM) to test the consistency of quantities before and after transferring. CRM was developed by NASA's Subsonic Fixed Wing(SFW) Aerodynamic Technical Work Group. This wing-body-horizontal (without nacelle-pylons) configuration is representative of a contemporary high-performance transonic transport detailed description[15].

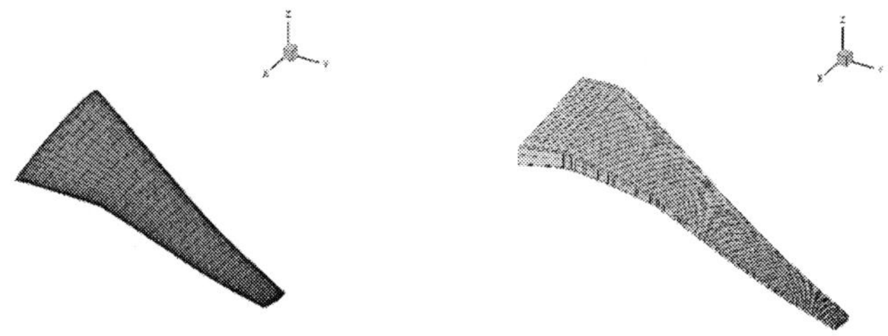

(a) Fluid surface mesh (b) Finite element model

Figure 4. Coupling surface of NASA Common Research Model

Figure 4 shows the outer wing region of the CFD and CSM coupling surface of NASA CRM model. The wing surface is discretized by 11522 quadrilateral elements and 12099 vertices. The structure part of CRM wing is represented by equivalent wing-box model which is provided by the official site of CRM from NASA and is discretized by 4626 nodes, 8476 quadrilateral elements and 14 triangular elements. Since the element distribution on CFD surface mesh and wing-box model differ a lot, point-wise method is expected to achieve poor results in transferring distribution of quantities from source mesh to target mesh. For static aeroelastic analysis, researchers tend to pay more attention to moment distribution along certain direction instead of force at a specific point so that they could obtain the static balance position of the wing. In this section, we will compare the moment distribution along span-wise direction between point wise method and adaptive refinement based method. Also the influence of threshold value on the error of moment distribution between surface mesh and structural mesh will be compared.

Firstly, we started from a RANS CFD simulation, which is computed on the fluid mesh. Our test case description is shown in table 1.

Table 1. Description of CRM test case.

Parameter	Value
Mach Number	0.85
Angle of Attack	2.5°
Reynolds Number	5×10^6
Reference Temperature	100°F

Then, we transferred the pressure distribution on fluid mesh to structural mesh using two different kind of methods. One is TPS interpolation, which is an RBF interpolation using $\phi(\|x\|) = \|x\|^2 log(\|x\|)$. But in practical problems, researchers tend to divide fluid mesh into several groups so that the force on fluid mesh could be transferred group by group to the corresponding structural grids using TPS interpolation and will greatly improve the conservation of force distribution before and after transferring. To compare the results of grouped TPS interpolation and original TPS interpolation, both interpolation methods were adopted. The other method is adaptive refinement based method. To count the moment distribution along specific direction, we divided the wing into 19 sections by 20 planes, construct a local coordinates and accumulate the moment distributed from tip to certain plane along span-wise direction. The divided CFD surface mesh is shown in figure 5. Each color represents a section divided by planes. GID is the ID of section and varies from 0 to 18.

Figure 5. Sections divided by planes.

Since our method determines whether to split the fluid cell or not by calculating the threshold ε, we set $\varepsilon = 0.1$ which is approximately 1/10 of the longer side of the largest cell in surface mesh and start splitting the surface mesh. Figure 6 shows the grid at the wing tip before and after splitting. The surface mesh is totally refined by 929583 cells and 1542594 points based on wing-box model so that the aerodynamic loads could be transferred conservatively.

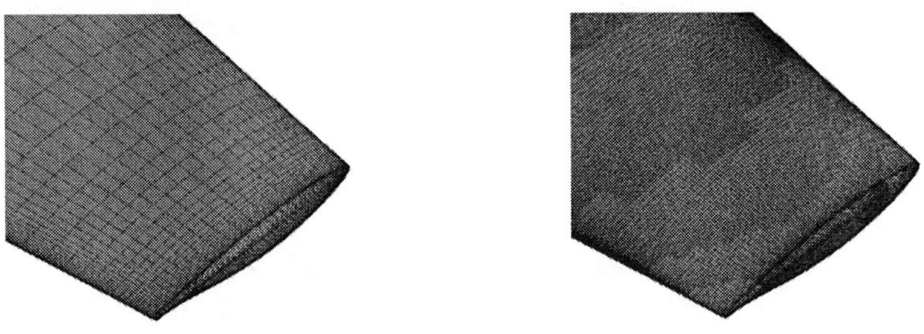

(a) Before (b) After

Figure 6. Fluid mesh of wing tip before and after splitting ($\varepsilon = 0.1$)

After splitting the surface mesh on fluid side, the software starts to transfer the aerodynamic loads on surface mesh to wing-box model and calculates the accumulated moment along the direction we gave. The moment distribution before and after transferring are shown in the figure 7. We could find out that the moment distribution on fluid mesh differs a lot from that transferred to structure grids by the original TPS interpolation. Grouped TPS interpolation improves conservation significantly comparing to the original TPS interpolation but is still inferior to the adaptive refinement based method. In general, the adaptive refinement based method is the most conservative method comparing to the point wise methods.

When the threshold value ε is rather large, none of the cell on fluid mesh need to be split. Then the adaptive refinement method degenerates to nearest neighbour search method. To determine how threshold value ε affects the error of moment distribution, we split the surface by two threshold value

0.1, 0.5 and compare the absolute error of moment distribution with the surface mesh without splitting. The absolute error of moment distribution is defined as $Error = |M_f - M_s|$ in which M_f represents the moment on fluid surface mesh and M_s represents the moment that transferred to structure. Results in figure 8 indicate that as the threshold ε decreases, the error of moment distribution is also minimized especially at the section around $X/L = 0.2$. In general, a relatively small threshold value helps to decrease the error of moment distribution and to some extent average the error along the direction we provided.

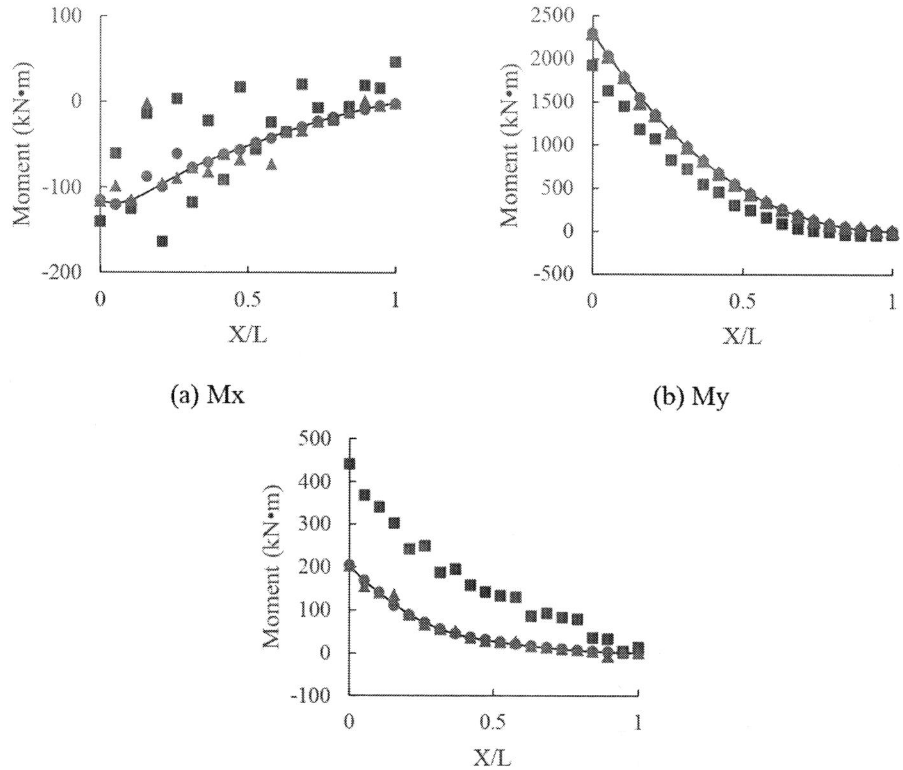

(c) Mz

Figure 7. Moment distribution along span-wise direction (solid line: moment distribution before transferring, circle: adaptive refinement based method, triangle: grouped TPS interpolation, square: TPS interpolation)

From the results above, we can draw out a conclusion that our method reconstructs the topology of surface mesh, enrich information in fluid cells and generate a new surface mesh that is suitable for a more conservative data transformation based on structural grids.

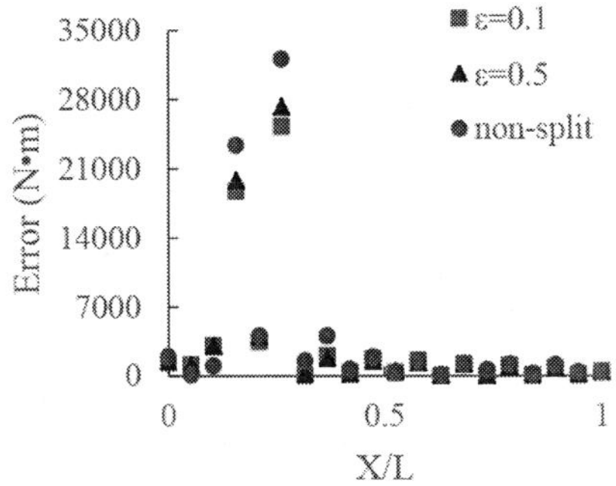

(a) Mx

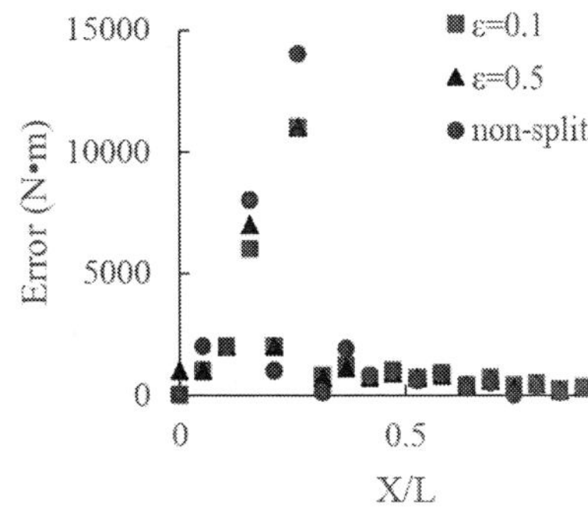

(b) My

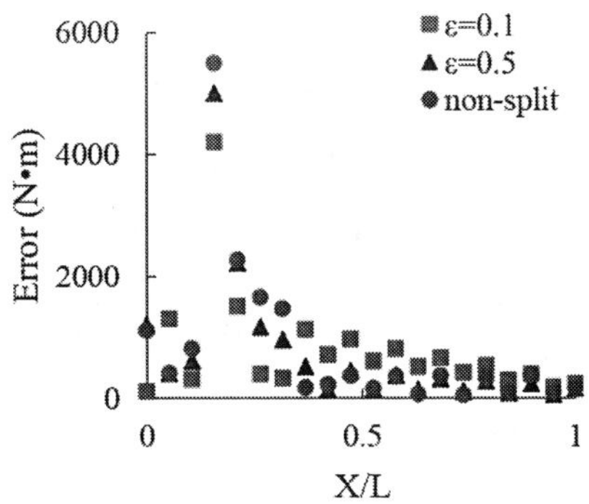

(c) Mz

Figure 8. Error of mement distribution along span-wise in three directions

5. Conclusions

In this paper, we overviewed a family of load transfer methods in fluid-structure interaction problems and presented a flexible, conservative method that only requires surface mesh on fluid side and grids on structure side to transfer force from fluid to structure. This method can be utilized as a subroutine in fluid-structure interaction problems especially the aeroelastic problems. When dealing with practical problems that fluid and structure do not share the same geometrical support such as 'fish-bone' and 'wing-box' model, point-wise method can be inaccurate in transferring quantity distribution. The method we presented in this paper combines the advantages of point wise methods and mapping methods. By reconstructing the topology of surface mesh based on structural grids, the error will be minimized and the distribution of quantities will be transferred consistently from source mesh to target mesh comparing to the point wise method. Moreover, surface mesh on structure side is not required which is likely to lack in practical problems. As a result, the adaptive refinement based method can be implemented more extensively than mapping methods. In addition, the threshold value can be adjusted to determine whether to pursuit efficiency or accuracy. Finally, our claims were validated by numerical results of Common Research Model.

References

[1] Hübner B, Walhorn E and Dinkler D 2004 A monolithic approach to fluid-structure interactionusing space-time finite elements Comput. *Methods Appl. Mech. Eng.* **193** 2087–2104

[2] Rendall T C S and Allen C B 2008 Unified fluid – structure interpolation and mesh motion using radial basis functions *Int. J. Numer. Meth. Engng* **74** 1519–1559

[3] Jiao X and Heath M T 2004 Common-refinement-based data transfer between non-matching meshes in multiphysics simulations *Int. J. Numer. Meth. Engng* **61** 2402–2427

[4] Farhat C, Lesoinnea M and Letallecb P 1998 Load and motion transfer algorithms for fluid / structure interaction problems with non-matching discrete interfaces: Momentum and energy

conservation, optimal discretization and application to aeroelasticity *Comput. Methods Appl. Mech. Engrg.* **157** 95-114

[5] Jiao X and Heath M T 2004 Overlaying Surface Meshes, Part I: Algorithms *Int. J. Comput. Geom. Appl.* **14** 379–402

[6] Jiao X and Health M T 2004 Overlaying Surface Meshes, Part II: Topology Preservation and Feature Matching Int. J. Comput. Geom. Appl. **14** 403–419

[7] M. D. Buhmann 1993 New developments in the theory of radial basis function interpolation *Multivar Approx. From CAGD to Wavelets, K. Jetter FI Utreras, eds World Sci.* pp 35–75

[8] Smith M J, Hodges D H and S. Cesnik C E 2008 Evaluation of Computational Algorithms Suitable for Fluid-Structure Interactions *J. Aircr.* **37** 282–294

[9] Dai K Y, Liu G R, Lim K M and Gu Y T 2003 Comparison between the radial point interpolation and the Kriging interpolation used in meshfree methods *Comput. Mech.* **32** 60–70

[10] Smith M J, Hodges D H and S. Cesnik C E 2008 Evaluation of Computational Algorithms Suitable for Fluid-Structure Interactions *J. Aircr.* **37** 282–294

[11] Harder R L and Desmarais R N 1972 Interpolation using surface splines *J. Aircr.* **9** 869–871

[12] Boer A De, Schoot M S Van Der and Bijl H 2007 Mesh deformation based on radial basis function interpolation *Comput. Struct.* **85** 784–795

[13] Rendall T C S and Allen C B 2009 Efficient mesh motion using radial basis functions with data reduction algorithms *J. Comput. Phys.* **228** 6231–6249

[14] Aurenhammer T 1991 Voronoi diagrams *ACM Comput. Surv.* **23** p 94

[15] Vassberg J 2011 A Unified Baseline Grid about the Common Research Model Wing/Body for the Fifth AIAA CFD Drag Prediction Workshop 29th AIAA Applied Aerodynamics Conference 2011

IOP Conf. Series: Materials Science and Engineering **685** (2019) 012012 doi:10.1088/1757-899X/685/1/012012

Distance-aware Data Forwarding Protocol for Named Data Mobile Ad-hoc Network (NDMANET)

Junyu Lai[1,2], Zhongwei Chen[1], Chengkang Li[1], Yingbing Sun[1] and Han Xiao[1]

[1]School of Aeronautics & Astronautics, University of Electronic Science and Technology of China, Chengdu 611731, China
[2]Science and Technology on Communication Networks Laboratory, China

laijy@uestc.edu.cn

Abstract. Information Centric Network (ICN) is a promising future networking architecture, while Named Data Networking (NDN) is regarded as ICN's representative and authoritative scheme. In the recent past, applying ICN/NDN concept in mobile ad-hoc network (MANET), i.e. Named Data Mobile Ad-hoc Network (NDMANET), has attracted a plenty of research efforts. Compared to legacy IP protocol, NDN scheme is considered to be more robust and profitable to the node mobility and the dynamically changed topologies in the MANET domain. This paper focuses on NDMANET's packet forwarding protocols, which to a large extent decides the efficiency of data dissemination and is a vital module of NDMANET. Firstly, the paper investigates and analyzes some existing packet forwarding protocols, which are widely accepted in NDMANET. On that basis, an innovative distance-aware data forwarding (DDF) protocol is elaborated for NDMANET, with the purpose of further enhancing its network performance. Comprehensive simulation experiments have been conducted and the results illustrate that the proposed DDF protocol can introduce considerable reduction in bandwidth and in energy consumption, with only slightly degrading the response ratio.

1. Introduction

MANET technology is designed to achieve data communication among nodes with mobility. Due to dynamic changing of network topology, and intermittent connections, MANET generally requires that there exists at least one complete end-to-end path between two nodes at some time for successful data transmission, and is highly vulnerable to packet loss and error in wireless environment. Therefore, an appropriate networking protocol is essential to MANET. TCP/IP protocol suite applied by the current Internet requires IP address to identify and locate a host, and then to maintain the end-to-end connection to realize reliable data delivery. That is however not suitable for MANET, considering the fact that MANET nodes are with mobility, and it is difficult to locate them by IP address.

At networking layer, legacy IP deliveries data in the sender-driven (push) mode. ICN/NDN as a clean-slate approach, adopts the receiver-driven (pull) mode, and can enable two communication nodes to interact asynchronously with no need to establishing and maintaining an end-to-end connection. In addition, ICN/NDN's in-networking caching technology enable each node to exploit full potentials of wireless channel's broadcast nature, so to lower redundant requests and decrease content retrieval latency. Compared with IP, the above merits of ICN/NDN are consistent with MANET requirements.

According to [1], naming, routing (forwarding), caching, transport and security currently are top five research topics in NDMANET. Most existing packet forwarding schemes are originally proposed for fixed network, not dedicatedly designed and optimized for MANET. Therefore, aiming at further enhancing network performance, this paper explores innovative data forwarding strategies and elaborates a distance-aware data forwarding protocol for NDMANET. Performance evaluation indicates that the proposed protocol is capable of reducing redundant packets to be transmitted, decreasing bandwidth and energy resource consumption, with only slightly degrading response ratio. To summarize, the major contribution of this paper includes:

- Elaborates a distance-aware data forwarding protocol to improve NDMANET's performance;
- Conducts comprehensive simulation experiments to evaluate the performance of the proposed distance-aware data forwarding protocol.

The remaining part of this paper proceeds as follows: First of all, Section 2 introduces the state-of-the-art followed by an in-depth background introduction in Section 3, which includes IP based MANET, NDN based MANET, as well as packet forwarding challenges in NDMANET. Then, in Section 4, as the major contribution, an innovative distance-aware data forwarding protocol is derived for NDMANET. Comprehensive performance evaluation by means of simulation for the proposed data forwarding protocol is conducted to demonstrate its effectivity and efficiency in Section 5. Finally, Section 6 summaries the paper and gives the outlook.

2. Related work

Research on NDN's application in MANET has recently gained much attention in academia. Packet forwarding protocol is one essential mechanism in NDN domain. There has been quite a lot of research efforts on it. The existing packet forwarding protocols can be classified into two types.

The first type is *proactive forwarding* based, which requires the content providers to periodically broadcast their own content name prefixes to other nodes, and consequently creates large overheads. Protocols of this type are only applicable to small-scaled and relatively stable networking scenarios. In [2], Yu et al. propose a protocol (BFR) using a bloom filter to advertise the name prefix of popular contents, which was applied in a typical VANET scenario. Kuang et al. in [3], propose a Content-Scent based architecture (CSAR), where different content is assumed to have its own content-scent. Each provider node periodically broadcasts a SCENTP that records the names of all the provided contents. However, the merit of low content transmission delay brought by the proactive forwarding protocols is offset by the overhead of maintaining routing states [4].

The second type is *reactive forwarding* based, which embodies the promising research direction in NDMANET. Protocols of this type do not need to maintain the routing table, while the flooding strategy is adopted to forward the interest packets. The blind flooding strategy is the simplest reactive forwarding protocol, which can well establish and maintain end-to-end route paths. However, it can cause broadcast storm problems [5] if not well controlled. Therefore, the x-aware based reactive forwarding protocols are proposed. Here, x-aware can be geographic-aware, neighbour-aware, distance-aware, etc. In [6], Lu et al. propose a direction-selective forwarding protocol, which can effectively reduce the redundant packets, but in the meanwhile can decrease the possibility of discovering potential content sources. In [7], Wang et al. increase geographic location information to content naming and only support location-related applications. BLOOGO (BLOOm filter based GOssip) [8] is a typical neighbour-aware protocol that determines whether to forward packets by comparing the containment relationships between neighbouring nodes. MADN (Multipath Ad-hoc Data Network) [9] uses a utility function to select the next hop forwarding node. RUFS (RobUst Forwarder Selection) [10] forwards neighbour information by prioritizing it. Neighbour-aware based protocols need to periodically exchange information with neighbouring nodes and need to make trade-offs between forwarding accuracy and forwarding overhead. LFBL (Listen First Broadcast Later) [11] is a distance-aware protocol. Each node in the network maintains a distance table. The distance information of the packets is equivalent. The distance measurement determines whether to forward the packets. CHANET (Content centric fasHion mANET) [12] introduces Int-Ack packet. When the producer receives the interest packet, it will return the data object packet according to the information in the PRT. BREB (Best Route Error Broadcast) [13] is based on the improved forwarding protocol of LFBL and CHANGE. It maintains the distance-based design concept of LFBL protocol, integrates the three data packets of CHANET protocol, and increases the path recovery function. CHANET and BRBE are only suitable for relatively low-speed mobility scenarios. Therefore, more efficient data forwarding protocol are desired for further improving the NDMANET performance.

3. Challenges in NDMANET

3.1. IP based MANET

MANET is extremely vulnerable in wireless environment featured of signal interference, fading and shadowing. In data link layer (L2), media access mechanisms such as CSMA/CA, TDMA, etc. are adopted to avoid access violation, and to make the most efficient use of the wireless broadcast channels. In networking layer (L3), legacy IP protocol is naturally employed due to its highly successful application in fixed networks. However, IP is ill-suited for MANET due to the following reasons: Firstly, IP based MANET belongs to the host-centric networking (HCN) domain. IP address is not only the node's identifier but also the locator representing its topological locations. Once a node moves out of the ad-hoc setting, it will fail to be located and then IP addresses aggregation will not be feasible [14], Secondly, nodes of highly mobility suffer from a large overhead to keep the binding between the node and its current address. Thirdly, the sender-driven transmission fails to utilize the other receivers to take over the forwarding when the intended receiver may not be expected any more.

3.2. NDN based MANET

It is an elemental demand for ad hoc networking to support nodes mobility. Appling NDN scheme in MANET can naturally satisfy this requirement. The major reasons are given below: Firstly, different from HCN, NDMANET employs unique name to identify with each Data unit, and it enables the applications to publish and consume Data directly through its name without being providing the network layer address. Secondly, in the HCN domain, TCP/IP session are required to be constantly reestablishments due to intermittent links between two communication moving nodes, which introduce considerably large overheads and delays. While NDMANET's receiver-driven mode makes sessions established for communication nodes unnecessary anymore. Thirdly, NDMANET separates data from physical node to make the data reusable, plus with the in-network caching mechanism, which further accelerates content transmission by means of local caching of previously retrieved data.

3.3. Challenges in content forwarding scheme of NDMANET

Receiver-driven mode based content forwarding scheme can fully use the broadcasting nature of wireless channels. However, there exist some challenges for broadcasting transmission, such as broadcast packets redundancy and collisions, which will eventually introduce bandwidth resource wastage, etc. Therefore, it is necessary to solve the broadcasting issues for enhancing the forwarding performance of typical NDMANETs.

4. An innovative distance-aware data forwarding protocol for NDMANET

4.1. Important factors of packet forwarding protocols in NDMANET

Due to the high dynamicity of MANET topology, designing packet forwarding protocols should minimize the dependence on fixed topology, and should pay attention to the logical process of the multi-faceted information obtained by networking nodes. Besides, other factors should also be considered includes: Firstly, avoiding to generate unnecessary packets. Since the energy and storage resources of mobile nodes are typically limited, introducing additional packets will increase energy and computation resource consumption; Secondly, protocol should be concise and robust. NDMANET nodes are divided into three roles, namely producer, consumer, and intermediate forwarding node. The interaction between nodes of the three types should be effective and efficient; Last but not least, forwarding decisions should be calculated locally at intermediate nodes based on local information obtained through inter-nodes packet interactions.

4.2. Principle of distance-aware data forwarding protocol

4.2.1. Packet modification and table description. To keep the protocol concise and to reduce extra traffic and processing cost introduced by new packets, the proposed DDF protocol does not introduce any new packet formats except the Interest and Data packet originally from NDN. However, some necessary modifications to the packet formats are made for the proposed protocol, and are shown in figure 1. Table 1 explains the meaning of each field.

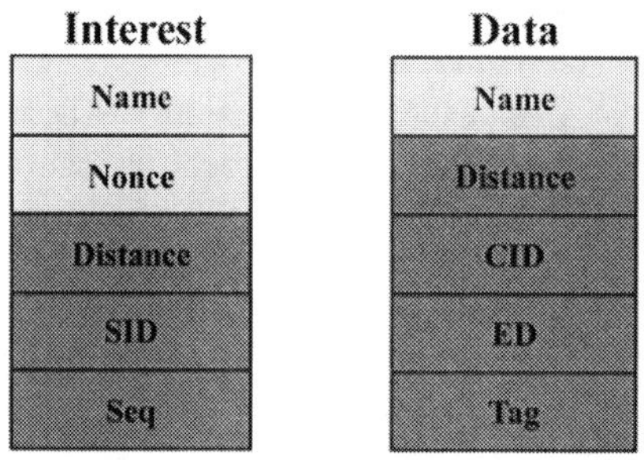

Figure 1. The modified packet formats for DDF protocol.

Table 1. Field description for the modified packet formats.

Interest Packet		**Data Packet**	
Name	Content name	**Name**	Content name
Nonce	Random number used to identify duplicate Interest packet	**Distance**	Distance between data packet and producer. When data packet is forwarded once, the value is incremented by one
Distance	Distance between Interest packet and consumer. If Interest is forwarded once, the value is added by one	**CID**	Record the target consumer identity information.
SID	Records unique identifier of source node (MAC addresses or application ID)	**ED**	Record the expected distance from the producer to the consumer
Seq	Monotonically increasing integer, which records the order of interest packets	**Tag**	Forwarding status of packet (0 represents the normal state; 1 indicates the data packet is in a state to be discarded)

In order to record the distance information used for making forwarding decisions, a Consumer Distance Table (CDT) is introduced to each forwarding node, and to record the distance between the nodes and the consumers. The structure of the table is shown in figure 2.

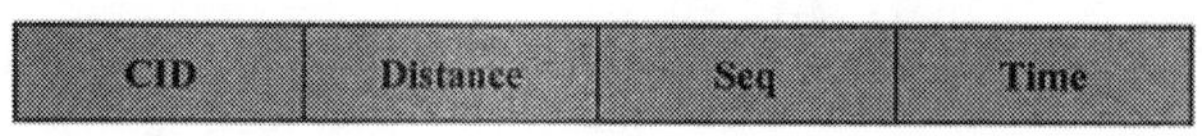

Figure 2. Format of the Consumer Distance Table (CDT).

4.2.2. Interest packet processing. The consumer assigns its identity information to the *SID* field and set the sequence number value for the *Seq* field of the Interest packet before sending it out. The

Distance field of the Interest packet is initialised as 0. In the process of flooding Interest packets, the Interest packet processing algorithm running on the forwarding nodes is shown in figure 3.

```
Interest packet processing
Interest packet preprocessing
if Name match CS then
    update CDT
    return Data packet
else
    if Name match PIT then
        update CDT
        drop Interest packet
    else
        update CDT
        add new PIT item, forward Interest packet
    end if
end if
```

Figure 3. Interest packet processing algorithm.

```
Data packet processing
Data packet preprocessing
if Name match PIT then
    Tag=0
    add new CS item
    delete PIT item
    if ED - D_pro >= D_con then
        forward Data packet
    else
        drop Data packet
    end if
else
    if Tag==0 then
        if Name match CS then
            drop Data packet
        end if
    else
        Tag=1
        add new CS item
        delete PIT item
        Distance = Distance-1
        forward Data packet
    end if
end if
```

Figure 4. Data packet processing algorithm.

CDT will create a new entry for the Interest packets that can not match the content name. If an Interest packet matches the CDT, the CDT will be updated based on the *Seq* field. If the *Seq* value in the latest received interest packet is smaller than the *Seq* in the CDT, the interest packet is considered to have expired, so the CDT is not changed and the Interest packet is discarded. If the *Seq* value is greater than the *Seq* in the CDT, the interest packet is modified to carry the newest distance information, namely the *Seq* field and the Distance field in the CDT are updated to the *Seq* value and the *Distance* value in the interest packet. If *Seq* is equaled, the smaller *Distance* value in Interest will be updated in CDT. In particular, the node will set a timer for each table item of the CDT. The timeout period is *T*. If the entry of CDT is updated, the timer will be reset; if the table item of the distance table is updated, the timer will be reset. If the entry has not been updated after time *T*, it is considered invalid, and the entry will be deleted immediately.

4.2.3. Data packet processing. After flooding Interest packets, CDT of all the nodes are updated. The qualified producer will match the related entry in the CDT and assign the Distance value to the ED field when Data packet is generated. The initial values of the Tag field and the Distance field in Data packet are both 0. The intermediate node will determine whether or not to forward the Data packet by equation (1).

$$ED - D_{pro} >= D_{con} \tag{1}$$

D_{con} represents the Distance value in the CDT, D_{pro} represents the Distance field in the packet. Equation (1) is adopted to determine whether the Data packet is in the shortest return path. If so, the returning delay can also be assumed to be shortest and the path is optimal. Thus, the node will forward the Data packet. Otherwise, the Data packet will be not forwarded. Equation (1) effectively filters the redundant forwarding packets caused by flooding, and consequently reduces the transmitted traffic.

In order to improve the response rate and to avoid the problem that node may discard the Data packet directly when it cannot match the PIT entry, in the proposed protocol, intermediate nodes will try to forward the Data packet once, and the *Tag* field of the Data packet will be set as 1. If PIT can be matched by the Data packet after the procedure mentioned above, the *Tag* field is set to be 0. If PIT still cannot be matched, the Data packet is then discarded. The above algorithm is given in figure 4.

4.3. Advantages and Disadvantages
The advantages of the derived DDF protocol lie in the fact that it can guarantee the stability and reliability of communications between consumers and producers in high dynamic conditions. At the same time, the proposed protocol retrieves Data packet over the shortest distance path, which can reduce the consumption of bandwidth, energy, cache resources, as well as chance of conflicts caused by flooding packets. Since the DDF protocol updates the CDT with flooding the Interest packets, which generates additional traffic, and broadcast storms may still occur in the case of dense consumer situations. This protocol does not take into account acquisitions of distance information from unintentionally monitored packets, which has little impact on network performance. The continuous requests of different consumers may reduce the retrieving efficiency of network distance information.

5. Performance evaluation by means of simulation

5.1. Performance metrics
The BF and BREB forwarding protocols are two most widely deployed forwarding protocols, which are compared with the elaborated DDF protocol. Five performance metrics (RR, BC, ND, EC, CRC) are chosen to be evaluated, as is defined in table 2.

Table 2. Description of evaluated performance metric.

RR (Response Ratio)	Ratio of the number of Data packets received by the consumer to the number of Interest packets sent by the same consumer
BC (Bandwidth Consumption)	Total amount of traffic transmitted by all the networking nodes
ND (Number of Data packets)	Total number of Data packets transmitted in the network
EC (Energy Consumption)	Total energy consumption for nodes to send interest & data packets
CRC (Computation Resource Consumption)	Total computation resource consumption for nodes to process interest and data packets

5.2. Simulation introduction
The proposed DDF protocol is implemented in ndnSIM / NS-3 network simulator. In order to investigate node mobility's impact on the performance of the proposed protocol, the simulated geometry scale is set to be a square of 1500m*1500m, and the maximum transmission range for each node is assumed to be 200m. A random rectangular positional allocator is used to assign the initial position of each node, while Random Waypoint Mobility Model [15] is used to simulate the motion trajectory of the nodes. The simulation time is 200 seconds.

5.3. Results analysis

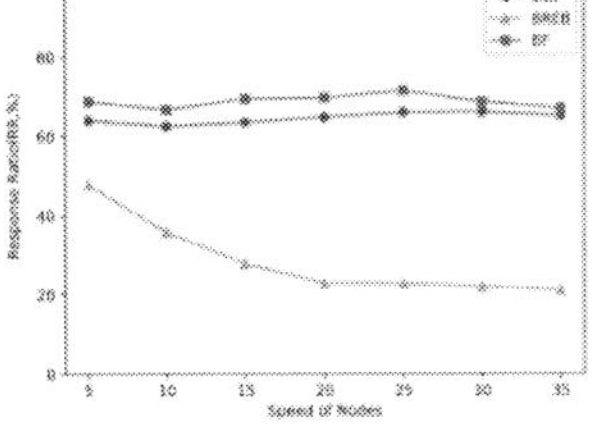

Figure 5. Result of Response Ratio

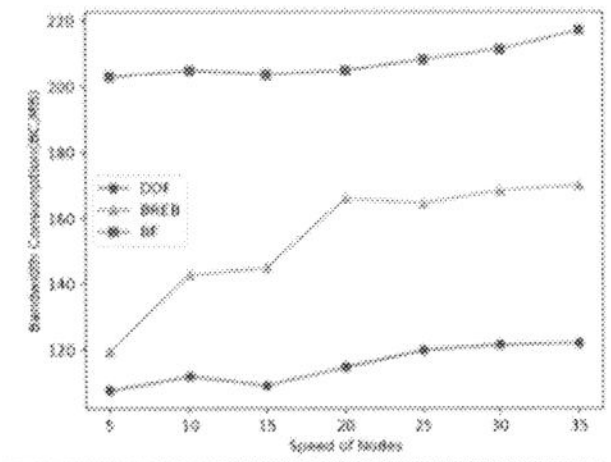

Figure 6. Result of Bandwidth Consumption

The aforementioned three packet forwarding strategies, namely BF, BREB and DDF forwarding protocols are simulated respectively. To ensure the three protocols are comparable, random numbers in the mobile model and the distribution model are generated based on the same random seeds. The simulation results of the *Response Rate* of the three protocols at different speeds are shown in figure 5. The Response Rate of DDF and BF are close at different speeds. This reflects the accuracy of the distance information generated by the DDF protocol, indicating that the ability to improve the distance perception can be improved by flooding interest packets. The Response Rate of BREB decreases with the increase of node moving speed. This is because BREB cannot respond to topology changes in time, resulting in untimely update of distance information. Figure 6 shows the bandwidth usage of the three protocols at different speeds. DDF has the lowest bandwidth consumption, which means that nodes occupy less bandwidth and means fewer collisions. This explains to some extent, in figure 5, why the response rate of BF will fall back when the speed is close to 35 m/s. BREB will trigger the adaptive routing adjustment algorithm after the topology changes, which will increase the bandwidth usage, thus the bandwidth growth of BREB is relatively volatile.

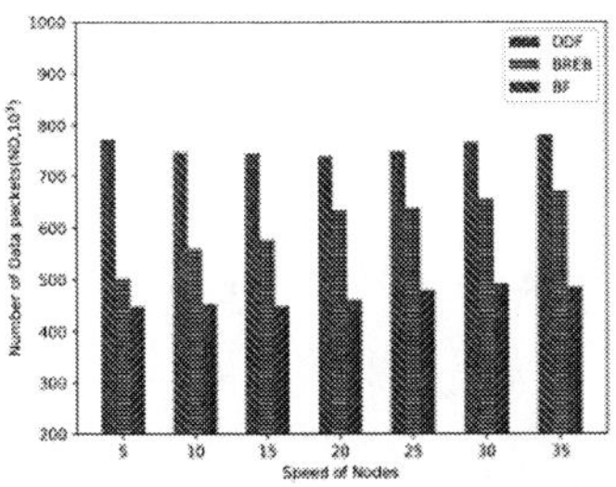

Figure 7. Result of ND

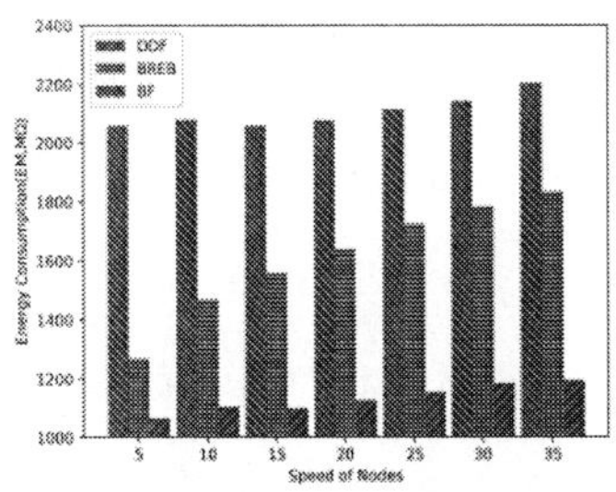

Figure 8. Result of EC

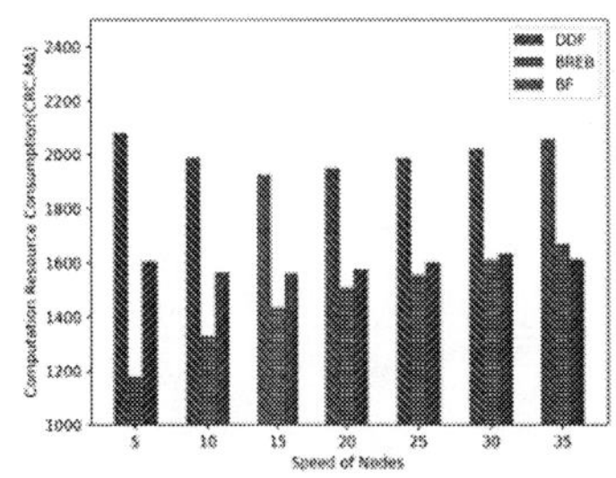

Figure 9. Result of CRC

Figure 7 shows the total number of Data packets sent by the three protocols at different speed. E experimental results intuitively indicate that distance perception can effectively reduce the redundant data packets brought by BF. The number of data packets sent by the BF and the DDF protocols does not change much as speed increases. BREB protocol can effectively reduce redundant data packets at low speeds, while the relatively slow distance update algorithm introduces negative influence on the performance of BREB as the speed continues to increase. Figure 8 and 9 show the energy and computation resource consumption of the three protocols at different speeds. The DDF protocol updates the distance information by flooding interest packets which consumes additional energy and computation resources. The simulation results show that the DDF protocol is superior to the BF and the BREB protocols on energy consumption. However, at low speed situations, the computation resource consumption of the DDF protocol is higher than that of the BREB protocol, although the difference is getting smaller as the speed increases. On the other hand, the DDF protocol always performs better than the BREB protocol regarding resource utilization.

To conclude, the proposed DDF protocol can effectively reduce redundant Data packets in the network, thereby reducing bandwidth and energy resource consumption. The fast update of the distance information by flooding Interest packets ensures the timeliness of the distance information, which greatly improves the response ratio of the distance sensing strategy. Although flooding Interest packet consumes additional computation resources, it is beneficial at high speeds.

6. Summary and outlook

NDMANET has attracted quite a lot of research efforts in the past few years. This paper investigates its data forwarding protocol, a vital factor influencing MANET performance. A novel DDF Protocol is elaborated aiming to further reducing redundant packets to be transmitted, decreasing bandwidth and energy resource consumption in MANET scenario. According to the results of comprehensive simulation experiments, the proposed DDF protocol outperforms the legacy BREB and BF approaches, which indicates the proposed protocol is effective and efficient.

The next step of this work could focus on the real-time video streaming in NDMANET, which is more challengeable compared with other applications. On the basis of the elaborated DDF protocol, it is valuable to explore some other innovate approaches to enhance the QoS & QoE of real-time video applications, including both video-on-demand and live broadcasting, in typical NDMANETs.

Acknowledgments

This work is partially supported by the National Natural Science Foundation of China for Young Scholars (Grants No. 61402085 & 61872051), the Science and Technology on Communication Networks Laboratory (Grant No. XX17641X011-03), and the 54th Research Institute of CETC.

References

[1] Liu X, Li Z, Yang P and Dong Y. 2016. Information-centric mobile ad hoc networks and content routing: a survey. *Ad Hoc Networks*

[2] Yu T, Thomas P, Mario G and Sanadidi M Y. 2013. Content routing in the Vehicle Cloud. *Military Communications Conference IEEE*

[3] Kuang, J and Shun, Z.Yu. 2015. CSAR: A Content-Scent based Architecture for Information-Centric mobile ad hoc networks. *Computer Communications*

[4] Matteo V, Ivica R, Uichin L, Lloyd G and Volker H. 2011. On the design of content-centric MANETs. *Eighth International Conference on Wireless On-demand Network Systems & Services IEEE*, 2011:1-8

[5] Tonguz O K, Wisitpongphan N, Parikht J S, Bait F, Mudaliget P and Sadekart V K. 2006. On the Broadcast Storm Problem in Ad hoc Wireless Networks. *International Conference on Broadband Communications. IEEE*

[6] Lu Y, Zhou B, Tung L C, Gerla M, Ramesh A and Nagaraja L. 2013. Energy-efficient content retrieval in mobile cloud. *Workshop on Sigcomm*

[7] Lucas W, Alexander A, Romain K, Rama V, Ryuji W and Lixia Z. 2012. Rapid traffic information dissemination using named data. *Proceedings of the 1st ACM workshop on Emerging Name-Oriented Mobile Networking Design - Architecture, Algorithms, and Applications* - NoM \"12

[8] Angius F, Gerla M and Pau G. 2012. BLOOGO: BLOOm filter based GOssip algorithm for wireless NDN. *Proceedings of the 1st ACM workshop on Emerging Name-Oriented Mobile Networking Design - Architecture, Algorithms, and Applications ACM*

[9] Angius F, Bhiday A, Gerla M and Pau G. 2013. MADN - Multipath Ad-hoc Data Network prototype and experiments. *Wireless Communications & Mobile Computing Conference, IEEE*

[10] Ahmed S H, Bouk S and Kim R D. 2015. RobUst Forwarder Selection in Vehicular Content-centric Networks. *IEEE Communications Letters* 19.9(2015):1-1

[11] Michael M, Vasileios P and Lixia Z. 2010. Listen First, Broadcast Later: topology-agnostic forwarding under high dynamics. *Army Research Laboratory and the U.K. Ministry of Defence*, W911NF-06-3-0001

[12] Amadeo M and Molinaro A. 2012. CHANET: A content-centric architecture for IEEE 802.11 MANETs. *2011 International Conference on the Network of the Future, IEEE*

[13] Bettstetter C, Resta G and Santi P. 2003. The Node Distribution of the Random Waypoint Mobility Model for Wireless Ad Hoc Networks. *IEEE Transactions on Mobile Computing* 2(3):257-269

[14] Han H, Wu M, Hu Q and Wang N. 2014. Best Route, Error Broadcast: A Content-Centric Forwarding Protocol for MANETs. *Vehicular Technology Conference, IEEE*

[15] Bettstetter C, Resta G and Santi P. 2003. The Node Distribution of the Random Waypoint Mobility Model for Wireless Ad Hoc Networks. *IEEE Transactions on Mobile Computing*, 2(3):257-269

IOP Conf. Series: Materials Science and Engineering **685** (2019) 012013 doi:10.1088/1757-899X/685/1/012013

Enabling Edge Computing Ability in Mobile Satellite Communication Networks

Junyu Lai[1,2], Yudi Zhang[1], Lei Zhong[1], Ying Qu[1]and Rui Liu[1]

[1]School of Aeronautics & Astronautics, University of Electronic Science and Technology of China, Chengdu 611731, China
[2]Science and Technology on Communication Networks Laboratory, China

laijy@uestc.edu.cn

Abstract. Mobile Satellite Service (MSS) can provide users reliable communication abilities wherever they are on earth surface, even in a high-speed motion status, which can either be stand-alone or a complementary service to the terrestrial communication systems. In the recent past, diverse MSS applications have been emerging rapidly. Some introduce unprecedentedly heavy traffic transmission pressure, and some others require real-time response, both of which cannot be satisfied by legacy mobile satellite communication networks anymore. This paper proposes an enhanced strategy to enable edge computing ability on the existing satellite networks. More precisely, this paper re-architects a GEO satellite node to become a cloudlet which possesses a set of computation, storage and networking resources, and can accomplish a certain amount of computing jobs which are typically happened on centralised cloud centre on the ground. A test-bed containing the emulated cloudlet-enabled GEO satellite has been built on a single PC. Functional test and performance evaluation are conducted for FTP file transmission and HTTP web browsing services. Experimental results show that the enhanced edge computing ability can effectively and efficiently decrease bandwidth consumption and lower the response latency in mobile satellite networks.

1. Introduction

Limited by the building costs and technology restrictions, ground base stations of terrestrial communication networks cover only 1/5 of the total land area. Mobile satellite communication system is a vital complement to terrestrial networks in desert, polar, ocean surface, and other geographical locations where terrestrial base stations are difficult to be installed. Mobile satellite communication service leverages microwave broadcasting feature to seamlessly cover the major part of earth surface, although the transmission bandwidth is relatively low and the latency is much higher than typical terrestrial networks. In the past decade, as the emerging of diverse novel applications such as HDTV, VR/AR, interactive online gaming, etc., increasingly more transmission bandwidth consumption and shorter responding latency are required in communication networks. Legacy mobile satellite communication networks are not able to satisfy these needs anymore.

This paper mainly focuses on alleviating the bandwidth shortage and long response delay problems. It innovatively integrates the edge computing framework to the traditional mobile satellite communication networks. More specifically, the paper leverages a set of promising ICT technologies such as SDN, NFV, light-weight virtualization technologies to re-architects a GEO satellite node to be a cloudlet which is equipped with a pool of computation, storage and networking resources, so to handle a number of computing tasks which originally run on the central cloud located behand the gateway station on the ground. A test-bed containing an emulated cloudlet-enabled GEO satellite is developed on a single physical machine. Functional test and performance evaluation are conducted for two typical applications, namely FTP based file transmission and HTTP based web browsing. The results illustrate that the introduced edge computing ability can effectively and efficiently decrease bandwidth consumption and lower the response latency in mobile satellite networks.

The remaining part of this paper proceeds as follows: First of all, Section 2 introduces the state-of-the-art followed by challenges faced by legacy mobile satellite networks in Section 3. Then, as the major contribution, Section 4 enhances mobile satellite communication network with edge computing ability by re-architect the GEO satellite nodes. Functional test and performance evaluation are conducted in Section 5. Finally, Section 6 summaries the paper and gives the outlook.

2. Related work

Research on edge computing's application in ground terrestrial networks has recently gained much attention in both academia and industry. In the meanwhile, a promising topic of enabling edge computing ability in satellite networking domain also start to attract increasingly more attentions. Denby et al. in [1] propose orbital edge computing. Within the framework of this concept, satellite clouds and corresponding TV and Radio programs available for users, as well as new opportunities and their potential influence on the information resources development have been discussed. Kanev et al. in [2] elaborate the concept of satellite-based cloud computing, which integrates virtualized information resources from satellites and the Internet. Within the framework, satellite clouds and corresponding television and radio programs available to users are discussed, as well as new opportunities and their potential impact on the development of information resources. Zhang et al. in [3] introduce the architecture and application scenarios of satellite-terrestrial network, and proposed to integrate the satellite mobile edge computing (SMEC) technique in STN-base mobile communication systems to improve QoS of STN. To improve the performance of 5G satellites, Yan et al. in [4] propose a 5G satellite edge computing framework (5GsatEC) to reduce latency and to expand network coverage. The framework consists of an embedded hardware platform and edge computing micro services in the satellite. To improve the flexibility of the framework in complex scenarios, authors unify the resource management of central processing units (CPUs), graphics processing units (GPUs) and field programmable gate arrays (FPGAs). Wang et al. in [5] present a new type of satellite terrestrial network with double edge computing to gain the benefits of providing computing services in remote areas. The strategy aims to solve the problem of effectively scheduling edge servers distributed in satellite terrestrial networks to provide more powerful edge computing services. Wang et al. in [6] describe edge computing to ease the pressure on the data center. A new approach called Edge Computation Based Differential Positioning (ECDP) does not send location information to the data center, but instead selects the nearest reference station to perform edge calculations and transmits the difference value directly to the mobile receiver.

The aforementioned research efforts on combining edge computing and satellite network are just staying in the concept designing stage, the concrete emulation or implementations are rarely to be discussed. Therefore, this paper not just focuses on the architecture design but also on prototype implementation leveraging on a bunch of promising ICT technologies.

3. Challenges in legacy mobile satellite communication networks

3.1. Typical mobile satellite communication network

Currently, mobile satellite communication systems can be either GEO satellite based (e.g. Inmarsat) or LEO satellite based (e.g. Iridium & GlobalStar). This paper focuses on the former one due to its commercial success. Figure 1 illustrates a typical mobile satellite system, which is consist of three major components, namely a number of terminal stations, a GEO satellite, as well as a gateway station. In particular, terminal stations can be small devices handheld by users or be larger ones mounted on vehicles, ships, and planes. GEO satellite serves as the signal relaying node either in digital mode or analogue mode, which is merely a networking node. As the name indicates, the gateway station is the gateway for terminal stations to the backbone network on the ground.

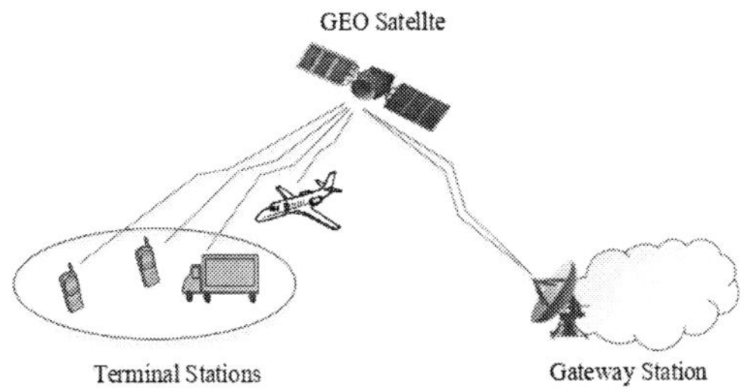

Figure 1. A typical mobile satellite system.

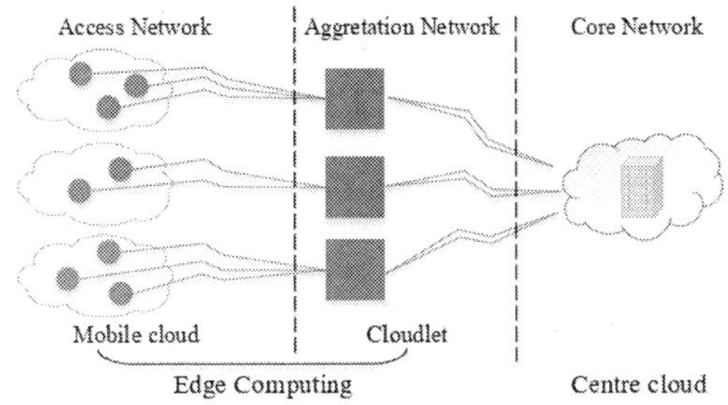

Figure 2. Concept of edge computing.

3.2. Analysis on Existing challenges
Originally, mobile satellite system shown in figure 1 are designed to support legacy voice, video and data transmission application for users. In the recent past, as the fast development of smart devices, diverse novel applications appear and develop rapidly which consequently introduce much higher bandwidth consumption, as well as need lower response latency. Legacy mobile satellite system cannot satisfy these increasingly difficult requirements anymore. Therefore, it is vital to decrease the consumed network bandwidth and to accelerate the task responding speed from the backend.

4. Enhancing mobile satellite communication network with edge computing ability

4.1. Concept of edge computing
Edge computing has attracted quite a lot of research interests in the past five years. It was originally motivated by IoT (Internet of Things) and telecommunication applications in terrestrial communication networks on the ground. The basic idea is to deploy computation resources on the networking nodes located at the networking edge. These edge nodes can either be the aggregation nodes (each of which connects and serves a number of terminal devices), or directly be the terminal nodes themselves. In this sense, some parts of the data or tasks generated by the terminal users or devices can be processed at the networking edge, which consequently has a potential to decrease the bandwidth consumption of the backhaul networks, and in the meanwhile to lower the response delay of the upper layer real-time applications. It is worth noting that edge computing technology is not proposed to completely replace the legacy centralized cloud computing framework. On the contrary, edge computing is a decent complement to the centralized computation architecture. Figure 2 presents the concepts of edge computing, as well as its integration with cloud computing centres located in the core network. At present, applying edge computing technology in existing satellite networks start to attract research attentions from both academia and industry, as indicated in Section 2.

4.2. Re-architect legacy mobile satellite communication network to support edge computing
A basic illustration of mobile satellite communication network has been given in figure 1. Aiming at enhancing the legacy mobile satellite communication network to support edge computing, this paper focuses on the GEO satellite node, which service a number of terminal stations distributed on a wide geographic area of the earth surface. The principle idea is to deploy a certain computation and storage resources on each GEO satellite node plus with its original networking functionalities, so some parts of the data and tasks generated at the ground terminal stations can be processed and accomplished on networking edge nodes just one hop away.

In order to re-architect the GEO satellite nodes, a set of promising ICT technologies are employed to transfer a legacy GEO satellite node to be a cloudlet, These ICT technologies mainly include software-defined networking (SDN), light-weight virtualization (i.e. Docker container), network functional virtualization (NFV), cloud and service orchestration technologies. More precisely, SDN divides the control plane from the data plane in the satellite networks, which makes the network to be programmable so to satisfy the flexibility requirements in cloud computing domain. Light-weighted virtualization (i.e. Docker container) can utilize the underlay physical computation and storage resources in a much more efficient manner compared with traditional Virtual Machine (VM) methods. All the edge computation tasks will happen on these light-weighted virtualization instances. NFV moves the data plane from hardware device to light-weighted virtual instances running on commodity servers, which permits more agile software-based resource and service orchestrations. Cloud technology implements best practices in scalable services, leveraging software-based solutions, micro-services, virtualized commodity platforms, elastic scaling, and service composition, which allows network operators to innovate more rapidly [7]. Service orchestration technology assembles and composes (micro) services. It unifies cloud infrastructure services, SDN control plane services, and NFV services. Figure 3 demonstrates the architecture of the edge computing framework in mobile satellite communication systems, integrating the above technologies.

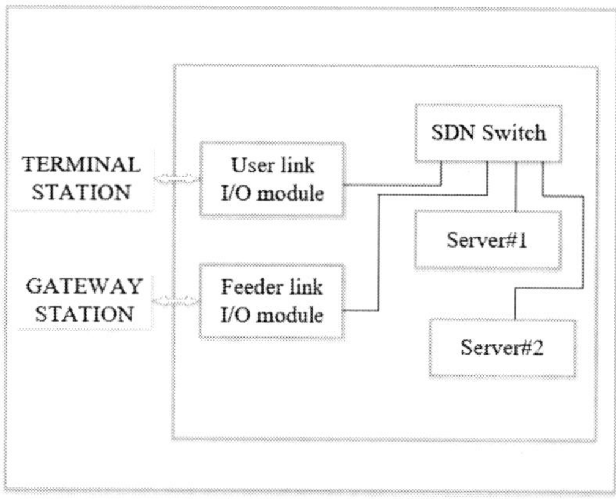

Figure 3. Edge Computing framework in Mobile Satellite Communication Systems.

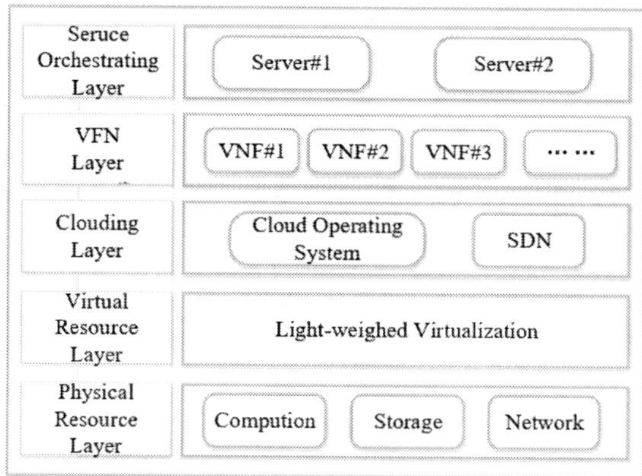

Figure 4. Reference payload

The advantage of re-architecting the GEO satellite node can fully utilize its feature of aggregating traffic of all the ground terminals, namely the GEO satellite naturally is a suitable node to deploy cloudlet and to conduct edge computing. On the other hand, only the satellite backhaul bandwidth consumption can be saved and due to the distance between a ground terminal and a GEO satellite is still long (at least ~35,786 kilometres), the decrease of response latency is limited to one half of the previous delay. However, it is believed that the exploration carried out in this paper is still valuable in mobile satellite system domain.

4.3. Implementation details

As already discussed, the key point of this design is to reconstruct the legacy GEO satellite node to be a cloudlet. Without loss of generality, this paper implements the above architecture on commodity hardwires. Figure 4 illustrates a reference payload implementation of edge computing ability enabled GEO satellite. It mainly includes:

- Physical servers: ARM architecture based computers with commodity RAMs and Discs, which are robust enough to be deployed in space environment;
- SDN switches: Ethernet based switch supporting OpenFlow protocols and can be configured remotely via southbound interface;
- User link I/O module: Interface module between the GEO satellite and the user terminal stations on the ground, realizing data transmission switch between wireless and Ethernet;

- Feeder link I/O module: Interface module between the GEO satellite and the Gateway stations on the ground, realizing data transmission switch between Ethernet and wireless.

In this reference implementations, all the adopted software are open source based. In particular. Operating systems on physical servers are Ubuntu 16.04; Light-weighted virtualization are using Docker container technology; Kubernetes is selected as the cloud OS; SDN controller adopts ONOS controller; Open vSwitch (OVS) is chosen as the software switch; Service orchestration uses XOS.

5. Functional Test and Performance Evaluation

5.1. Test-bed Setting up

An emulation test-bed is developed on a single physical machine for functional tests and performance evaluation. The architecture of the test-bed is demonstrated in figure 5, where only one terminal station (TERM), one GEO satellite and one gateway (GW) station are considered. More precisely, Kubernetes & Docker are installed on this physical machine, a POD is generated to emulate the TERM node. Mininet emulator is adopted to construct the network connecting GEO and GW nodes. In fact, GEO and GW nodes are LXCs (Linux Containers) generated by the mininet emulator, which again is a POD in Kubernetes. By configuring the link between the GEO and the GW nodes, as well as the link between two PODs connecting the GEO and the TERM nodes, the emulation test-bed is established. Detailed configurations are summarised in Table 1.

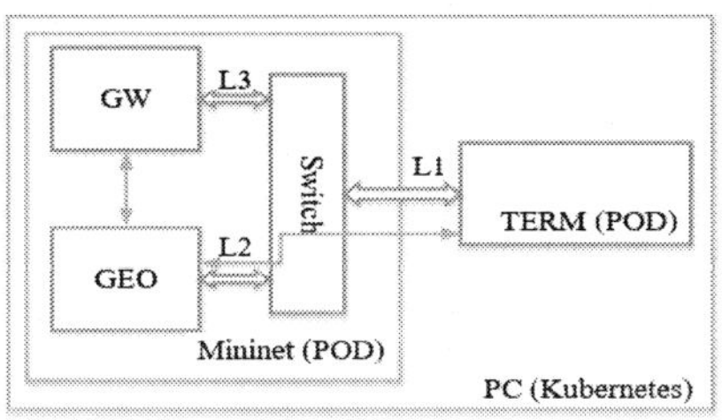

Figure 5. Architecture of Emulation Test-bed.

Table 1. Configuration of the emulation test-bed.

Item	Value
Bandwidth(L1)	1Mbps
Bandwidth(L2)	4Mbps
Bandwidth(L3)	4Mbps
Delay(L1)	40ms
Delay(L2)	270ms
Delay(L3)	135ms

5.2. Functional Test

The key point of the re-architected network is the edge computing ability in the GEO node. To test this enhanced architecture, this paper chose two most common applications: FTP file transfer and HTTP web browsing. A file server and a web server are installed on the emulated GEO node in the above test-bed. FTP and HTTP Client programs running on the TERM node are tested to fetch files from the GEO node. The snapshots of the testing results are given in figures 6 and 7, respectively. It is indicated that by enabling the edge computing ability in GEO nodes, applications can be conducted in a position more closing to users, which will definitely introduce potential advantages.

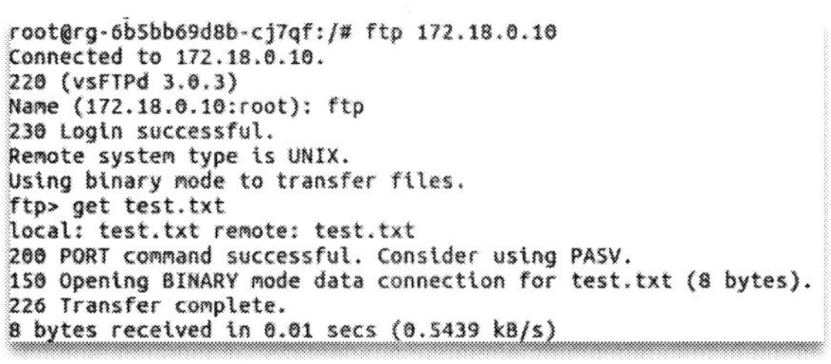

Figure 6. FTP file transfer.

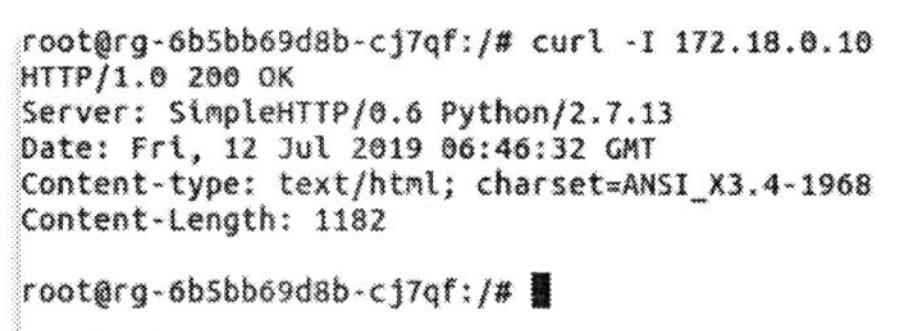

Figure 7. HTTP web browsing.

5.3. Performance evaluation

On one hand, bandwidth is precious in mobile satellite network. On the other hand, transmission delay is unavoidable due to long distance between GEO node and Earth. Particularly, bandwidth consumption (BC) is the sum of data delivered via all links, and transmission delay (TD) refers to the E2E latency, which are two most important performance metrics. This paper carried out comparative evaluation experiments for legacy and edge computing ability enabled mobile satellite network.

For the bandwidth consumption metric, FTP file server program is installed in the GEO node and the GW node, respectively. 5 files of different sizes ranging from 0.5MB to 2MB are transferred from the FTP server program to the TERM node. The results are present in figure 8, which shows that around half bandwidth can be saved by the edge computing enhanced network architecture.

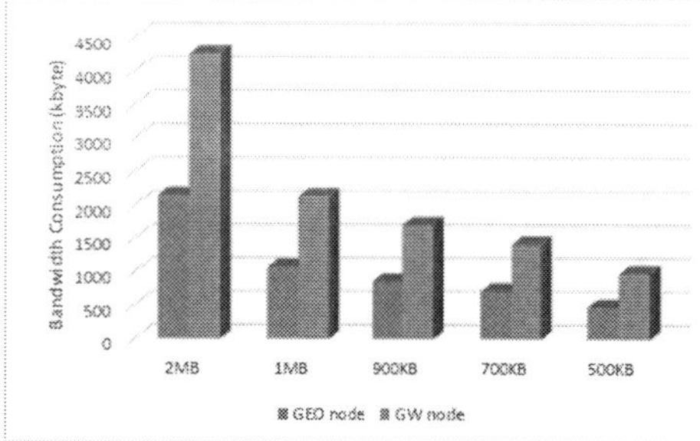

Figure 8. FTP test results.

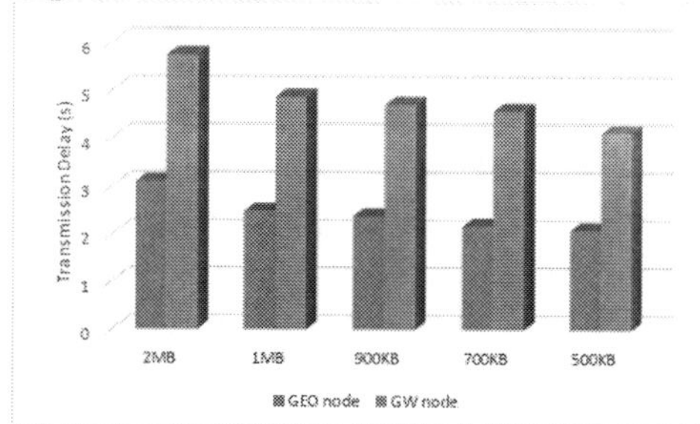

Figure 9. HTTP test results.

For the transmission delay metric, HTTP Web server is deployed in GEO and GW nodes, respectively. 5 HTML files of different size ranging from 0.5MB to 2MB are delivered from the Web server to the TERM node. The transmission delay is the latency from the moment a HTML file request sent by a TERM node to the instance the TERM node receives the requested file. The results in figure 9 illustrated that significant latency decreasing can be achieved by the enhanced network architecture.

6. Conclusions

Edge computing has attracted many research efforts in the past few years. This paper investigates its application in mobile satellite communication network domain to decrease bandwidth consumption and lower the response time. A set of promising ICT technologies are employed to reconstruct the legacy GEO satellite node to be a cloudlet so to accomplish some real-time tasks locally. Experimental results demonstrate the proposed edge computing framework's effectivity and efficiency. The next step of this work could focus on more complicated GEO-LEO double layer satellite networks, deploying cloudlets on both GEO and LEO layers, which is technically more challengeable.

Acknowledgments

This work is partially supported by the National Natural Science Foundation of China for Young Scholars (Grants No. 61402085 & 61872051), the Science and Technology on Communication Networks Laboratory (Grant No. XX17641X011-03), and the 54th Research Institute of CETC.

References

[1] Denby B and Lucia B. 2019. Orbital Edge Computing: Machine Inference in Space. *IEEE Computer Architecture Letters*

[2] Kanev K and Mirenkov N. 2011. Satellite Cloud Computing. 2011 IEEE Workshops of International Conference on IEEE

[3] Zhang Z, Zhang W and Tseng F H. 2019. Satellite Mobile Edge Computing: Improving QoS of High-Speed Satellite-Terrestrial Networks Using Edge Computing Techniques. IEEE Network, 33(1):70-76

[4] Yan L, Cao S, Gong Y, Han H, Wei J, Zhao Y and Yang S. 2019. SatEC: A 5G Satellite Edge Computing Framework Based on Microservice Architecture. Sensors

[5] Wang Y, Zhang J, Zhang X, Wang P and Liu L. 2018. A Computation Offloading Strategy in Satellite Terrestrial Networks with Double Edge Computing. IEEE International Conference on Communication Systems (ICCS)
[6] Wang LN, Li LL and Qiu R. 2019. Edge Computing-based Differential Positioning Method for BeiDou Navigation Satellite System. KSII Transactions on Internet and Information Systems
[7] Peterson L, Al-Shabibi A, Anshutz T, et al. 2016. Central office re-architected as a data center [J]. IEEE Communications Magazine, 54(10):96-101

IOP Conf. Series: Materials Science and Engineering **685** (2019) 012014 doi:10.1088/1757-899X/685/1/012014

Study on Linear and Nonlinear Coupling Damage Model of BGA Solder Joint

Dongyang Lei[1], Yufeng Sun[1], Chaojie Qi[1] and Weiwei Hu[1]

[1] School of reliability and system engineering, Beihang University, Beijing China 100191

Email: hww@buaa.edu.cn

Abstract. Due to the complex and changeable environment in space, the ball grid array (BGA) has become one of the most important form of the electronic package in aerospace industry. As the main factors which affect the reliability of electronic products, temperature and vibration will lead to creep and fatigue respectively. Based on the damage mechanism, taking SnAgCu solder joints as the research object, the relationship between the resistance and the damage index are established by using the micro-conductive model of resistance. The mathematical model between stress cycle and damage value is established by damage mechanics. Then, the linear and nonlinear interaction models are established to simulate the initial and mature damage of solder joints respectively based on the principle of area interference. Finally, Monte Carlo method is used to simulate the growth of the crack, concluding that it's suitable for the linear model when the damage index is less than 0.155 and the nonlinear model is suitable when the damage index is more than 0.155.

1. Introduction

The stress in space is complex and diverse, which makes the ball grid array (BGA) be wildly used in rockets and satellites. The reliability of BGA also become more important than before. In the aspects of physics of failure (PoF), the failure of BGA can be regard as the crack or tiny-hole in the joints. Up to now, the damage caused by thermal stress, vibratory stress and other stresses alone has been studied widely [1]. However, the solder joints are often exposed to the thermal stress, vibratory stress and voltage stress at the same time during service. This coupling effect caused by different physical stresses will affect the failure mechanism and failure mode of the joints, making the failure of joints more complex.

The life prediction of solder joint under multiphysical interaction is usually inaccurate up to now [2]. On the one hand, there is no universal interaction model included all kinds of stresses, on the other hand, the stress in model can't fit the actual situation well. Because the BGA solder joints are mainly subject to thermal stress and vibratory stress during mission profile [3,4]. This paper will explore models to describe the damage of BGA based on damage mechanics, then explore the probabilistic failure models of BGA caused by creep and fatigue. Finally, the using situations of the linear model and nonlinear model will be simulated by Monte-Carlo sampling.

2. Damage model of creep and fatigue of BGA

2.1 Damage

Damage is the fundamental cause of deterioration of the material's mechanical properties, which eventually reduces the stiffness, strength or toughness of the material and reduces the life of the devices. The solder joints are often subject to the combined action of multiphysical stresses at the same time, causing internal damage of the solder joints, reducing the macroscopic mechanical properties, and shorten the remaining life ultimately [5]. Due to the micro-size and high-dense arrangement of solder joints, it is impossible to reflect the actual damage inside the solder joints by measuring the change of mechanical properties of the solder joints. Therefore, the resistance of solder joints is used to characterize the damage. The simplified model of the solder joint is shown below in Figure. 1 [6,7].

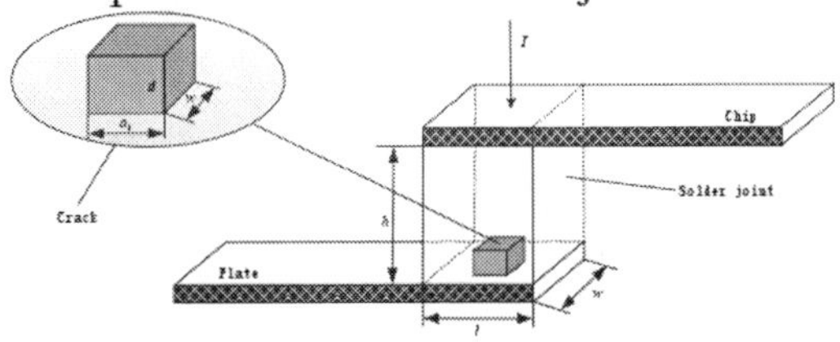

Figure 1. Simplified damage model of BGA joint

The current can't pass the damage area due to the atomic separation at the crack, so the resistance could get increased. No matter the low frequency or high frequency, the damage can be measured as the increase of resistance. According to the damage mechanics [8], the relationship between damage and resistance can be represented as:

$$D = 1 - \frac{R_0}{R_t} \tag{1}$$

Where R_0 is the primary resistance of the solder joint, R_t is the resistance after the crack occurs.

2.2 Creep failure mechanism of BGA solder joints

Creep refers to the phenomenon that the material produces the deformation after the temperature reach or exceed the 1/3 of the melting temperature. The melting temperature of solder joint is around 200°C, so when the components heat up, creep deformation occurs in BGA solder joints during long service. The tensile stress produces after each cycle of temperature, the creep deformation will accumulate until the failure. In theory of energy, the creep damage of solder joint increases with the increase of dissipated inelastic strain energy, and eventually leads to creep failure [9,10].

In theory of continuum damage mechanics, based on the dissipative function put forward by Lemaitre [4], the relationship between creep damage D_c and the cycle of temperature N_c can be deduced as:

$$D_c = 1 - \left(1 - \frac{N_c}{n_c}\right)^{\beta} \tag{2}$$

$$n_c = \beta\sqrt{2ES_0}(w_{acc})^{-1} \tag{3}$$

Where β is the index of creep, n_c is the upper limit of temperature cycle, E is the modulus of elasticity, S_0 is the constant related to the dissipation.

2.3 Fatigue failure mechanism of BGA solder joints

Fatigue refers to the accumulation of damage after stress cycles. Under the stress cycles, the fatigue crack will initiate, expand and destruct [11]. As for electronic component, the fatigue usually comes from the vibration of devices. Taking the cycle hardening into account, the relationship between fatigue damage D_c and the cycle of stress N_c can be deduced as:

$$D_f = 1 - \left(1 - \frac{N_f}{n_f}\right)^{\gamma} \tag{4}$$

$$n_f = \frac{2K'ES_0\gamma}{\sigma_a^2}\left(\frac{2K'}{\sigma_a}\right)^{\frac{1}{n'}-1} \tag{5}$$

Where γ is the parameter of fatigue, n_f is the upper limit of stress cycle, K' is the cycle strengthening factor, n' is the cyclic hardening index.

3. Coupling mechanism of creep and fatigue

The coupling mechanism can be defined as two kinds of form: the linear coupling and the nonlinear coupling. In engineering, the linear coupling is wildly used for it's easy to calculate [12]. But the total damage is not the sum of creep and fatigue actually. There is an overlap of damage area when the two damages meet.

3.1 Linear coupling

In the early time of the crack, there is no interaction between the creep damage and the fatigue damage because of the tiny area of damage. So there is no overlap area of the two damage area, and the total damage of the creep-fatigue coupling is $S_D = S_c + S_f$, the damage can be regard as the sum of the creep damage and the fatigue damage:

$$D = \frac{S_D}{S} = \mathrm{D_c} + D_f \tag{6}$$

Both of the thermal stress and stress exist in the mission profile, so regard the cycle of the temperature and vibration are same: $N_c = N_f = N$, combined with the (2)(4), the relationship between damage D and cycle N can be deduced as:

$$D = 2 - \left[1 - \frac{N}{\beta\sqrt{2ES_0}(w_{acc})^{-1}}\right]^{\beta} - \left[1 - \frac{N}{\frac{2K'ES_0\gamma}{\sigma_a^2}\left(\frac{2K'}{\sigma_a}\right)^{\frac{1}{n'}-1}}\right]^{\gamma} \tag{7}$$

3.2 Nonlinear coupling

When arrive the mid-age of the damage, the damage area of creep and fatigue start to overlap. So the total area of damage is:

$$S_D = S_c + \frac{S_f}{S}(S - S_c)$$

The damage can be deduced as:

$$D = \frac{S_D}{S} = \mathrm{D_c} + D_f - \mathrm{D_c}D_f$$

This nonlinear coupling model suggest that the damage area can be calculated as the sum of creep area and fatigue area, and subtract the common area of these two. The model can be reshaped as:

$$D = 1 - (1 - D_c)(1 - D_f) \tag{8}$$

Same to 3.1, combined with (2)(4), the relationship between damage D and cycle N can be deduced as:

$$D = 1 - \left[1 - \frac{N}{\beta\sqrt{2ES_0}(w_{acc})^{-1}}\right]^{\beta}\left[1 - \frac{N}{\frac{2K'ES_0\gamma}{\sigma_a^2}\left(\frac{2K'}{\sigma_a}\right)^{\frac{1}{n'}-1}}\right]^{\gamma} \tag{9}$$

3.3 Environmental experiment and solving of parameters

The progress of experiment and result data are referred from [11]. The parameters in (7) and (9) can be solved by the data from the experiment, which constituted by marble frame, invar alloy, aluminum alloy, and BGA joint. The thermal expansivity of invar alloy resembles the silicon chips', and the thermal expansivity of aluminum alloy resembles the FR4 substrates.

The cycles of temperature profile and the vibration profile loads in same steps. The temperature ranges from -40°C to 125°C, and the stress range from -40MPa to 40MPa. The hold time of max and min in each cycle is 15min, and the resistance are tested for each 15 cycles after back to room temperature. The model of the experiment is shown in Figure. 2.

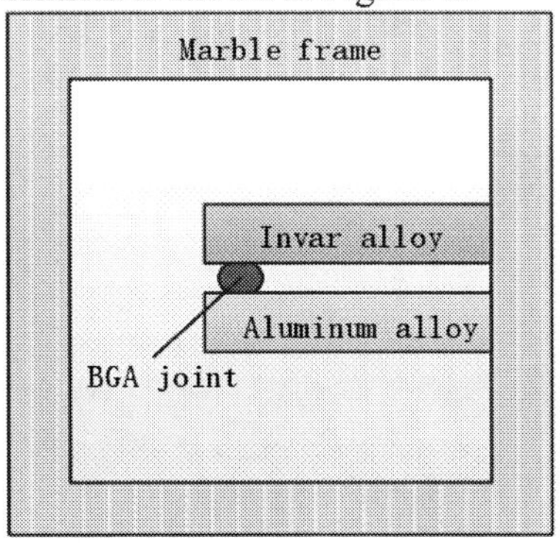

Figure 2. Life test of BGA joint

According the method in [12], the parameters in n_f and n_c can be deduced. Taking these parameters into (2) and (4), the creep damage and the fatigue damage can be calculated by (10) and (11). The Figure. 3 shows the model is consistent with the experimental data mostly [13].

$$D_c = 1 - (1 - \frac{N_c}{605.57})^{0.2029} \quad (10)$$

$$D_f = 1 - (1 - \frac{N_f}{387.82})^{0.2383} \quad (11)$$

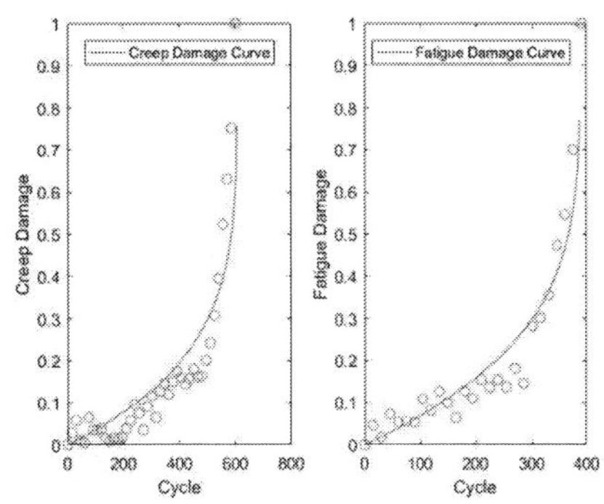

Figure 3. Simplified damage model of BGA

4. Simulation of sampling and result

Combined the (12) (13) with the result in formula (6) (8), the linear model D_l and nonlinear model D_n can be deduced as follow:

$$D_l = 2 - \left[1 - \frac{N}{605.57}\right]^{0.2029} - \left[1 - \frac{N}{387.82}\right]^{0.2383} \quad (12)$$

$$D_n = 1 - \left[1 - \frac{N}{605.57}\right]^{0.2029} \left[1 - \frac{N}{387.82}\right]^{0.2383} \tag{13}$$

From the micro perspective, these two crack show up at random locations. The point of creep and the point fatigue can hardly meet in the initial time of crack, which can be considered as linear coupling. As time goes by, the crack would get growth and the overlap of damage area occurs, after the overlap, the damage can be considered as nonlinear coupling. The cycle of environmental stress where the linear comes to an end and the nonlinear gets start can be deduced as:

$$N_0 = D_l^{-1}(D_0) \tag{14}$$

The boundary between the linear model D_l and nonlinear model D_n depends on the initial location and the speed of growth. The closer they are, the earlier they overlap. According to the law of large number, the conclusion will get convergence with the increase of the simulation times. The mechanism can be simplified as Figure. 4.

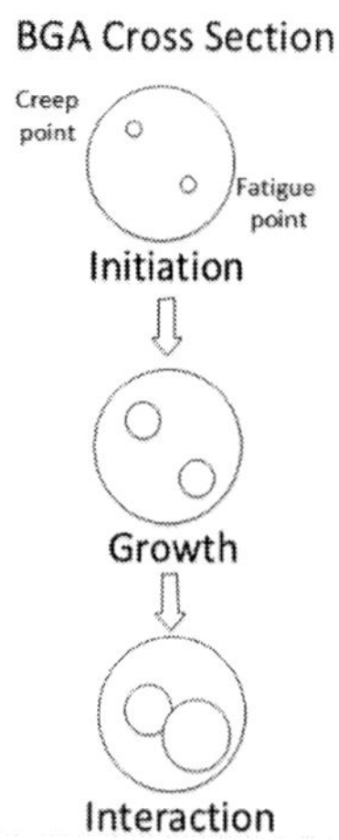

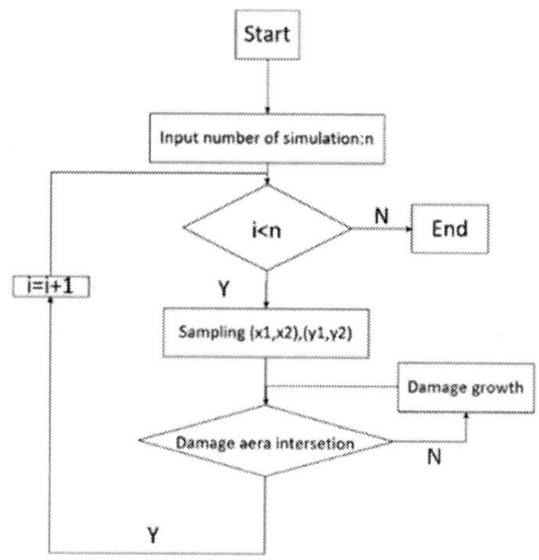

Figure 4. The extension of crack **Figure 5.** Program flowchart of simulation

Monte Carlo method is an efficient way to solve this problem. In each simulation cycle, assuming that there are only one creep crack and one crack point when the damage occurs and their location of the creep crack and damage crack are both random variables. With the environment stresses load on, the cracks extend along by their own speed [14]. When the two damage area overlaps, the simulation stops, returns the D_0 and starts the next cycle. The program flowchart and the simulation result are shown in Figure. 5.

After 1000 times of simulation, the mean of D_0 converges to 0.155. And the load cycle $N_0 = 143$ can be get from formula (16). The result of Mont Carlo simulation and then damage curve are shown in Figure. 6 and Figure. 7:

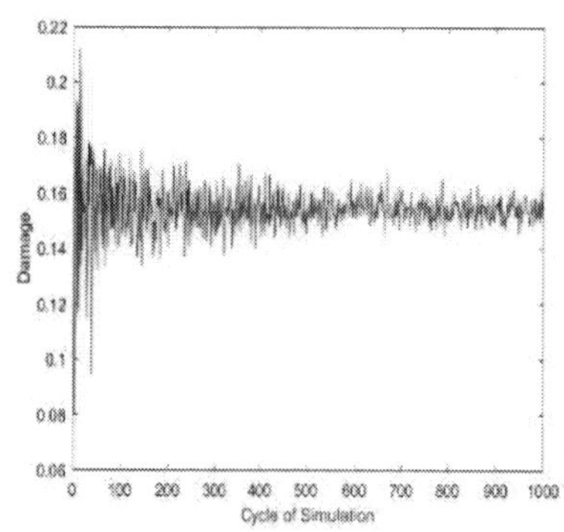

Figure 6. Simulation result

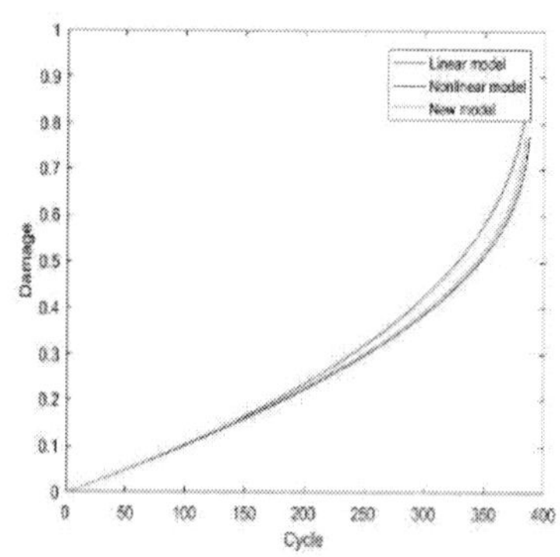

Figure 7. Comparison of models

IOP Conf. Series: Materials Science and Engineering **685** (2019) 012014 doi:10.1088/1757-899X/685/1/012014

5. Conclusion

Probabilistically, there is hardly overlap between the creep damage area and the fatigue damage area at the initial time of the crack, and the total damage can be calculated by: $D = \mathrm{D_c} + D_f$. As the stresses go on, the damage area get larger until the overlap occurs, and the damage can be calculated by: $D = \mathrm{D_c} + D_f - \mathrm{D_c}D_f$. According to the results of simulation. The D=0.155 can be regarded as the boundary of linear model and nonlinear model. So when the proportion of damage area is lower than 0.155, the coupling can be regarded as linear model, and when the proportion of damage area is higher than 0.155, the coupling can be regarded as nonlinear model. The life of BGA made of other materials can be calculated by this method too.

References

[1] Mattila, T. T., & Kivilahti, J. K., 2010, *The role of recrystallization in the failure of snagcu solder interconnections under thermomechanical loading*. (IEEE Transactions on Components and Packaging Technologies, vol 33,issue 3), p 629-635.

[2] Chen, L., Jiang, J., Fan, Z., Chen, X., & Yang, T., 2007, *A new model for life prediction of fatigue–creep interaction*. (International Journal of Fatigue, vol 29, issue 4), p 615-619.

[3] J. Lamaitre, 1996, *Damage mechanics*

[4] Hui, X., Li, X, Yan, Y., Liu, N., & Shi, Y. 2013, *Damage behavior of snagcu solder under thermal cycling*. (Rare Metal Materials and Engineering, vol 42,issue 2), 221-226.

[5] Schubert, A. , Dudek, R. , Auerswald, E. , Gollhardt, A. , Michel, B. , & Reichl, H. . (2003). *Fatigue Life Models for SnAgCu and SnPb Solder Joints Evaluated by Experiments and Simulation*. (Electronic components and technology conference. IEEE Xplore.)

[6] C.Basaran, 2002, *Thermomechanical analysis of solder joints under thermal and vibrational loading*. (Trans. ASME, , Vol 124) p. 60–66

[7] Ghaffarian, R., & Kim, N. P.. 2000. *Reliability and failure analyses of thermally cycled ball grid array assemblies*. (IEEE Transactions on Components and Packaging Technologies), vol 23 issue 3),) p. 528-534.

[8] Zuo, Y., Ma, L., Liu, S., Wang, T., Guo, F., & Wang, X., 2013, *The coupling effects of thermal cycling and high current density on sn58bi solder joints*. (Journal of Materials Science, vol 48 issue 6), p. 2318-2325.

[9] Pang, J. H. L., Wong, F. L., Heng, K. T., Chua, Y. S., & Long, C. E. 2013. *Combined vibration and thermal cycling fatigue analysis for SAC305 lead free solder assemblies*. (Electronic Components & Technology Conference. IEEE.)

[10] Xiao, H., Li, X. Y. Hu, Y. , Guo, F. , & Shi, Y. W. , 2013, *Damage behavior of snagcu/cu solder joints subjected to thermomechanical cycling*. (Journal of Alloys and Compounds), p.110-117.

[11] Qi C., 2018, Study on the probabilistic failure physical model of BGA solder joint with multi-field coupling (Beihang University) p. 30

[12] Y. Hu, 2006, *Reliability study of SMT solder joint and finite element analysis of CBGA solder joint*. (Nanjing university of aeronautics and astronautics) p. 46

[13] Q. Xu, Y. Xu, W. Chen, 2016, *Analyze of the effect of micro hole on stress based on MC method and FEM*. (Materials for Mechanical Engineering) p. 36

[14] A. Todorut, N. Cordos, M.D. Burdea, M. Balcau, 2015, *The evaluation of normal load redistribution on the static axles and on the wheels, when the vehicle is in motion* (Acta Technica Napocensis, Series: Applied mathematics, mechanics and engineering, vol 58, issue 3) p. 349

IOP Conf. Series: Materials Science and Engineering **685** (2019) 012015 doi:10.1088/1757-899X/685/1/012015

Deep Learning for Aircraft Wake Vortex Identification

Pan Weijun, Duan Yingjie*, **Zhang Qiang, Tang Jiahao and Zhou Jun**

Civil Aviation Flight University of China, Air Traffic Management College

*Dwanyj@sina.com

Abstract. Aircraft wake vortex refers to a pair of closed vortices due to the air pressure difference between the upper and lower surfaces of the wing when the aircraft is flying. During the take-off and landing of the aircraft, the wake vortex of the preceding aircraft threatened the safety of the aircraft following its approach. Especially in the near-field phase, due to the low altitude of the aircraft, the wake vortex is more harmful under the influence of the ground effect. Therefore, the wake vortex of the front aircraft can not be ignored. In order to maintain air traffic safety and accelerate the smooth flow of air traffic, this paper used a YOLO v3 model based on the principle of deep learning to identify the wake vortex generated by the aircraft. Through laser radar scanning, data processing, and deep neural network simulation experiments, the results show that the model can achieve high confidence recognition of the aircraft wake vortex, and can provide air traffic controllers with auxiliary decision information in actual work to ensure aircraft safety.

1. Introduction

The wake vortex formed by the aircraft in flight will form a wake vortex field with weakening intensity under the influence of gravity, atmospheric turbulence, wind speed and atmospheric stability. In the early 1970s, the first aircraft crash which results of wake vortex was reported, when the aircraft crashed into the Boeing 747 (the largest civil aircraft at the time) and encountered a wake vortex[1]. On November 12, 2001, the crash of American Airlines Flight 587 was also due to the wake vortex[2].

In the near-field flight phase, the International Civil Aviation Organization (ICAO) and the Federal Aviation Administration (FAA) are in order to prevent the aircraft from entering the forward wake vortex region, resulting in pitching, rolling, and stalling, meteorological conditions (instrument meteorological condition, visual meteorological condition), means of control (procedural control, radar control), use of runway conditions (single runway, parallel or crossover runway) have developed a relatively complete set of wake vortex spacing standards [3]. China's National Air Traffic Control Committee and the Civil Aviation Administration have also promulgated China's aircraft wake spacing standards based on ICAO standards, and have developed relevant operational specifications[4]. Based on the above problems, the identification of aircraft wake vortex has become a key issue in aviation research. At present, the research on wake vortex detection and identification of civil aviation aircraft mainly includes theoretical analysis, numerical simulation and radar test[5]. Among them, radar test has been recognized as one of the most effective tools for wake vortex detection and identification. At present, the detection tool for the aircraft wake is mainly Doppler lidar. In 2001, Holzapfel F et al. used Doppler lidar to conduct a number of airport wake vortex field detection experiments in Germany, and gave a method for estimating the vortex ring volume of the wake vortex[6]. In 2006, France's Thales used X-band radar to perform a wake vortex detection experiment on an airliner at an airport near Paris[7]. In 2008, Frederic et al. conducted an X-band laser radar wake vortex detection experiment at the Paris airport in France to evaluate the wake vortex monitoring capability under various atmospheric conditions[8]. In recent years, with the hot research of artificial intelligence, deep learning (DL) has become a new method for lidar target recognition. The concept of Deep Learning was proposed by Hinton et al. in 2006. In addition, the convolutional neural network proposed by Lecun et al. is the first real multi-layer structure learning algorithm, which uses spatial relativity to reduce the number of parameters to improve training performance. Deep learning is a representation-based learning method in machine learning, which can simulate the neural structure of human brain. The concept of deep learning originates from the study of artificial neural networks. Artificial Neural Network (ANN) abstracts the human brain neuron network from the perspective of information processing, establishes a simple model, and forms different networks according to different connection modes, referred to as neural network or similar neural network. Therefore, in-depth learning is also called Deep Neural Networks (DNN), which is developed from the previous ANN model of artificial neural networks. With the development of deep learning, especially the excellent performance of convolutional neural network in image recognition, convolutional neural network is widely used in target recognition of lidar. Because the convolutional neural network mimics the human visual mechanism and can more fully express the image information, it is more suitable for practical applications. Therefore, based on the Doppler lidar detection principle, this paper proposes an algorithm based on deep learning theory and uses convolution. The neural network extracts the image characteristics of the tail vortex cloud image to identify the wake vortex, and verifies its effectiveness by accuracy and false alarm rate to assist the decision-making control in the configuration of the wake vortex interval.

2. Design of Data Acquisition Method

The main principle of lidar detection is to emit a laser beam of a specific wavelength to scan the target airspace, and receive the backscattered signal of the laser by atmospheric molecules or aerosol particles through the receiver, in the detection space, due to atmospheric molecules and aerosol particles. Brownian motion causes the captured scatter signal to produce Doppler broadening, while the overall average velocity of the particle results in Doppler shift of the atmospheric echo signal, and the Doppler shift and the detected wake vortex radial velocity. And the following relationship exists between the Doppler shift Δf_D and the detected wake vortex radial velocity V_R:

$$\Delta f_D = 2f_0\frac{V}{C} = \frac{2}{\lambda_0}V_R \tag{1}$$

This paper chooses the mode of transverse detection of wake vortex, i.e. Range-Height-Indication (RHI). As shown in Figure 1 below, the laser beam emitted by radar scans the cross section perpendicular to the flight direction of the aircraft. After processing, the velocity field of the wake vortex detected in the target airspace is obtained, and convolution is used. Meanwhile, the recognition of aircraft wake by lidar is realized by feature extraction layer processing method in the network.

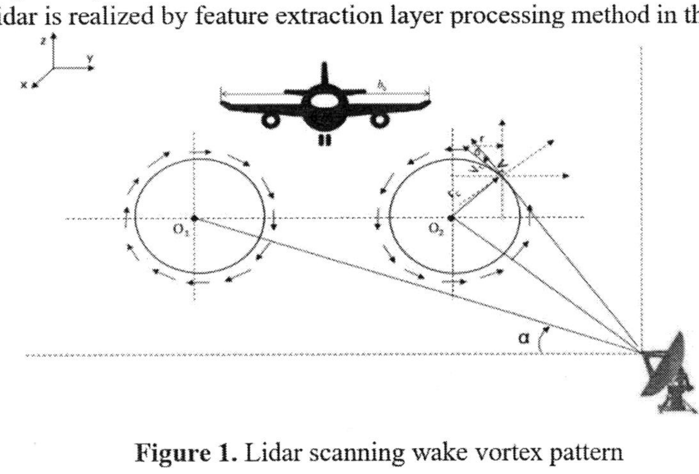

Figure 1. Lidar scanning wake vortex pattern

3. Convolutional Neural Network and Its Detection Implementation

3.1. Convolutional Neural Network

In recent years, convolution neural network has been applied to the field of target recognition, and remarkable research results have been achieved. The convolution neural network is introduced into the target detection of aircraft wake vortices in atmospheric wind field, which provides a new idea and direction for the recognition of wake vortices detected by lidar[9]. Convolution Neural Networks (CNN) is a deep feedforward neural network widely used in computer vision and pattern classification. The training process of convolutional neural network is essentially a process of continuously learning and updating parameters. Training gives an initial random parameter and obtains the output value. When the output value differs from the standard value, the error is calculated, and then the back propagation algorithm is used. The parameters are updated layer by layer from the output layer to the opposite direction until the training accuracy is reached. With the continuous development of deep learning research at home and abroad, CNN has achieved great success in the fields of image recognition and target detection. In the aspect of lidar image recognition, convolutional neural network has its unique advantages[10]. Therefore, CNN is widely used in the field of target recognition of laser radar. Deep learning applications A wide range of target detection algorithms can be divided into two categories. The first category is based on regional nomination algorithms, also known as two-step target detection methods, such as Fast R-CNN, Faster R-CNN, Mask R-CNN, etc. These algorithms divide the process of target detection into two steps. First, the Region Proposal Network(RPN) is used to extract the candidate target information, and then the detection network is used to predict and identify the location and category of the candidate target. The second kind is the end-to-end algorithm, also known as the single-step target detection method, such as SSD, YOLO, YOLO 9000, YOLO V3. This kind of algorithm does not need RPN. It generates the location and category information of the target directly through the network and completes in one step. Therefore, the single-step target detection algorithm has faster detection speed. Because the end-to-end detection algorithm can complete the training quickly, it is hopeful to realize the function of real-time identification and early warning in aircraft wake vortex recognition. Therefore, this paper uses YOLO to realize the recognition of aircraft wake vortex.

3.2. Network Model Construction

YOLO v3 uses the newly designed Darknet53 residual network and combines the FPN network structure. After sampling on the two characteristic maps of the network, the corresponding feature maps are aggregated in the early stage of the network, and then the convolutional network is used to obtain the prediction result. Figure 2 shows YOLO v3 network structure diagram[11].

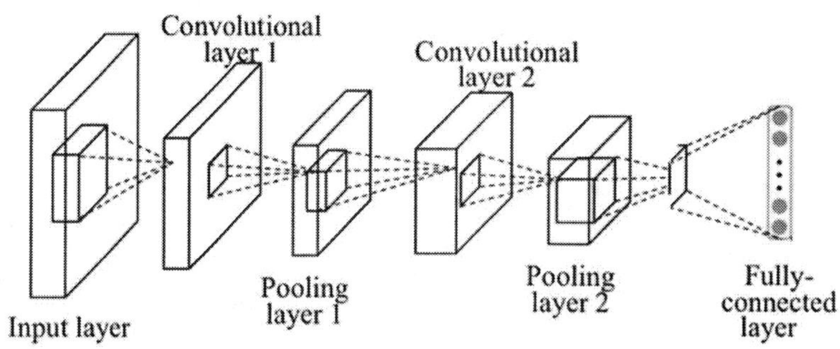

Figure 2. YOLO v3 algorithm flow chart

A convolutional neural network usually consists of an input layer, a convolutional layer, a pooling layer, and an output layer. The network inputs a two-dimensional image, and the convolution layer extracts and maps the detailed features of the image through the form of a sliding window; the pooling layer downsamples the input feature image, thereby reducing the computational complexity and extracting the main features on the one hand. The image feature information is extracted by convolving each convolution kernel with the previous feature map, and then the feature map of the current layer is generated by the activation function, and the extracted feature maps are combined at a higher layer to obtain global features. By convolving the same feature map by setting different convolution kernels, different feature information is obtained, and the purpose of extracting features is achieved.

In this paper, according to the actual wake vortex identification situation, the network structure is adjusted, the network parameters are changed, and the appropriate convolution kernel and activation function parameters are selected to design a network model suitable for actual air traffic control requirements.

3.3. Network Settings

The network modular structure of YOLO v3 is shown in Figure 3. The network structure of YOLO v3 builds the network structure of YOLO improved algorithm based on Darknet-53 network, and realizes the detection of moving targets. Darknet-53 is similar to ResNet and is more powerful than Dacknet-19, with a similar accuracy to ResNet-101 or ResNet-152, but at a faster rate[12].

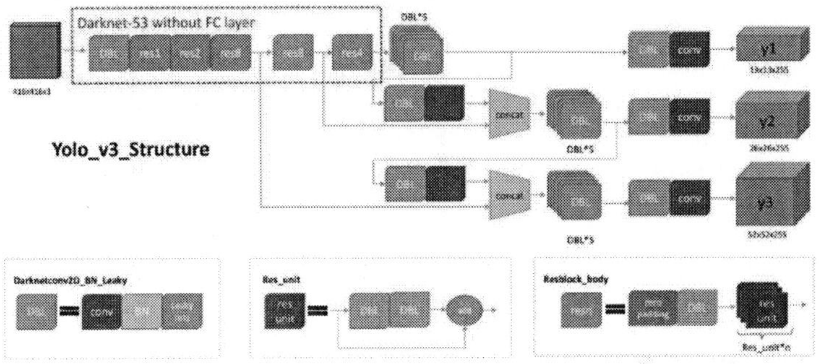

Figure 3. YOLO v3 network modular structure

As shown in Figure 3, the DBL block is the basic component of YOLO v3, namely the convolutional layer BN layer Leaky relu, corresponding to the Darknet conv2 d_BN_Leaky in the code. For YOLO v3, the convolutional layer, the BN layer, and the loss layer are inseparable parts that together constitute the smallest component. The n in the Resn block represents a number, and there are res1, res2, ..., res8, etc., indicating how many res_units are contained in this res_block. This is a large component of YOLO v3, YOLO v3 began to draw on ResNet's residual structure, using this structure can make the network structure deeper. The concat block is a tensor splicing that stitches up the upper layer of the darknet and one of the subsequent layers. The splicing operation is different from the operation of the residual layer add block superposition. The splicing expands the tensor dimension, and the add block just adds directly without causing a change in the tensor dimension.

The entire YOLO v3_body contains 252 layers, which are organized as shown in Table 1 below:

Table 1. Yolo_v3_body network layer

Total	**252**
Add	23
BatchNormalization	72
Concatenate	2
Conv2D	75
InputLayer	1
LeakyReLU	72
UpSampling2D	2
ZeroPadding2D	5

The network features in this article: 1) Multi-scale target prediction. There are 3 scales, and each anchor scale predicts 3 anchor frames. A total of 9 anchor frames can be obtained by K-means clustering. The YOLO v3 model is much more complex than the previous version, and you can weigh the speed and accuracy by changing the size of the model structure.(2) Adopting a better basic network structure Darknet53 residual network.(3) Classification loss The Softmax loss function is replaced by a binary cross entropy loss function. Since Softmax only assigns one target category to each border, and for open data such data sets, the target may have overlapping category labels, so Softmax is not suitable for multi-label classification; Softmax will select the category with the highest score to determine that the current box belongs to The category, while in reality one target may belong to multiple category labels; Softmax can be replaced by multiple independent logistic classifiers, and the accuracy will not decrease.

4. Experiment and analysis

4.1. Border Prediction

Like YOLO, the YOLO v3 algorithm divides the input image into a grid. When there is a center point of a target in the grid, the grid predicts the target. During training and testing, each grid predicts B bounding boxes, each of which has (x y, w, h, c) 5 prediction parameters here (x, y, w, h) is the information of the bounding box, (x, y) is the center point position of the object, w and h are the width and height of the bounding box, respectively, and c is the confidence of the target. The difference is that the YOLO v3 algorithm uses the offset of x and y as the linear transformation of the length and width of the border. The expression is as shown in equation (2),which effectively improves the problem that the boundary of the moving target is not accurate due to the fixed frame size of the YOLO algorithm.

$$\begin{cases} \hat{G}_x = P_w t_x(P) + P_x \\ \hat{G}_y = P_h t_h(P) + P_y \\ \hat{G}_w = P_w e^{t_w(P)} \\ \hat{G}_h = P_h e^{t_h(P)} \end{cases} \tag{2}$$

$\hat{G}_x, \hat{G}_y, \hat{G}_w, \hat{G}_h$ A is the position coordinate of the real frame on the feature map. The estimated value P_x, P_y is the center point of the preset anchor frame on the feature map. The coordinate P_w, P_h is the width and height of the preset anchor frame on the feature map. The relevant information value of the e-learning target is t_x, t_y, t_w, t_h.

The solution of the border information (x, y, w, h) is given by equation (3):

$$\begin{cases} \mathrm{b_x} = \sigma(t_x) + c_x \\ \mathrm{b_y} = \sigma(t_y) + c_y \\ \mathrm{b_w} = p_w^{\,e^{t_x}} \\ \mathrm{b_h} = p_h^{\,e^{t_h}} \end{cases} \tag{3}$$

The final frame information value is $\mathrm{b_x}, \mathrm{b_y}, \mathrm{b_w}, \mathrm{b_h}$, and the network learning target information value is $t_x, t_y, t_w, t_h . c_x, c_y$. , is the coordinate offset of the grid p_w, p_h is the side length of the preset anchor border.

The confidence score can be expressed as formula (4):

$$Pr(object) * IoU_{pred}^{truth} , \quad \Pr(object) \in (0,1) \tag{4}$$

$\Pr(object)$ reflects the confidence of the target in the current bounding box, and IoU_{pred}^{truth} reflects the accuracy of the current bounding box's predicted target position. If there is no object in the bounding box, then $Pr(object)=0$, if there is an object, then $Pr(object)=1$

The IOU value is calculated according to the ratio of the predicted bounding box and the real bounding box data, and the conditional probability of predicting a common class C object, each grid only predicting the class C object, $Pr(Class_i / object)$,i=1,2,...,C. Each grid predicts the position of the B bounding boxes. That is, the B bounding boxes share a set of conditional class probabilities $Pr(Class_i / object)$, i = 1, 2, ..., C. Based on the calculated $\Pr(Class_i / object)$, the class-related confidence within a bounding box can be calculated during the test, as in formula (5):

$$\Pr(Class_i/object)*\Pr(object)*\mathrm{IoU}_{pred}^{truth} = \Pr(Class_i)*\mathrm{IoU}_{pred}^{truth} \tag{5}$$

$\Pr(Class_i/object)$ is the category information for each grid prediction, and $\Pr(object)*\mathrm{IoU}_{pred}^{truth}$ is the confidence level for each bounding box prediction. Equation (5) can get the probability that the predicted border belongs to a certain class, and the accuracy information of the border prediction can also be obtained.

At the time of inspection, the classification confidence of each category bounding box is equal to the product of the confidence of each target bounding box and the category information for each grid prediction. After obtaining the classification confidence of each border, by selecting a reasonable threshold, the lower bounding box is removed, and the remaining borders are normalized to obtain the final detection result.

4.2. Loss Function

During the training process, YOLO uses the mean square and error as the loss function. It consists of three parts: coordinate error, IOU error and classification error, as shown in equation (6):

$$Loss = \sum_{i=0}^{s^2} coodErr + iouErr + clsErr \tag{6}$$

Taking into account the contribution rate and relative error of each loss, the loss of the YOLO v3 algorithm during the training process is calculated as (7):

$$\begin{aligned}
&\lambda_{coord}\sum_{i=0}^{S^2}\sum_{j=0}^{B}\mathbb{Z}_{ij}^{obj}\left[(x_i-\hat{x}_i)^2+(y_i-\hat{y}_i)^2\right]+\lambda_{coord}\sum_{i=0}^{S^2}\sum_{j=0}^{B}\mathbb{Z}_{ij}^{obj}\left[\left(\sqrt{w_i}-\sqrt{\hat{w}_i}\right)^2+\left(\sqrt{h_i}-\sqrt{\hat{h}_i}\right)^2\right]\\
&+\sum_{i=0}^{S^2}\sum_{j=0}^{B}\mathbb{Z}_{ij}^{obj}\left(C_i-\hat{C}_i\right)^2+\lambda_{nobj}\sum_{i=0}^{S^2}\sum_{j=0}^{B}\mathbb{Z}_{ij}^{obj}\left(C_i-\hat{C}_i\right)^2\\
&+\sum_{i=0}^{S^2}\mathbb{Z}_{i}^{obj}\sum_{C\in classes}\mathbb{Z}_{ij}^{obj}\left(p_i(C)-\hat{p}_i(C)\right)^2
\end{aligned} \tag{7}$$

Where $\hat{x}_i, \hat{y}_i, \hat{w}_i, \hat{C}_i, \hat{p}_i(C)$ is the predicted value and $x_i, y_i, w_i, c_i, p_i(c)$ is the training mark value. $\mathbb{Z}_{ij}^{obj}$ indicates that the object falls within the jth bounding box of the lattice i.

If there is no target in a cell, the classification error is not back-propagated; in the B bounding box, the back-propagation of the coordinate error is performed only for the predicted value with the highest IOU, and the rest are not performed.

4.3. Experiment result

In the depth learning of images, in order to enrich the image training set, better extract image features, generalize the model, and usually enhance the sample image data[13]. This experiment uses rotation processing, multi-scale scaling and changing contrast and three kinds of algorithms to enhance the training data, increase the amount of data in the training data set, and enable the trained model to learn more pixel information of the image, thereby improving the model. Classification performance and generalization ability[14]. This paper trains 5,000 samples, 4,600 training sets, 400 test machines, and a total of 1000 training sessions. The learning rate (LR) controls the learning progress of the model. If the learning rate is set too small, the network convergence will be slow and the learning efficiency will be significantly reduced. If the learning rate is set too large, the parameter update will be intensified, causing the network to converge to the local The best, or the loss value, starts to increase directly. Therefore, after many experiments, the basic learning rate is set to 0.001, the momentum coefficient is 0.9, and the attenuation coefficient is 0.005. Some experimental results of the tail vortex identification test are shown in Figure 4.

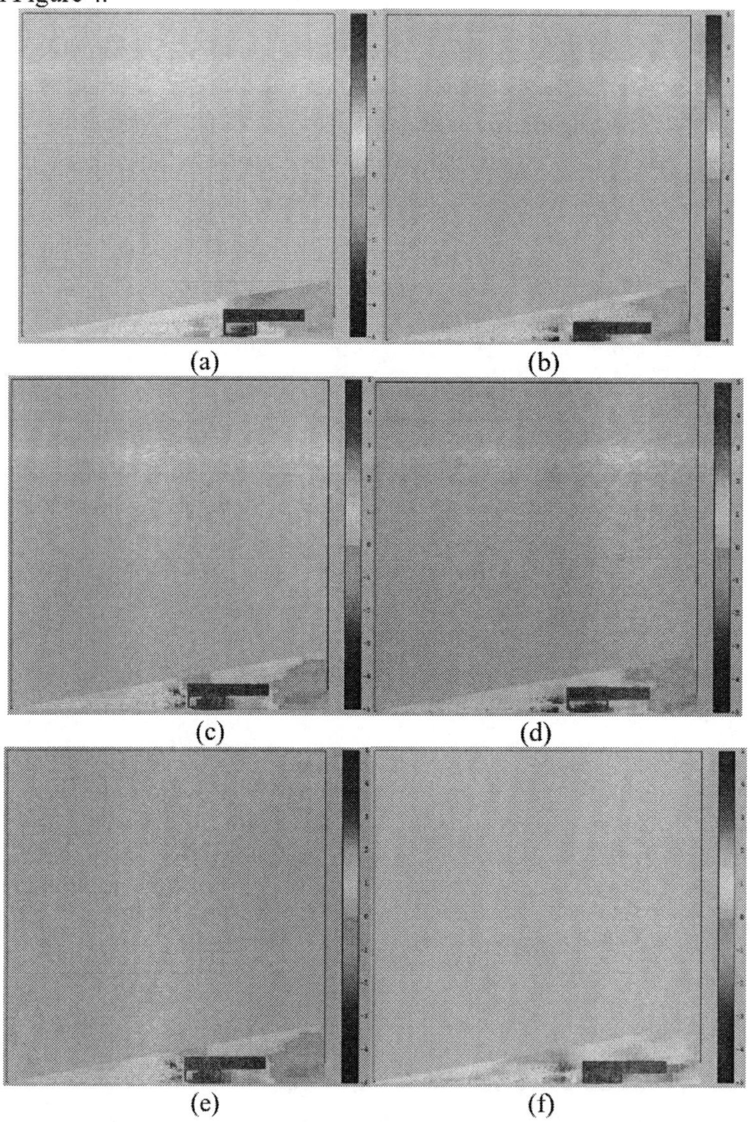

Figure 4. Part of experient results — identification of the wake vortex by YOLO v3

The experimental results show that when a wake vortex is detected, the network model displays 'Deteted' and gives a predicted value. Its accuracy rate reaches to 94.46%, and the recall rate is 91.27%.In order to effectively identify and avoid the danger of aircraft wake vortex, a YOLO v3 based algorithm is proposed, which can predict the target from multiple scales. The experimental results show that the proposed algorithm can accurately identify the wake vortex detected by the lidar and accurately predict the frame return to the target, which effectively improves the airport capacity. The algorithm can better detect the wale vortex of the approaching stage and provide technical support for the new civil aviation technology.

Acknowledgments
At the point of finishing this paper, I'd like to express my sincere thanks to all those who have lent me hands in the course of my writing this paper. First, I'd like to take this opportunity to show my great gratitude to my supervisor, Mr.Pan,who has given me so much useful advices on my writing. Secondly, I'd like to thanks my classmates who offered me references on time. Last,I'd like to thanks National Natural Science Foundation of China(U1733203),National key R&D Program Funding Project(2016YFB0502403),Technology Program(Grant No.2018GZ0167,2018JY0394),the Safety Capacity Building Program of Civil Aviation Administration of China(Grant No.0241928) and Innovation and Entrepreneurship Program of CAFUC(S201910624014)

Reference
[1] Crichley J B,Foot P B.Analysis of incidents reported between 1972 and 1990[C]//Proceedings of the Aircraft Wake Vortices Conference 1992 London:Civil Aviation Authority Wake Vortex Database
[2] Murphy B,Callaghan O,Fox M 2005 *Overview of the structures investigation for the American airline flight 587 investigation*(46th AIAA/ASME/ASCE/AHS/ASC Structures,Structural Dynamics&Materials Conference) (Austin,Texas:American Institute of Aeronautics and Astronautics)
[3] ICAO,Air regulations and air traffic services 2000 *Air Traffic Management Bureau of CAAC*
[4] Air traffic service manual(ICAO9426-AN/924) 1999 *Air Traffic Management Bureau of CAAC*
[5] Kopp F,Rahm S,Samlikho I, 2004 vol 21. *Characterization of aircraft wake vortices by 2μm pulsed Doppler lidar*(Journal of Atmospheric and Ocean Technology) p194-206
[6] Holzapfel F, Gera T, Kopp F, et al 2004 vol 20 *Strategies for circulation evaluation of aircraft wake vortex measured by lidar* (Journal of Atmospheric and Oceanic Technology) p 1183-1195
[7] Barbaresco F, Jeanntet A, Meier U 2007 *Wake vortex detection &monitoring by X-band Doppler radar: Paris Orly radar campaign results* (IET International Conference on Radar) p1-5
[8] FREDERIC B, UWE M 2010 vol 11 *Radar monitorin g of a wake vortex: electromagnetic reflection of wake turbulence in clear air*(Comptes Rendus Physique) p54-67
[9] Arel I, Rose D C, Karnowski T P 2010 *Deep Machine Learning - A New Frontier in Artificial Intelligence Research [Research Frontier]*(Computational Intelligence Magazine IEEE) p 13-18
[10] Yin B C, Wang W T 2015 vol 41 *Review of Deep Learning*(China: Journal of Beijing University of Technology*)* p 48-59
[11] Xu G B, 2019 *Moving target detection of remove tower based on improved YOLO algorithm*(Science Technology and Engineering) p 377-383
[12] Redmon J , Farhadi A 2018 *An Incremental Improvement*
[13] Duan M, Wang G P, Niu C Y 2008 *Method of small sample size image recognition based on convolution neural network*(Computer Engineering and Design) p 224-229
[14] Pan WJ, Duan YJ 2019 *Research on aircraft wake vortex recognition using AlexNet* (Opto-Electronic Engineering)

IOP Conf. Series: Materials Science and Engineering **685** (2019) 012016 doi:10.1088/1757-899X/685/1/012016

Study on the Influence of Swept Angle on the Aerodynamic Characteristics of the Cross-Section Airfoil of a Variable Swept-Wing Aircraft

Su Xinbing*, **Jiang Wen and Zhao Xiwei**

College of Aeronautical Engineering, Air Force Engineering University, Xi'an 710038, China

*2356093308@qq.com

Abstract. The variable forward swept-wing aircraft is a typical variant aircraft. The airfoil is a variable airfoil during the wing swept, and its aerodynamic characteristics have a huge impact on the flight performance of the variable swept-wing aircraft. It is also an aircraft. One of the difficult problems to be solved first in aerodynamic layout design. In order to explore the unsteady aerodynamic characteristics of the variable airfoil, a variable airfoil model with different relative thicknesses is established. Then the S-A turbulence model is used and the CFD simulation is used to calculate the variable airfoil unsteady. The variation of the angle of attack and the Mach number when the aerodynamic parameters change in the forward sweep angle. The results show that the airfoil lift coefficient and moment coefficient are negatively correlated with the relative thickness, while the drag coefficient is positively correlated with the relative thickness. These aerodynamic characteristics can provide a useful reference for the aerodynamic layout design and dynamic analysis of the variant aircraft.

1. Introduction

The variable swept-wing layout aircraft is a typical variant aircraft, which can change the wing grazing angle according to the needs, but the grazing will bring serious changes in aerodynamic characteristics, which have an impact on flight performance and safety. It is necessary for the aircraft to achieve the evolution of the configuration with changes in flight conditions.

In recent years, model design and numerical simulation and experimental verification have been carried out in the field of variant wings at home and abroad. The research focuses on the layout of the forward-forward wing or the aerodynamic characteristics of the whole wing [1-7], and there are few studies on the two-dimensional state.

With the change of the grazing angle of the wing, the relative thickness of the cross-section airfoil will change with the parameters of the airfoil, which will have a great influence on the aerodynamic performance of the overall wing. Therefore, it is necessary to study the variation of the section airfoil and its influence on the aerodynamic characteristics during the process of wing swept.

2. Modeling and feature parameter analysis

In order to make the wing of the variable swept-wing aircraft change smoothly in the state of flat wing, forward swept wing and delta wing, this paper selects NACA0006 thin symmetrical airfoil as the basic airfoil of the variable forward swept-wing aircraft wing, and makes It is perpendicular to the leading edge of the wing.

The wing is composed of an outer wing and an inner wing hidden in the rear of the fuselage in a flat wing state. The geometry and contour of the wing are shown in Figure 1. According to the principle of relative motion, this paper studies the different planes intercepting the wing around the leading edge point and the trailing edge point of the wing, and obtains 12 kinds of airfoil.

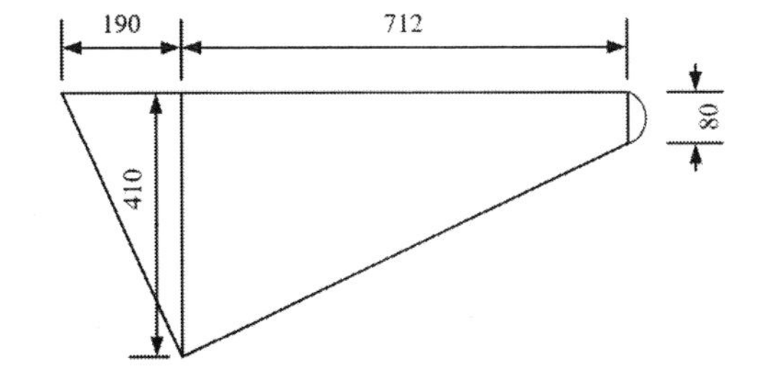

Figure 1. Geometric dimensions of the wing

In order to fully consider the convenience and representativeness of airfoil interception, the airfoil section of group A represents the rotation of the leading edge of the wing root, and the airfoil section of group B represents the rotation of the trailing edge of the wing root. The relative size is the largest. During the forward-swept deformation of the wing, the process of the profile of each airfoil with the forward sweep angle is shown in Figure 2.

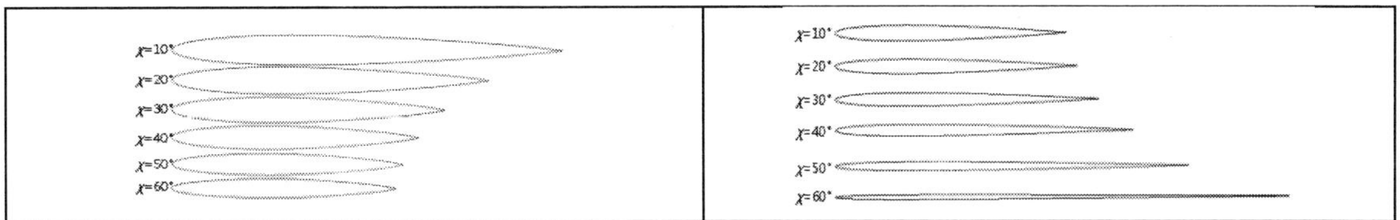

(a) Group A airfoil (b) Group B airfoil

Figure 2. Schematic diagram of the variation of the airfoil of each group with the swept angle

The group A airfoil is obtained by rotating around the leading edge of the wing root and is located in the triangular portion on the left side of the wing, and its chord length and relative thickness are changed. When the current swept angle of the B-set airfoil is large, its relative thickness will decrease sharply. Since the aerodynamic characteristics of the airfoil are greatly affected by the relative thickness, in order to facilitate the analysis, only the influence of the relative thickness variation on the aerodynamic characteristics of the section airfoil is considered. The relative thickness of the airfoil of Group A and Group B as a function of the forward sweep angle is shown in Table 1.

Table 1. Relative thickness of each airfoil group at different forward sweep angles

Forward sweep angle / Airfoil group	0°	10°	20°	30°	40°	50°	60°
A	6%	7.45%	8.45%	9.06%	9.26%	9.01%	8.29%
B	6%	5.61%	4.99%	4.21%	3.25%	2.16%	0.88%

It can be seen from Table 1 that the relative thickness of the airfoil of Group A increases first and then decreases, and the thickness of the airfoil of Group B increases while the thickness decreases gradually, so the relative thickness gradually decreases.

3. Pneumatic analysis

In the section, the corresponding airfoil is used to perform structural meshing. The mesh is of the order of 270,000, and different solvers are used to increase the convergence speed by using the Mach number of 0.3. The S-A turbulence model is used to select the ideal temperature of 288.15K free flow, calculate the variation law of each aerodynamic parameter.

3.1. Influence of angle of attack on lift characteristics when the sweep angle changes

Figure 3 shows the relationship between the lift coefficient of each cross-section airfoil at different forward sweep angles under the conditions of, T = 288.15K .

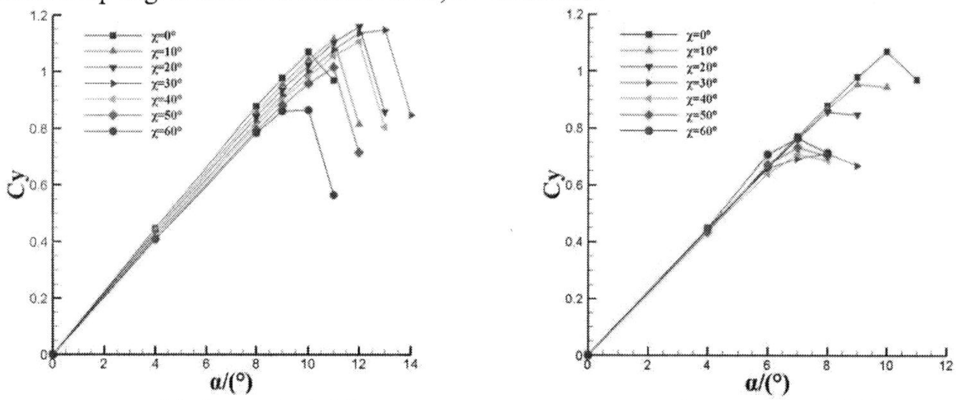

(a) Group A airfoil (b) Group B airfoil

Figure 3. The variation of the lift coefficient of each airfoil group with the angle of attack under the conditions of $Ma = 0.2$.

From the curve, the following law can be obtained: as the angle of attack increases, the lift coefficient first increases, and then decreases after reaching the stall angle of attack.

For the A-set airfoil, in the state of $\alpha < 8°$,the lift coefficient increases linearly with the angle of attack at the angle of attack, up to the maximum lift coefficient, and the lift coefficient begins to decrease after the stall angle of attack; As the forward sweep angle increases, the maximum lift first increases and then decreases, and the maximum lift is reached in the state of χ=20° . Moreover, the increase of the forward sweep angle also causes the stall angle to increase first and then decrease, and stall angle of attack is the largest in the state of χ=30° . The airfoil of the forward sweep angle is used to illustrate the flow mechanism[8]. In the state of $\alpha = 4°$, the upper surface of the airfoil maintains an attached flow, and the lift coefficient increases linearly with increasing angle of attack. When the angle of attack is further increased, the negative pressure zone of the leading edge of the airfoil increases, the suction increases, and at this time the airflow separation begins to occur at the trailing edge of the upper surface of the airfoil. In the state of $\alpha = 12°$, the area separating the upper surface of the airfoil further increases, and the separation vortex of the airflow occurs at the trailing edge of the airfoil, the leading edge negative pressure zone decreases, the suction force decreases, the lift coefficient decreases rapidly, and the airfoil stalls.

For the relationship between the lift and the forward sweep angle, it can be seen from the curve in the figure that when the relative thickness of the cross-section airfoil is large, the lift coefficient decreases as the forward sweep angle increases. This is because as the forward sweep angle increases, the leading edge of the airfoil has a negative pressure region with a smaller absolute value, and the area of this negative pressure region also decreases as the forward sweep angle increases. Therefore, its lift coefficient decreases as the forward sweep angle increases.

For the B-set airfoil, in the state of $\chi \le 20°$,the stall angle of attack and the maximum lift coefficient decrease with the increase of the forward sweep angle, while the maximum lift coefficient increases with the advance sweep angle in the state of $\chi > 20°$. When the thickness of the section airfoil is small, the lift coefficient will increase with the increase of the forward sweep angle. For example, the airfoil thickness of the B group airfoil is less than 6%. It can also be seen from the lift coefficient curve of Figure 3. As the forward sweep angle increases, the airfoil lift coefficient gradually increases.

3.2. Influence of Mach number on lift characteristics when the sweep angle changes

Figure 4 shows the variation of the lift coefficient of each airfoil group with the Mach number under the conditions of $\alpha = 4°$.

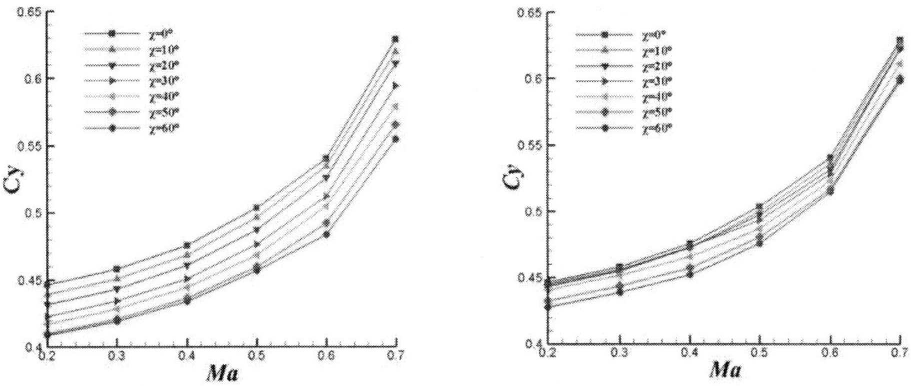

(a) Group A airfoil (b) Group B airfoil

Figure 4. The variation of lift coefficient with Mach number for each airfoil group under the conditions of $\alpha = 4°$

It can be seen from the figure that the lift coefficient of each airfoil group increases with the increase of the Mach number, and the rate of increase of the lift coefficient increases with the increase of the Mach number. At the same Mach number, as the forward sweep angle increases, the lift coefficient gradually decreases.

With the increase of the Mach number, the pressure on the upper surface of the airfoil has a negative pressure region with a larger absolute value, and as the Mach number increases, the negative pressure region also gradually increases, so the lift coefficient of the airfoil increases. When the Mach number is increased to 0.6, the leading edge of the airfoil gradually generates a shock wave. The appearance of the shock wave makes the absolute value of the leading edge pressure of the airfoil larger, and the area of the negative pressure zone is further increased, so the airfoil lift coefficient has increased dramatically.

3.3. Influence of angle of attack on resistance characteristics when the sweep angle changes

Figure 5 shows under the conditions of $Ma = 0.2$, the variation of the airfoil drag coefficient with the angle of attack at different forward sweep angles.

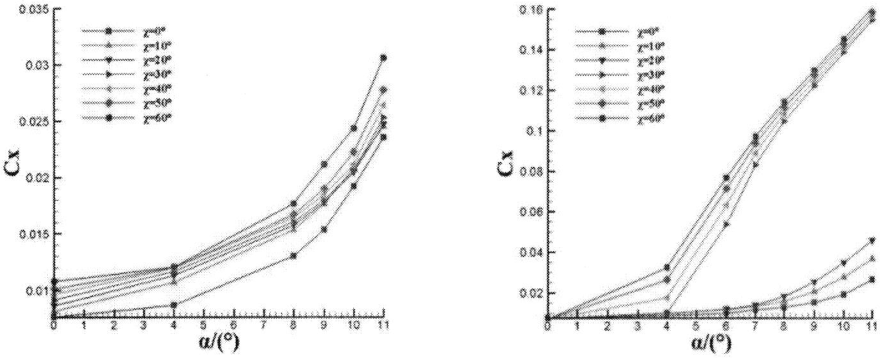

(a) Group A airfoil (b) Group B airfoil

Figure 5. The variation of the drag coefficient of each airfoil group with the angle of attack under the conditions of $Ma = 0.2$.

It can be seen from the figure that the drag coefficient increases with the increase of the angle of attack as the forward sweep angle is constant, and the slope of the drag coefficient increases with the increase of the angle of attack; at the same angle of attack, the drag coefficient As the plunging angle increases, it increases. For group A airfoil, the increase of the forward sweep angle causes the relative thickness of the airfoil to gradually increase, so the drag coefficient gradually decreases. For the group B airfoil, the increase of the forward sweep angle causes the relative thickness of the airfoil to decrease, when the relative thickness is reduced. When reduced to a certain value, its drag coefficient will rise sharply at a small angle of attack, mainly because the thin-wing type stall angle of attack is small, and the airfoil drag coefficient will increase sharply after divergence.

3.4. Influence of Mach number on resistance characteristics when the sweep angle changes

Figure 6 shows the variation of the lift coefficient of each airfoil group with the Mach number under the conditions of $\alpha = 4°$. It can be seen from the airfoil of Group A that the change of the drag coefficient of each airfoil at each forward sweep angle is relatively gentle during the process from 0.2 to 0.5, and the drag coefficient increases sharply during the process from 0.5 to 0.7. Mainly due to the generation of local shock waves around 0.7 Ma, the drag coefficient of each airfoil increases sharply [9]; from the B-group airfoil, the larger the forward sweep angle, the smaller the Mach number of the resistance divergence, and the Mach number. The effect on the resistance is also more pronounced, mainly because the relative thickness of each airfoil decreases as the forward sweep angle increases.

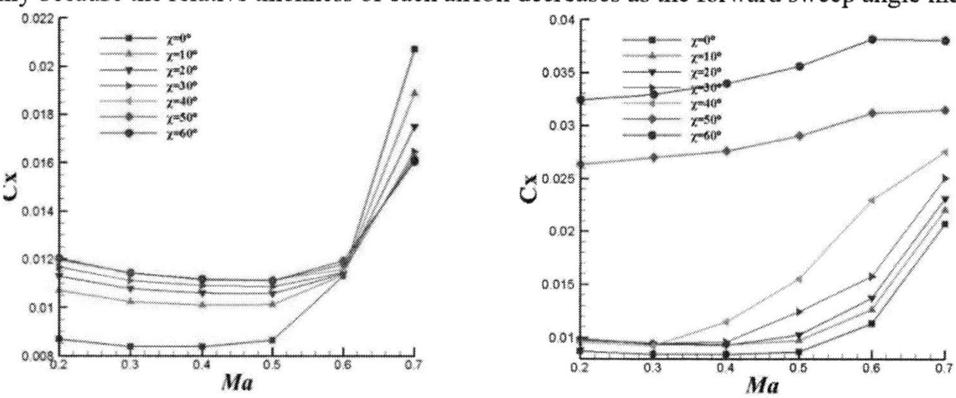

(a) Group A airfoil (b) Group B airfoil

Figure 6. The variation of the drag coefficient of each airfoil group with the Mach number under the conditions of $\alpha = 4°$

3.5. Influence of angle of attack on torque characteristics when the sweep angle changes

Figure 7 shows the variation of the pitching moment of each airfoil with the angle of attack at different forward sweep angles under the conditions of $Ma = 0.2$.

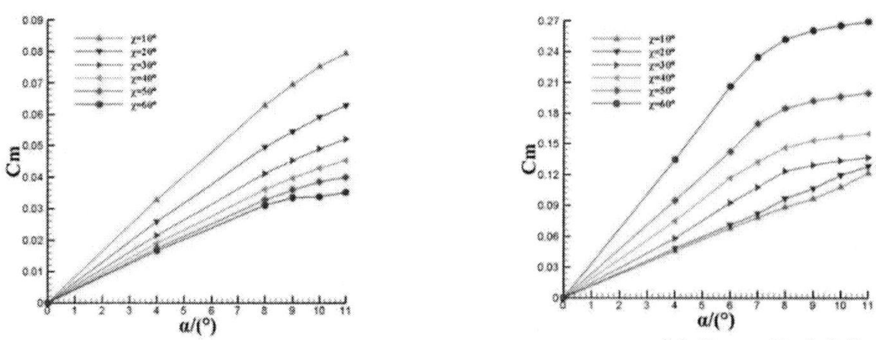

(a) Group A airfoil (b) Group B airfoil

Figure 7. The variation of the pitching moment coefficient of each airfoil group with the angle of attack under the conditions of $Ma = 0.2$

It can be seen from the figure that the torque coefficient of the leading edge point of the airfoil gradually increases with the increase of the angle of attack at the same forward sweep angle. At the small angle of attack, the moment has a linear relationship with the change of the angle of attack. Then, as the angle of attack increases, the moment changes gradually. When the angle of attack remains unchanged, the airfoil torque coefficient of group A decreases with the increase of the forward sweep angle, while the airfoil torque coefficient of group B increases with the increase of the forward sweep angle. The reason is that the change of the forward sweep angle causes a change in the relative thickness of the section airfoil, and the moment coefficient of the airfoil is negatively correlated with the relative thickness, that is, the larger the relative thickness is, the smaller the moment coefficient is, and the smaller the relative thickness is, the larger the moment coefficient is.

4. Conclusion

For the change of the cross-section airfoil during the swept process, the following conclusions are obtained.

(1) When the grazing angle of the wing changes, the relative thickness of the section airfoil and other parameters will also change, and the aerodynamic characteristics will change accordingly;

(2) When the angle of attack changes, the lift coefficient decreases with the increase of the forward sweep angle for the cross-sectional airfoil with larger relative thickness; the lift coefficient with the forward sweep angle is the relative airfoil with the relative thickness less than 6%. The increase is slightly increased; when the Mach number changes, the lift coefficient of the cross-section airfoil decreases as the forward sweep angle increases.

(3) When the angle of attack changes, the drag coefficient of each airfoil increases with the increase of the forward sweep angle; when the Mach number changes, the drag coefficient of the cross-section airfoil also appears with the increase of the forward sweep angle. Increase the trend.

(4) When the angle of attack changes, the pitching moment coefficient decreases with the increase of the forward sweep angle for the cross-sectional airfoil with larger relative thickness; the pitching moment coefficient with the relative thickness is less than 6% of the cross-section airfoil The angle increases and increases.

References

[1] Neal D A, Good M G, et al. Design and wind tunnel analysis of fully adaptive aircraft configuration [R], AIAA Paper 2004-1727.

[2] Secanell M, Suleman A. Gamboa P. Design of a morphing aircraft for a light unmanned aerial vehicle using high-fidelity aerodynamic shape optimization [R]. AIAA Paper 2005-1891.

[3] Chen Qian, Bai Peng, et al. Analysis of aerodynamic characteristics of variable camber airfoil with continuous smooth deflection trailing edge[J]. Journal of Aerodynamics, 2010, 28 (1): 46-53.

[4] Brett A N. Robert L S. Classical Flight Dynamics of a Variable Forward-Sweep Wing Aircraft [J]. Journal of Guidance, 1986, 9(2): 352-362.

[5] Northrop Grumman Corporation. Aircraft with Variable Forward-Sweep Wing: USA, US5984231[P]. 1999-11-16.

[6] Liu Wenfa, Wang Xu. A new concept of aerodynamic layout of variable swept-wing UAV[J]. Acta Aeronautica Sinica, 2009, 30(5):832-836.

[7] Wang Xu, Su Xinbing, Feng Haoyang. Simulation of Unsteady Aerodynamic Characteristics of Variable Airfoil of Variable Aircraft[J]. Computer Simulation. 2015.10,32(10):54-58.

[8] Chen Tingnan. Applied Fluid Mechanics [M]. Beijing: Aviation Industry Press, 2000

[9] Chen Yuankai, Dong Yanfei et al. Analysis of Resistance Characteristics of Variable Swept Wing Body Assembly[J]. Journal of Flight Mechanics, 2014, 32(4): 308-311.

IOP Conf. Series: Materials Science and Engineering **685** (2019) 012017 doi:10.1088/1757-899X/685/1/012017

Modeling and Simulation of Bow Wave in Aerial Refueling

Kunxian Zhai, Suiping Zhou*, Xiaoping Li and Yanming Liu

School of Aerospace Science and Technology, Xidian University, 710126, China

*Email: spzhou@xidian.edu.cn

Abstract. In a probe-and-drogue aerial refueling system, the tanker docking system consists of a hose and a drogue. When the receiver approaches the drogue, the dynamics of the drogue may be seriously disturbed by the aerodynamic effect from the receiver which is commonly known as the bow wave effect. In this paper, a numerical simulation method based on potential flow theory is proposed to investigate the bow wave effect. The proposed method is efficient and can be conveniently used in the modelling and simulation of autonomous aerial refuelling system. The results of this method are highly consistent with the CFD results, which demonstrates the effectiveness of the proposed method.

1. Introduction

The aerial refueling technology can greatly improve the endurance of the aircraft and is highly valued in the airline industry. Although the probe-and-drogue aerial refueling technology has made significant progress during the past few decades, it still remains as a challenging task for most pilots. In addition, as more and more UAVs (Unmanned Aerial Vehicles) will used in the aviation industry, the autonomous refueling of the UAVs has attracted many researchers in recent years, which has a very high requirement on the accuracy of the refueling system.

In the probe-and-drogue aerial refueling process, the bow wave effect is one of the main factors that contribute to the docking failure. Due to the complex aerodynamics involved, modeling the bow wave effect is still a challenging research problem. In [1], the effects of the bow wave is analyzed based on look-up tables obtained from CFD simulations. Dai and Wei used a series of line doublets to simulate the nose and cockpit of the receiver aircraft, and developed a correction function is to reduce the modeling error [2]. In 2018, Zhang and Hu proposed a modeling method based on potential flow theory, where the fuselage is treated as a Rankine body [3]. Zhong and Li select several velocity field data and model the bow wave by fitting and interpolation [4].

Considering the modeling convenience and accuracy, this paper proposes a method based on potential flow theory to model the bow wave effect. The main contributions of this paper are: (1) the method considers the geometrical characteristics of the head of receiver, so it is suitable for different types of aircraft (2) the obtained model can be easily applied to the controller design and the real-time simulations due to its computational efficiency and modeling accuracy.

The remainder of this paper is organized as follows. Section 2 presents the bow wave effect modeling assumptions and major methods involved. Section 3 presents the results of the proposed method and also compares the results the proposed method and CFD simulation, which indicates that the modeling method is effective. Conclusions and future work are summarized in Section 4.

2. Bow wave effect modeling

2.1. Modeling assumptions
Figure 1 (a) shows the flow field around the receiver. Many parts of the receiver, such as the head, the cockpit, the wing, the probe, may affect the bow wave effect. For some aircraft with special configurations, the wings, air inlet or the propeller may also have an effect on aerodynamics of the drogue [2]. In general, the bow wave effect can be approximated based on the principle of flow field superimposition as shown in (1):

$$W_{bow} = W_{nose} + W_{cockpit} + W_{wings} + W_{probe} + W_{others} \tag{1}$$

(a) (b)

Figure 1. CFD modeling of velocity distribution of F-16 in flow field.

Since the flow field attenuates rapidly with increasing distance from the drogue [2], the nose of receiver and the cockpit are the main causes of the bow wave, as shown in figure 1(b). In this paper the equation (1) is simplified as

$$W_{bow} = W_{nose} + W_{cockpit} \tag{2}$$

2.2. 2-D Bow wave effect modeling method
First, we present the panel method to model the 2-D bow wave effect. To solve aerodynamic problems, the panel method discretizes the surface of an object, and places singularities (such as source, vortex, doublets or their combinations) on the panel.

Assuming that a body is placed in a uniform flow of speed U [5], the velocity potential of the flow field is superimposed by the uniform flow and the source which placed on panels to obtain a total velocity potential of the form

$$\phi = Ux + \varphi \tag{3}$$

φ represents the velocity potential due to the sources on the panels.

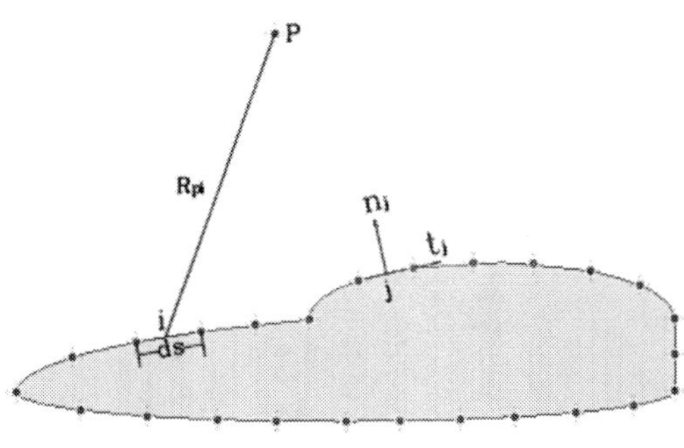

Figure 2. Surface distribution of singularities.

According to the panel method, the potential flow around a body of any given shape can be modeled by a distribution of sources over its surface.

As shown in Figure 2, the first step is to discretize the 2D contour and divide it into N panels. The velocity potential at P due to sources on panel i is given by

$$\Delta\Phi_{Pi} = \sigma_i \int_{-\Delta s/2}^{\Delta s/2} lnR_{Pi} ds \tag{4}$$

$$R_{Pi} = \sqrt{(x_P - x_i)^2 + (y_P - y_i)^2} \tag{5}$$

where σ_i is the source strength, R_{Pi} is the distance from P to center of the panel i, (x_p, y_p) is the coordinates of P; (x_i, y_i) is the coordinates of the center of panel i, ds is the length of the panel.

The velocity potential at P due to all sources is

$$\Phi_P = \sum_{i=1}^{N} \Delta\Phi_i = \sum_{i=1}^{N} \sigma_i \int_{-\Delta s/2}^{\Delta s/2} lnR_{Pi} ds \tag{6}$$

Thus, following equation (3), the total velocity potential at P can be written as

$$\phi_P = Ux + \Phi_P = Ux + \sum_{i=1}^{N} \sigma_i \int_{-\Delta s/2}^{\Delta s/2} lnR_{Pi} ds \tag{7}$$

In order to use equation (7) for numerical modeling, the source strength on panels is required. Next, we discuss how to obtain source strength.

Similar to equation (6), the velocity potential at panel i due to sources on other panels is given by

$$\Phi_i = \sum_{j=1}^{N} \sigma_j \int_{-\Delta s/2}^{\Delta s/2} \ln R_{ij} ds \tag{8}$$

The normal velocity of the source panel is

$$V_{ni} = \frac{\partial \Phi_i}{\partial \hat{n}_i} \quad \left(when \;\; i = j, \;\; V_{ni} = \pi\sigma_i\right) \tag{9}$$

Superimposed with the uniform, and because of the boundary condition which the normal velocity is 0, we can get a set of equations with N unknowns and N equations as shown in (10). By solving these equations, the source strength of each panel can be obtained.

$$V_n = \mathbf{U} \cdot \hat{n}_i + \pi\sigma_i + \sum_{\substack{j=1 \\ j \neq i}}^{N} \sigma_j \int_{-\Delta s/2}^{\Delta s/2} \frac{\partial \ln R_{ij}}{\partial \hat{n}_i} ds = 0 \tag{10}$$

Once the source strength of each panel is obtained, the velocity potential at P can be obtained through equation (7).

The velocity at P is then given by

$$\begin{cases} u = \dfrac{\partial \phi_P}{\partial x} = U + \displaystyle\sum_{i=1}^{N} \sigma_i \int_{-\Delta s/2}^{\Delta s/2} \frac{\partial lnR_{Pi}}{\partial x} ds \\ v = \dfrac{\partial \phi_P}{\partial y} = \displaystyle\sum_{i=1}^{N} \sigma_i \int_{-\Delta s/2}^{\Delta s/2} \frac{\partial lnR_{Pi}}{\partial y} ds \end{cases} \tag{11}$$

where u and v are the velocity components in the x direction and y direction.

2.3. 3-D Bow wave effect modeling method

For 3-D objects, the solution method is similar to the 2-D case, where the 3-D surface is discretized into a number of 2-D planes. The distance in equation (5) is changed to

$$R_{Pi} = \sqrt{(x_P - x_i)^2 + (y_P - y_i)^2 + (z_P - z_i)^2} \tag{12}$$

where (x_p, y_p, z_p) is the coordinates of P, (x_i, y_i, z_i) is the coordinates of the center of panel i. The major process involved is similar to the 2-D case, however, the difficulty in the 3-D case lies in how to discretize the 3-D surface.

As most aircraft has a symmetric body along the vertical plane of the body coordinate system, we can simplify the computation procedure as described blow.

$$x_p = x, y_p = \sqrt{y^2 + z^2} \tag{13}$$

By substituting (x_p, y_p) into equation (8), the 2-D velocity can be obtained, and the 3-D velocity vector can be obtained by

$$\begin{cases} v_x' = u(x_P, y_P) \\ v_y' = \dfrac{y}{\sqrt{y^2 + z^2}} v(x_P, y_P) \\ v_z' = \dfrac{z}{\sqrt{y^2 + z^2}} v(x_P, y_P) \end{cases} \tag{14}$$

However, since the cross section of most receivers is not circular but elliptic, coordinate transformation is required, as shown in Figure 3, and the new velocity component is closer to the actual situation.

$$v_x = v_x', v_y = \frac{b}{a} v_y', v_z = v_z' \tag{15}$$

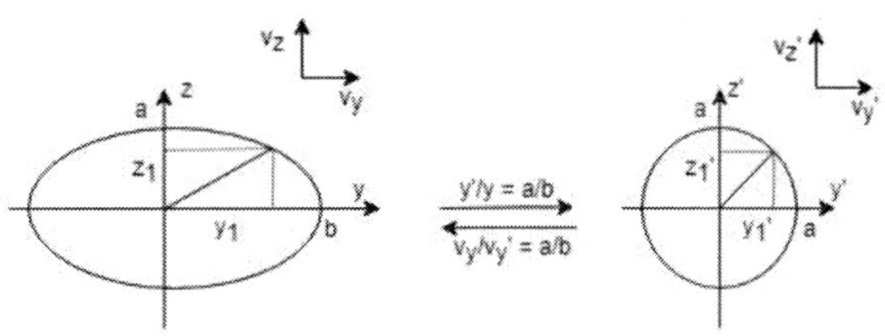

Figure 3. Coordinate transformation.

3. Simulation results and analysis

To analyze the effectiveness of the proposed modeling and computation method as described in Section 2, we have implemented the proposed method in the MATLAB on a laptop computer. For comparison, we generate the flow field results using the commercial software Fluent. We use Tecplot360 to display the results.

3.1. 2-D bow wave simulation

Figure 4 shows the contour plots of the velocity field. To generate the results, we discretize the 2-D model into 364 panels. In Figure 4, (a) and (c) are the x-direction and y-direction velocity distribution generated from our proposed method, (b) and (d) are the x-direction and y-direction velocity distribution generated from the CFD method. During the docking process, we pay more attention to the flow field near the receiver nose and the cockpit. The obtained velocity distribution characteristics are consistent well with the CFD results, the error is less than 3%. The average calculation time of our method is 0.1s.

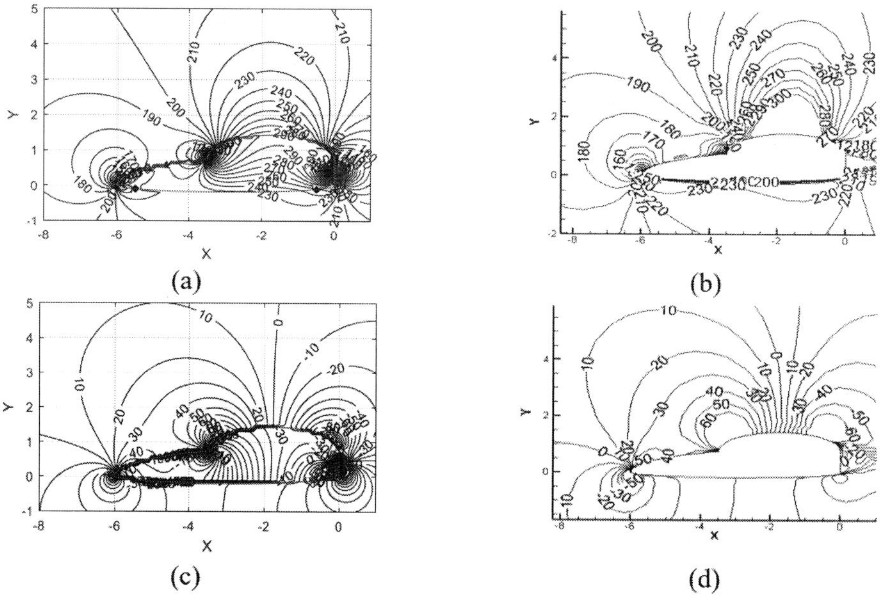

Figure 4. 2-D bow wave model velocity distribution from proposed method((a), (c)) and CFD ((b), (d)).

Next, we observe the velocity characteristics with the change of the angle between the axis of the fuselage and the relative airflow, and observe the cases where the angle is 5°, 10°, and 15°, as shown in Figure 5, 6, and 7. In the figures, (a) and (c) are the x-direction and y-direction velocity distribution generated from our proposed method, and (b) and (d) are the x-direction and y-direction velocity distribution generated from the CFD method. It can be seen that the velocity distribution of our method and the CFD result are consistent. As the angle increases, the accuracy decreases, and the error increases to about 10%. The average calculation time of our method is 0.1s.

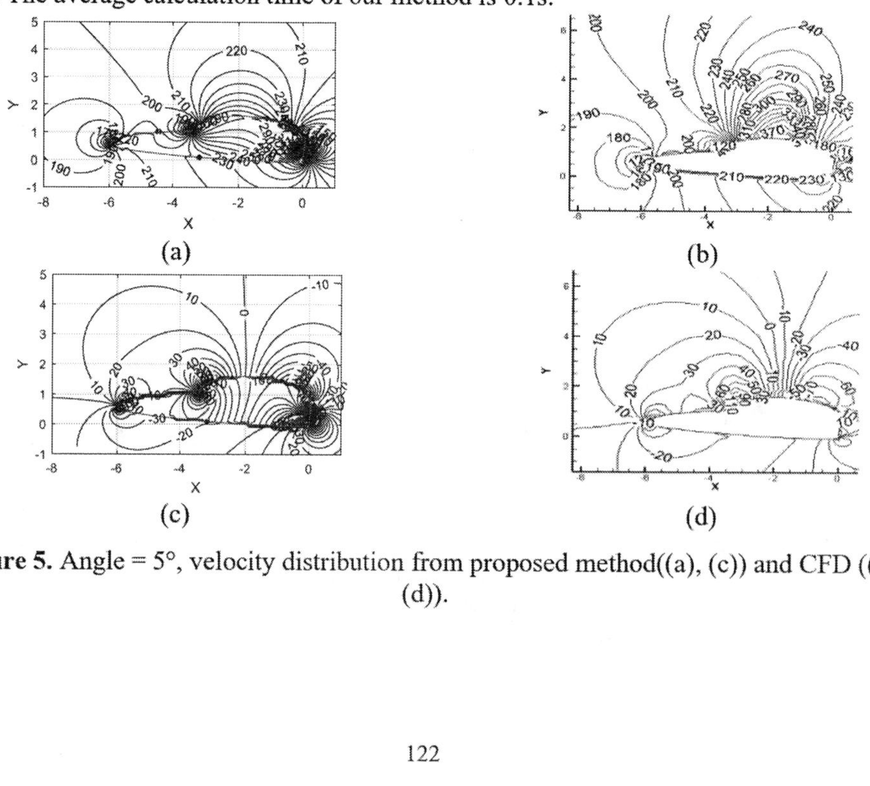

Figure 5. Angle = 5°, velocity distribution from proposed method((a), (c)) and CFD ((b), (d)).

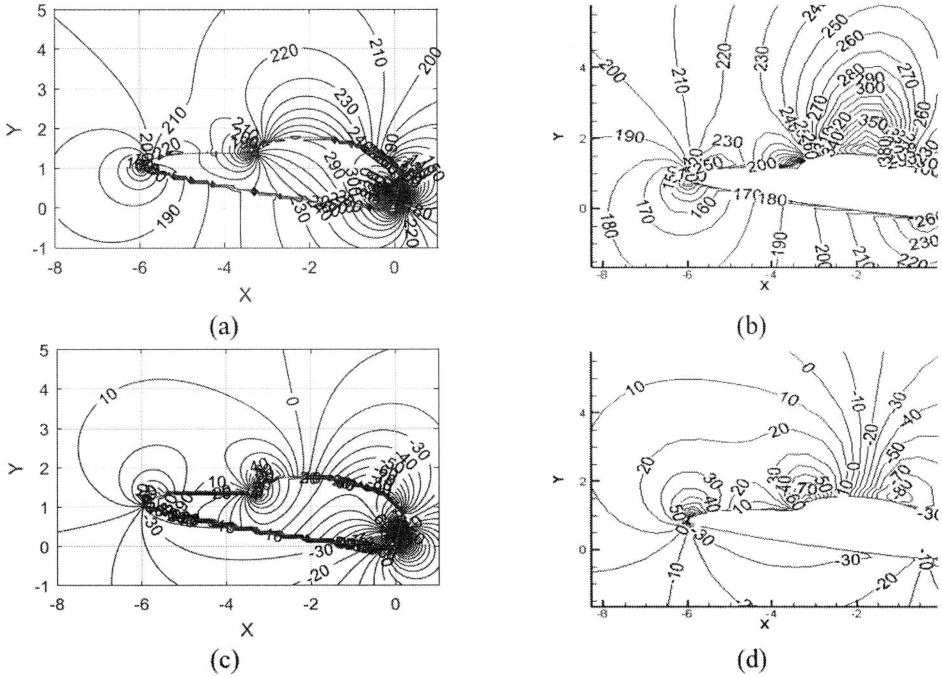

Figure 6. Angle = 10°, velocity distribution from proposed method((a), (c)) and CFD ((b), (d)).

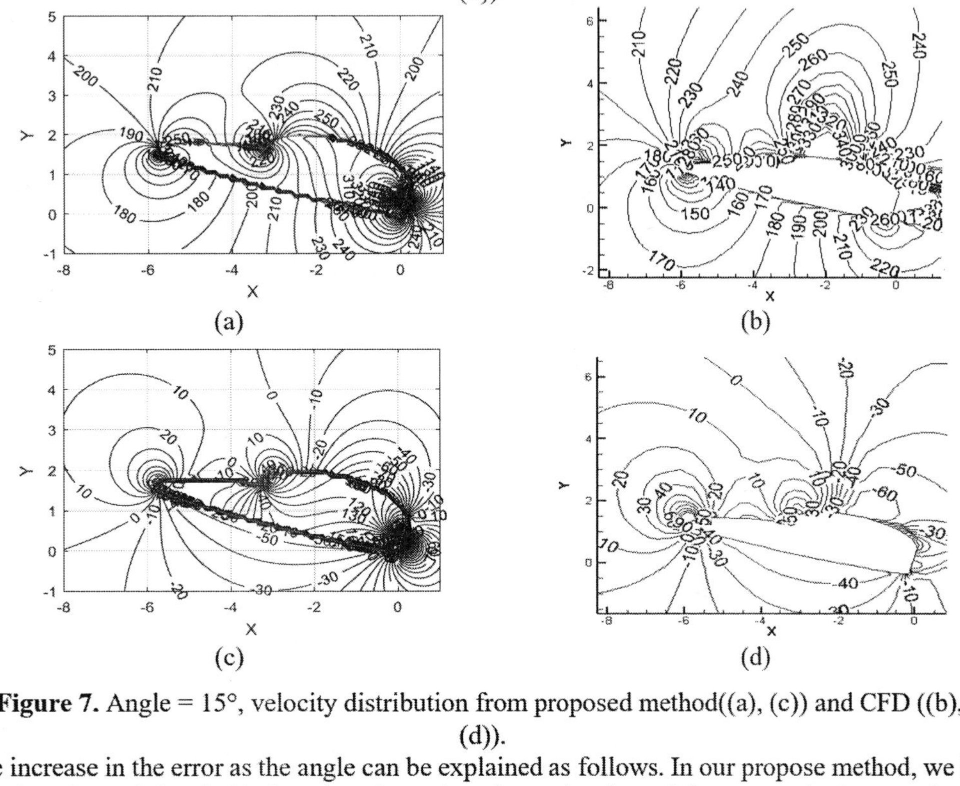

Figure 7. Angle = 15°, velocity distribution from proposed method((a), (c)) and CFD ((b), (d)).

The increase in the error as the angle can be explained as follows. In our propose method, we ignore the air viscosity and the air friction, therefore when the angle of attack increases, the increase in the lift is simply ignored. This error can be reduced if the effect of air viscosity, friction, and attack of angles is included in our model by introducing a correction factor based on some CFD results or wind tunnel test results.

3.2. 3-D bow wave model simulation

The 3D receiver head model as shown in Figure 8. Figure 9 and Figure 10 show the bow wave velocity distribution. In the figures, (a)(c)(d) are the x-velocity, y-velocity, z-velocity distribution generated from the proposed method, and (b)(d)(f) are the x-velocity, y-velocity, z-velocity distribution generated from CFD method. It can be seen that in the important place near the aircraft nose and the cockpit, the simulation results and the CFD results are still consistent with each other. The average calculation time of our method is 0.13s.

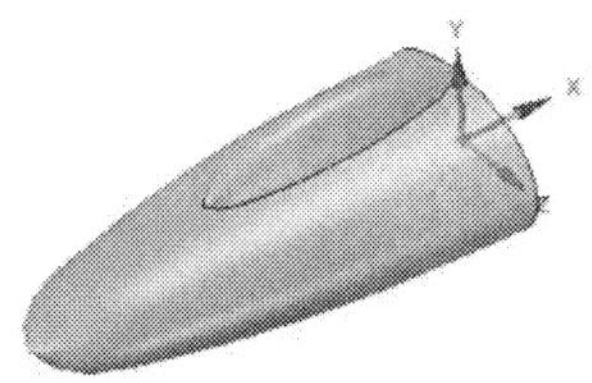

Figure 8. The 3D receiver head model.

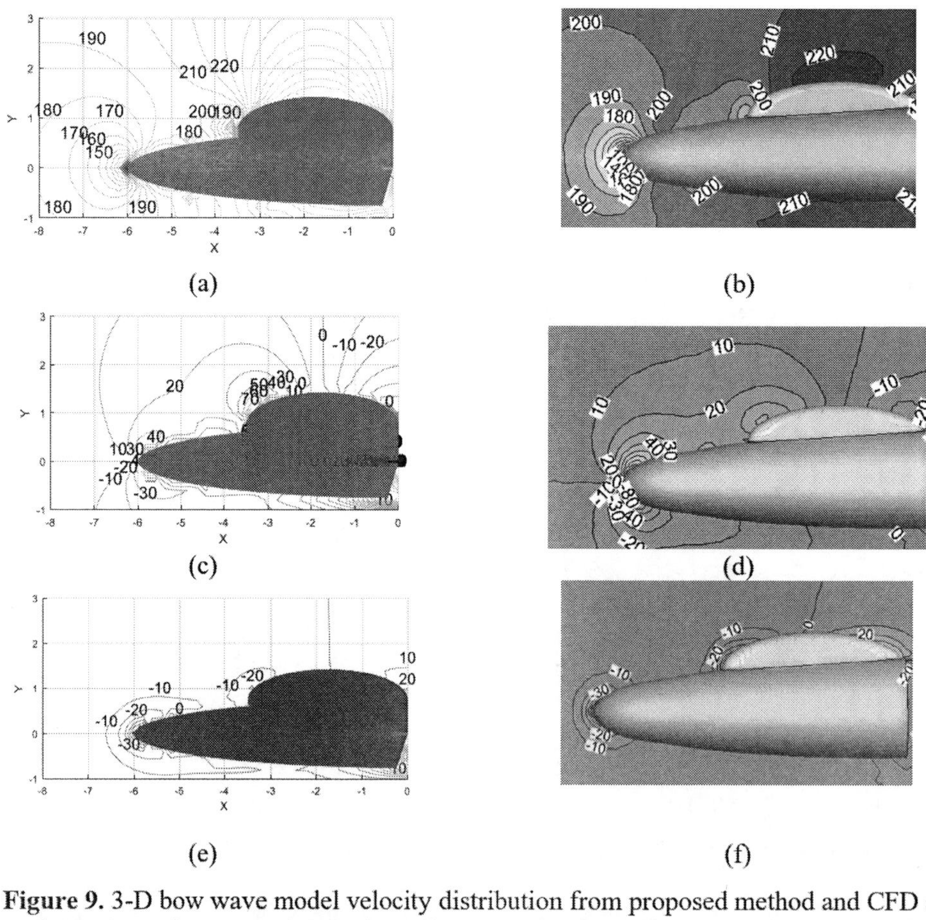

Figure 9. 3-D bow wave model velocity distribution from proposed method and CFD at panel Z=0 (note that at panel Z=0, the z-direction velocity is almost equal to 0 everywhere, so in figure (e) and (f), the results shown are on the panel Z = -0.25).

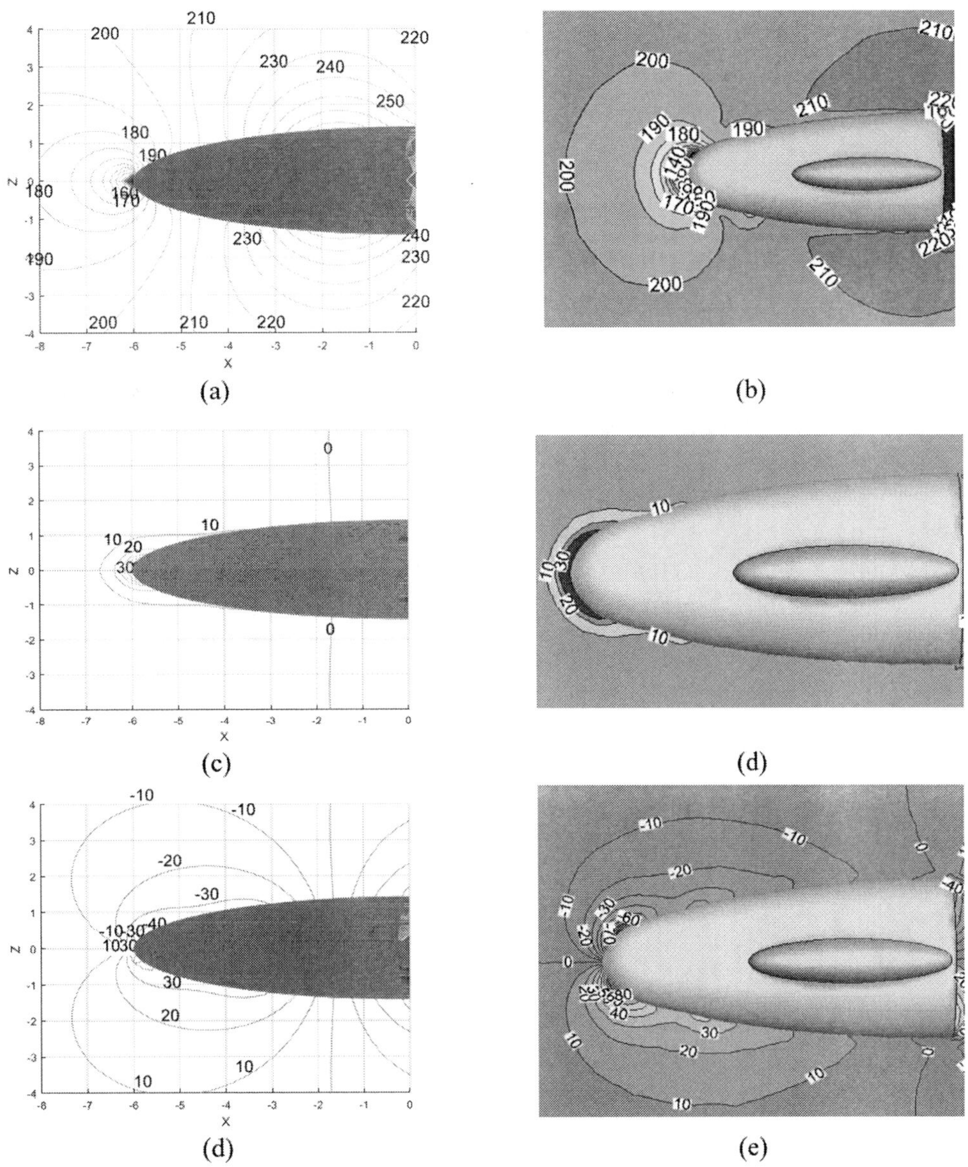

Figure 10. 3-D bow wave model velocity distribution from proposed method and CFD at panel Y=0.

The small computation error of our method is mainly due to:

- the inherent defects of the panel method lead to some errors in the 2-D model, and the 3-D model is based on the 2-D model, which result in the accumulation of errors;
- the nose and cockpit are simulated by two ellipsoids. The overlap between them leads to a superposition of calculations, thus the result is slightly larger than the the CFD result;
- the entirety of object geometry in the panel method. The nose and the cockpit can only be deformed at the same time, although the eccentricity of the two cross sections is different.

4. Conclusions and future work

This paper proposes a method based on potential flow theory to model the bow wave. The method considers the geometrical characteristics of the head of receiver, which can be used for different types of aircrafts. Experimental results from the proposed method are consistent with the CFD results, and the average computation time for the 3-D cases used in the simulation experiments is around 0.13s using MATLAB on a laptop computer. Therefore, the propose method could be used in the controller design for UAVs autonomous aerial refueling and some real-time simulations.

Our future work will includes: (1) refining the model to improve the accuracy; and (2) applying the proposed model to the autonomous air refueling system.

References

[1] Bhandari U, Thomas P R, Bullock S and Richardson T S 2013 Bow wave effect in probe and drogue aerial refuelling *AIAA* 2013-4695

[2] Dai X H, Wei Z B and Quan Q 2016 Modeling and simulation of bow wave effect in probe and drogue aerial refueling *Chinese Journal of Aeronautics* **29**(2) pp 448-461

[3] Zhang X F *et al* 2018 *J. Phys.: Conf. Ser.* **1064** 012045

[4] Zhong Z W, Li D W, Wang H L and Su Z K. Modeling and analysis of receiver-aircraft bow wave based on fitting and interpolation. *Electronics Optics & Control* 2018 25(2) pp 69-73.

[5] Houghton E L *et al* 2012 *Aerodynamics for Engineering Students 6th ed* (Waltham: Butterworth-Heinemann) chapter 3 pp 195-203

Research on Method of Phantom Track Generation Based on Cooperative Control of UAVs in Uniform Linear Motion

Zhiping Ouyang[1], Chen Xu[1*], Jiqiang Feng[1] and Bo Wang[2]

[1] College of Mathematics and Statistics, Shenzhen University, Shenzhen, China
[2] College of Science, Inner Mongolia University of Technology, Huhhot, China

*xuchen@szu.edu.cn

Abstract. In order to effectively interfere with the networking radar, this paper establishes a reverse-pyramid model firstly based on the uniform linear motion of UAVs and the characteristic of phantom track. Furthermore, the multi-UAV cooperative track deception model is built to analyze the locations and motions of UAVs. Finally, we design a simulation experiment to obtain the motion parameters of each UAV and analyze the minimum number and speed of UAVs.

1. Introduction

With the rapid development of the networking radar, how to effectively interfere with the networking radar and generate valid phantom track has become a benchmark in the field of electronic interference. [1] establishes the motion model of the false target and UAVs in a set-height flight mode based on the geometrical relationship; the one-one deception model and the theory of track deception was proposed by [2]; [3] builds a dynamics model of phantom and team of aircrafts and carries out the simulation based on the constraint conditions and cost function; the principle of distance deception jamming and the traditional track deception jamming model was combined by [4] to established a track deception jamming model, which not only achieves track deception but also effectively increases the degree of freedom of the jammers' invasion route.

In the above researches, it is often assumed that drones make curve motion, they can bend and accelerate at will. In fact, the drone is often automatically controlled by the system. For safety and convenience of control, it is hoped that the drone will do less maneuvers such as turning, climbing, and diving in the real life. Therefore, this paper proposes that the drones maintain a uniform linear motion during the execution activities. Furthermore, the above researches require all radars in the networking radars must be deceived during the process of generating the phantom track. But the requirement is not necessary in reality. In the case of radar network penetration, the effective false trajectories should interfere with a certain amount of radar, i.e., if the false target can be detected by a certain amount of radar, then the false target generated by the drones can be accepted by the networking radar. In this paper, 20 consecutive false targets accepted by the networking radar will be regarded as a real target track by the networked radar system. Since the RCS of the UAV is small and several stealth techniques are also adopted so that the drones can not be effectively detected by the radar when they fly at a certain distance from the radar. Assume that the networked radar system checks spatial state of all objects 10 seconds once. Now given a real target track that contains the position coordinate information of 20 moments, and the time interval is 10 seconds. The flight speed of the drones are controlled from 120km/h to 180km/h(i.e.from 33.334m/s to 50m/s) and the flight altitude is controlled from 2000m to 2500m. The problem is how to control the drones to generate the real target track in uniform linear motion.

2. The reverse-pyramid model

We start by introducing the proposition about pyramid: Any straight line can not pass through the three edges of the pyramid unless it pass through the vertex of the pyramid. The quadrilateral pyramid was shown in figure 1.

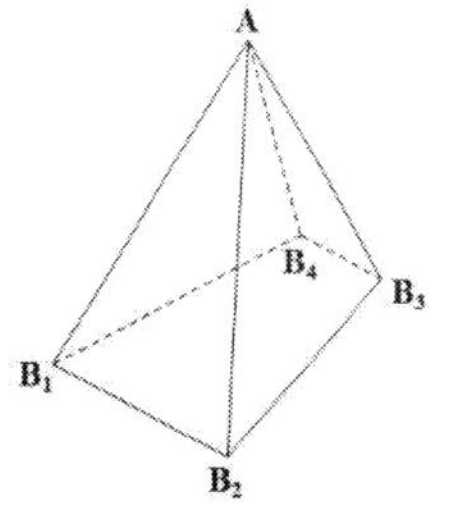

Figure 1. The Pyramid model.

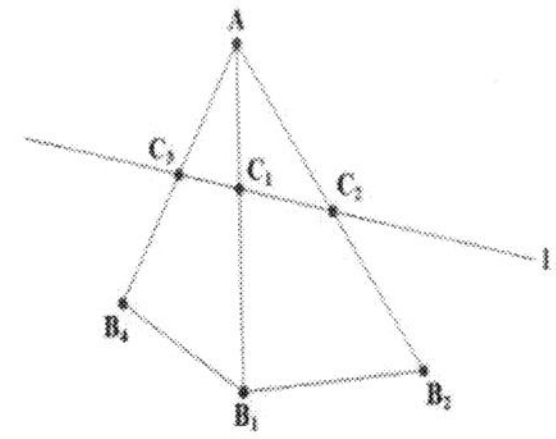

Figure 2. The proof of proposition.

2.1. The proof of the proposition

Now we prove the proposition by contradiction.

Proof: Let $P_{AB_1B_2B_3B_4}$ be the quadrilateral pyramid in figure 1. Taking the quadrilateral pyramid as an example, if there exists one straight line intersects the edges AB_4, AB_1, AB_2 at C_3, C_1, C_2 respectively. As shown in figure 2, C_3, C_1 and C_2 are collinear, then , which contradicted the definition of the pyramid. So any straight line can not pass t$C_3 \in$ Plane AB_1B_2hrough the three edges of the pyramid $P_{AB_1B_2B_3B_4}$ unless it pass through the vertex A of the pyramid $P_{AB_1B_2B_3B_4}$. Similarly, for any pyramid, any straight line can not pass through the three edges of the pyramid unless it pass through the vertex of the pyramid.

2.2. The data of the real target track and radar network

The real target track contains the position coordinate information of 20 moments. They are shown in table 1. The first and fourth columns *i* represent the serial number of each position. The unit of the table is meter(m). The radar network system consists of five radars. The maximum range of the radar is 150 km, which means that it can only effectively detect targets within 150 km of the radar. And the geographic coordinates of the five radars are radar 1: $R_1(80000,0,0)$, radar2: $R_2(30000,60000,0)$, radar3: $R_3(55000,110000,0)$, radar4: $R_4(105000,110000,0)$, radar5: $R_5(130000,60000,0)$.

Table 1. The coordinates of the real target track.

i	x_i	y_i	z_i	i	x_i	y_i	z_i	i	x_i	y_i	z_i	i	x_i	y_i	z_i
1	60600	69982	7995	**6**	63523	69359	7820	**11**	66131	67900	7395	**16**	68134	65817	6720
2	61197	69928	7980	**7**	64078	69131	7755	**12**	66594	67518	7280	**17**	68442	65361	6555
3	61790	69838	7955	**8**	64618	68870	7680	**13**	67026	67116	7155	**18**	68719	64897	6380
4	62377	69713	7920	**9**	65141	68577	7595	**14**	67426	66697	7020	**19**	68966	64429	6195
5	62955	69553	7875	**10**	65646	68253	7500	**15**	67796	66263	6875	**20**	69184	63957	6000

2.3. The possible lines of the positions

To effectively interfere with the radar, it is necessary to make sure that the drones on the connection between the radar and the false target. Let $H_i(x_i, y_i, z_i)$ be the i-th false target and $R_j(x_j, y_j, z_j)$ be the j-th radar. The line between radar R_j and false target H_i can be expressed as

$$\frac{x-x_j}{x_i-x_j}=\frac{y-y_j}{y_i-y_j}=\frac{z-z_j}{z_i-z_j}. \quad (1)$$

The flight speed of the drone is controlled from 120 km/h to 180 km/h(i.e.from 120/3.6 m/s to 50 m/s) and the flight altitude is controlled from 2000m to 2500m. Denote the possible positions line as $L_{i,j}$, which i, j represents the serial number of the false target and radar respectively. With the constraint of the altitude, the possible positions line of the drones can be formatted as

$$L_{i,j}:\frac{x-x_j}{x_i-x_j}=\frac{y-y_j}{y_i-y_j}=\frac{z-z_j}{z_i-z_j}\left(2000\leq z\leq 2500\right). \quad (2)$$

Here, the symbol $L_{i,j}$ means that the i-th false target belongs to the field of the j-th radar. Since the quantity of radar and false targets are 5,20 respectively, we can get all the possible positions lines: $L_{1,1}$, $L_{1,2}$, $L_{1,3}$, $L_{1,4}$, $L_{1,5}$, $L_{2,1}$, $L_{2,2}$, $L_{2,3}$, $L_{2,4}$, $L_{2,5}$, …, $L_{20,1}$, $L_{20,2}$, $L_{20,3}$, $L_{20,4}$, $L_{20,5}$. We find out that one drone can only affect exactly one radar because of the long pairwise distance among the five radars and the constraint conditions of the drone. Now classify these 100 lines into 5 sets by the field of radar, the first set is $S_1=\{L_{1,1},L_{2,1},L_{3,1},\ldots,L_{19,1},L_{20,1}\}$, the second set is $S_2=\{L_{1,2},L_{2,2},L_{3,2},\ldots,L_{19,2},L_{20,2}\}$, the third set is $S_3=\{L_{1,3},L_{2,3},L_{3,3},\ldots,L_{19,3},L_{20,3}\}$, the fourth set is $S_4=\{L_{1,4},L_{2,4},L_{3,4},\ldots,L_{19,4},L_{20,4}\}$ and the fifth set $S_5=\{L_{1,5},L_{2,5},L_{3,5},\ldots,L_{19,5},L_{20,5}\}$.

2.4. The minimum or maximum distance of the pairwise possible lines

The minimum or maximum distance of the pairwise possible lines are the most important determinant for giving the solution for UAVs.

2.4.1. The minimum distance of the pairwise possible lines. Given two arbitrary possible lines L_{i_1,j_1}, L_{i_2,j_2}, take two points P_{i_1,j_1}, P_{i_2,j_2} from the two possible lines respectively, and then connect the two points as a new line $L=P_{i_1,j_1}P_{i_2,j_2}$. Now we calculate the minimum distance $\min(L)$ and maximum distance $\max(L)$ of the new line L. The endpoints of line L_{i_1,j_1} are H_{i_1} and R_{j_1}. The endpoints of line L_{i_2,j_2} are H_{i_2} and R_{j_2}. $\forall P_{i_1,j_1}\in L_{i_1,j_1}$, $\forall P_{i_2,j_2}\in L_{i_2,j_2}$, we have

$$P_{i_1,j_1}=R_{j_1}+\alpha_{i_1,j_1}\left(H_{i_1}-R_{j_1}\right)\left(0\leq\alpha_{i_1,j_1}\leq 1\right), P_{i_2,j_2}=R_{j_2}+\alpha_{i_2,j_2}\left(H_{i_2}-R_{j_2}\right)\left(0\leq\alpha_{i_2,j_2}\leq 1\right). \quad (3)$$

Let $\vec{a}=H_{i_1}-R_{j_1}$, $\vec{b}=H_{i_2}-R_{j_2}$, $\vec{c}=R_{j_1}-R_{j_2}$, $\vec{d}=P_{i_1,j_1}-P_{i_2,j_2}$. According to equation (3), we can get the following formula.

$$\vec{d}=\vec{c}+\alpha_{i_1,j_1}\vec{a}-\alpha_{i_2,j_2}\vec{b}\left(0\leq\alpha_{i_1,j_1},\alpha_{i_2,j_2}\leq 1\right). \quad (4)$$

Given the two lines L_{i_1,j_1}, L_{i_2,j_2}, L may be the shortest connection when L is perpendicular to both lines L_{i_1,j_1} and L_{i_2,j_2} or $L\in\left\{H_{i_1}R_{j_1},H_{i_1}R_{j_2},H_{i_2}R_{j_1},H_{i_2}R_{j_2}\right\}$, which mean

$$\vec{d} \perp L_{i_1,j_1}, \vec{d} \perp L_{i_2,j_2} \Rightarrow \vec{d} \perp \vec{a}, \vec{d} \perp \vec{b} \Rightarrow \vec{d}\cdot\vec{a}=0, \vec{d}\cdot\vec{b}=0 \Rightarrow \begin{cases} \left(\vec{c}+\alpha_{i_1,j_1}\vec{a}-\alpha_{i_2,j_2}\vec{b}\right)\cdot\vec{a}=0 \\ \left(\vec{c}+\alpha_{i_1,j_1}\vec{a}-\alpha_{i_2,j_2}\vec{b}\right)\cdot\vec{b}=0 \end{cases}. \tag{5}$$

where $(\cdot)$ is the vector inner product between two vectors. If α_{i_1,j_1} and α_{i_2,j_2} solved by equation (5) are in the internal $[0,1]$, then we can get the value of $\min(L)$. If not, $\min(L)$ can be achieved int the set $L \in \{H_{i_1}R_{j_1}, H_{i_1}R_{j_2}, H_{i_2}R_{j_1}, H_{i_2}R_{j_2}\}$.

2.4.2. The maximum distance of the pairwise possible lines. Suppose there are two spatial lines AB and CD, $\forall \mathrm{P} \in \mathrm{AB}$, P and CD are coplanar. Denote $|\mathrm{AB}|$ as the length of AB. As shown in figure 3(b), figure 4(a) and figure 4(b), according to the geometric relationship of the triangle, we can know that

$$\begin{cases} \max\left(\{|\mathrm{PC}|,|\mathrm{PP}_1|,|\mathrm{PP}_2|,\ldots,|\mathrm{PP}_{n-1}|,|\mathrm{PP}_n|,|\mathrm{PD}|\}\right) \le \max\left(\{|\mathrm{PC}|,|\mathrm{PD}|\}\right), \forall \mathrm{P} \in \mathrm{AB} \\ \max\left(\{|\mathrm{BC}|,|Q_1\mathrm{C}|,|Q_2\mathrm{C}|,\ldots,|Q_{n-1}\mathrm{C}|,|Q_n\mathrm{C}|,|\mathrm{AC}|\}\right) \le \max\left(\{|\mathrm{BC}|,|\mathrm{AD}|\}\right) \\ \max\left(\{|\mathrm{BD}|,|Q_1\mathrm{D}|,|Q_2\mathrm{D}|,\ldots,|Q_{n-1}\mathrm{D}|,|Q_n\mathrm{D}|,|\mathrm{AD}|\}\right) \le \max\left(\{|\mathrm{BD}|,|\mathrm{AD}|\}\right) \end{cases}. \tag{6}$$

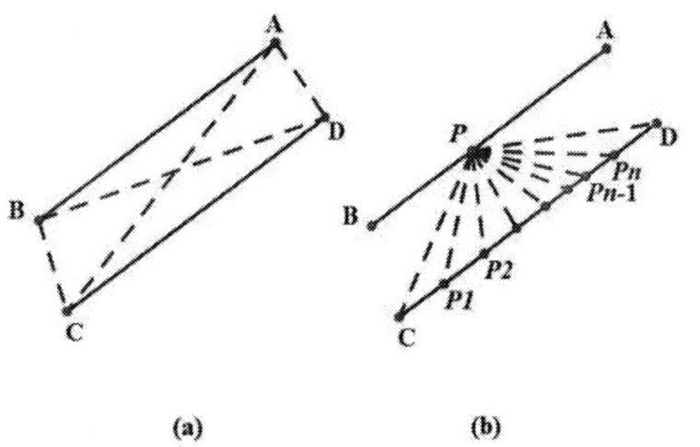

Figure 3. The connection lines between the P in AB and arbitrary points in CD.

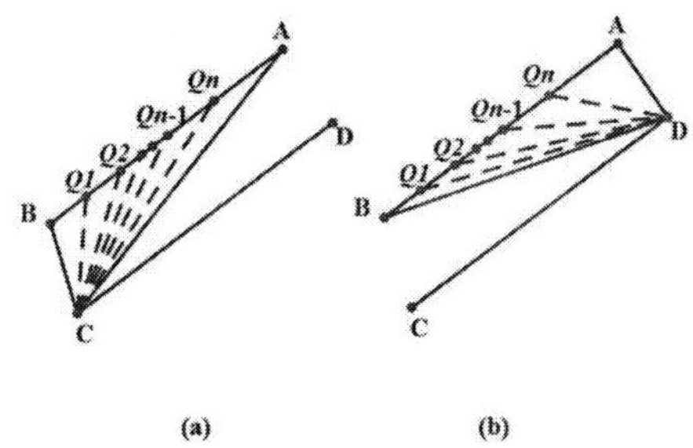

Figure 4. The connection lines between the AB and CD.

As a result, which is shown in figure 3(a), it can be expressed as

$$\max(L) = \max\left(|\mathrm{AC}|,|\mathrm{AD}|,|\mathrm{BC}|,|\mathrm{BD}|\right). \tag{7}$$

2.5. The reverse-pyramid model

In the real life, we find that one drone can only affect exactly one radar because of the long pairwise distance among the five radars and the constraint conditions of the drone. The connected lines between the false targets and radar are always the reverse-pyramid, which are shown in figure 5 and figure 6. Obviously, all pairwise lines in the same set can not be overlapped and all intersect at the location of radar. The drones can only fly through the lines in the same set to generate the phantom track. We can

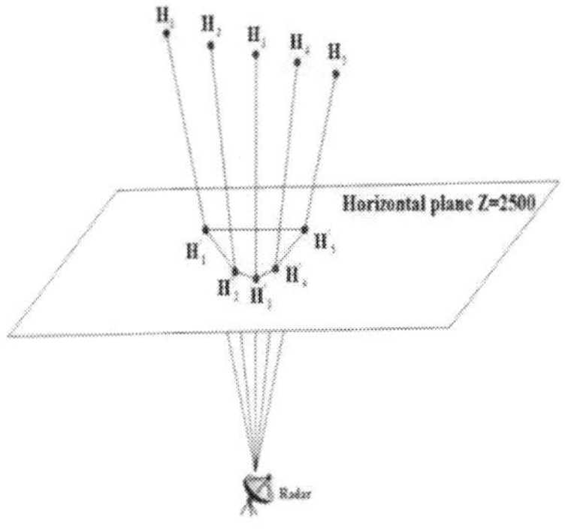

Figure 5. The lines connect the false targets and radar.

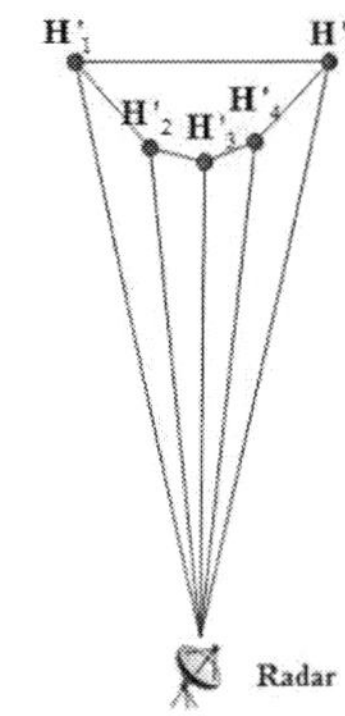

Figure 6. The reverse-pyramid model.

calculate all $\min(L)$ and $\max(L)$ for all pairwise lines. Let $s_{\min}$ and $s_{\max}$ be the shortest distance the longest distance in a 10 seconds' motion. If the $\min(L)$ and $\max(L)$ of pairwise lines satisfy the constraint $\left[\left[(\min(L)/(i_1-i_2))^2\right]^{1/2},\left[(\max(L)/(i_1-i_2))^2\right]^{1/2}\right]\cap[s_{\min},s_{\max}]\neq\varnothing$, then pairwise lines satisfy the distance constraints of drones. The drone may pass through the pairwise lines in the process of penetration. To simplify the problem, we just discuss the discrete time when the networking radar system check the spatial state, which is 10 seconds once and start from the any straight line can not pass through the three edges of the pyramid unless it pass through the vertex of the pyramid. That means if the drone is in uniform linear motion, it can effectively interfere with one radar twice at most in the whole process. Since all drones are in uniform linear motion and a false target needs at least three drones to make it be accepted by the networking radar system, the number of drones must be more than 3 × 20 ÷ 2 = 30. Thus, we get the lower bound of the number of drones N_l, $N_l = 30$. Let N be the minimum number of drones in the process of penetration, we can get $N \geq N_l$. But the precise minimum number of drones is unknown and how to distribute the drones is still not solved. In the subsection 2.3, there are five sets for the possible lines of the positions for the drones. One drone pass through the L_{i_1,j_1} at the i_1-th unit time(one unit time is 10 seconds), which means the drone effectively interferes with the j_1-radar and makes a "false information" at the i_1-th false target given by problem. If there are 3 drones do it at the same time, then the "false information" changes into the false target given by problem and the task is perfectly finished by drones.

3. The cooperative control of UAVs

The elements of the set $S_k, k=1,2,3,4,5$ are: $L_{1,k}$, $,L_{2,k}$, ..., $L_{19,k}$, $L_{20,k}$. Let $x^k_{i_1,i_2}$ be 1 if drone from $L_{i_1,k}$ at the i_1-th unit time to the line $L_{i_2,k}$ at the i_2-th unit time in uniform liear motion with satisfying the distance constraints in the filed of first radar(which means the drone interferes with the k-th radar and the drone passes through two edges of the reverse-pyramid); otherwise, $x^k_{i_1,i_2}$ is 0. Denote vertexes v^k_i as the elements $L_{i,k}$ of the set S_k. The $x^k_{i_1 i_2}, \forall i_1,i_2 \in 1,2,\ldots,19,20, k=1,2,3,4,5$ constructs the matrices $X^k_{20\times 20}, k=1,2,3,4,5$. Denote $X(i,j)$ as element of X in the i-th row *j-th* column, d as the distance and v as the speed of UAVs. These matrices describe the possible routes of the drones and each matrix of them is equivalent to a graph, which means that the drone fly from $v^k_{i_1}$ to $v^k_{i_2}$ if $x^k_{i_1,i_2}$ is equal to 1. Thus the phantom generating problem can change into following optimization problem. the distribution problem changes into a matrix problem: given the matrices $X^k_{20\times20}$, find the matrix X^k with same size make them satisfy the following constraints.

$$\min N = \sum_{k=1}^{5}\sum_{i_1}^{20}\sum_{i_2}^{20} X^k(i_1,i_2)$$

$$s.t.\begin{cases} X^k(i_1,i_2) \le X^k_{20\times20}(i_1,i_2), \forall i_1,i_2,k\ ; X^k(i_1,i_2)=0\ or\ 1, \forall i_1,i_2,k; \\ \sum_{k=1}^{5}\sum_{j=1}^{20}\left(X^k(i_1,j)+X^k(j,i_1)\right) \ge 3, \forall i_1,k\ ; \sum_{j=1}^{20}\left(X^k(i_1,j)+X^k(j,i_1)\right) \le 1, \forall i_1,k; \end{cases} \quad (8)$$

4. Numerical simulation

We use minimum-element method in [5] and the improved variable closed-loop method proposed by [6] to solve the problem. The results are shown in table 2. The minimum of N is 27 and another 6 drones should be added in the process of phantom track generation, which they only pass through the edge of pyramid just once and make one 'false information' in the process. So the minimum of the number of drones N is 33 and their motion parameters are shown in table 2.

Table 2. The solution of the distribution of drones.

i_1	i_2	k	$v(m/s)$	i_1	i_2	k	$v(m/s)$	i_1	i_2	k	$v(m/s)$	i_1	i_2	k	$v(m/s)$
2	20	2	37.30	9	13	2	35.27	7	15	3	40.37	5	17	4	35.22
3	19	2	36.75	10	12	2	35.27	8	14	3	40.16	6	16	4	34.74
4	18	2	36.31	2	20	3	42.93	9	13	3	40.05	7	15	4	34.38
5	17	2	35.95	3	19	3	42.18	10	12	3	40.00	8	14	4	34.10
6	16	2	35.67	4	18	3	41.56	2	20	4	37.44	9	13	4	33.93
7	15	2	35.47	5	17	3	41.06	3	19	4	36.55	10	12	4	33.83
8	14	2	35.33	6	16	3	40.67	4	18	4	35.82				

5. Conclusion

Uniform linear motion is a simple and easy motion and it can be easily and automatically controlled in cooperation control of UAVs. This paper proposes the reverse-pyramid model based on the characteristic of uniform linear motion and then get the lower bound of the number of drones. Combining with the distance constraints, the problem can be formulated as a graph problem or a integer linear programming problem. Finally, the problem is solved. We can get the motion parameters of each drone and the minimum number of drones.

References

[1] Shufen G, Guowen Y, Xin X and Peng X 2018 Academy Research on method of track deception jamming based on cooperative control of UAVs *J. Air Force Early Warning Acad.* **32** 44-7.

[2] Fei L, Zhongliang Z, Xinyu G and Zhongqi C 2013 Technology for penetrating radar net based on multiple combat air vehicles cooperation track deception *Syst. Eng. Electron.* **35** 2309-13.

[3] Tian Y, Jianfeng T and Xingcheng Li 2017 Simulationresearch of phantom track generation based on cooperativecontrol of a team of aircrafts *Comput. Simul.* **34** 100-4.

[4] Zhouyan S, Min W, Lu Z, Linzhi M, Lu C, Pei X and Su J 2018 Research on track deception technology based on jamming of flexible lines by jammers *MATEC Web Conferences* **176** 01039.

[5] Kai W, Huijun C, Bin G 2016 Application of minimum-element method in optimization of pipeline arrangement for dredging & reclamation works *Port & Waterway Eng.* **522** 188-92.

[6] Zhufang W, Wenqing M 2012 An improved algorithm to solve the transportation problems of relief materials *Opera. Res. Manage. Sci.* **21** 142-46.

IOP Conf. Series: Materials Science and Engineering **685** (2019) 012019 doi:10.1088/1757-899X/685/1/012019

Based on MATLAB Simulation to Analyze the Online Source Interaction of the Photovoltaic Power Grid

Lin Qing LUO[1,a] and Zhao Xing MA[2,b]

[1]Qingdao university of technology, School of information and control engineering, Qingdao 266520, China
[2]Qingdao University of Technology, Innovation institute for sustainable martime architechture research and technology 266520, China

Email: [a] 1339450991@qq.com; [b] 379071859@qq.com

Abstract. Firstly, the influence of voltage fluctuation and power loss on power flow distribution is analyzed by theoretical formula. Then, under the simulation experiment of MATLAB, the related problems arising from the power quality of the distribution network when the distributed photovoltaic power generation is connected to the power grid are studied, and the influence of voltage fluctuations and harmonics on it is analyzed.Through simulation test, the corresponding voltage, current and power waveforms are obtained. To study the effects of the waveform, and incorporate FFT analysis tools to observe harmonic content.The simulation results show that the node voltage will increase with the reverse current phenomenon of pv and netease. In the process of grid-connection, a large number of higher harmonics are difficult to filter completely.

1. Introduction

With the development of economy, energy sources are becoming increasingly scarce. As a clean and renewable energy, photovoltaic power generation has gradually entered people's life[1].Distributed photovoltaic power generation is one of the main utilization methods, so to study the effect of distributed grid-connected photovoltaic is imminent[2].Distributed generation through the inverter connected to the electricity grid which affect the power quality[3].In literature [4], the windowed interpolation basis 2FFT algorithm is used to detect and analyze various parameters of harmonics more clearly, and the harmonics of the grid-connected photovoltaic model are analyzed in the same way.Similarly,voltage distribution and fluctuation are also a manifestation of the influence of distributed photovoltaic grid connection on power quality[5].In literature [6], the forward and backward generation algorithm was used to calculate the power flow of the distributed photovoltaic integrated into the 10kV distribution network, and the dynamic voltage change of the distributed photovoltaic connected to the grid was simulated by MATLAB.Therefore, it is necessary to study the influence of distributed photovoltaic power grid connection. The relevant influence studied in this paper will provide data reference and relevant Suggestions for future distributed photovoltaic power grid connection technology.

Based on MATLAB simulation experiment, this paper studies the power flow related to distributed photovoltaic grid-connection, and adopts kirchhoff's first law calculation method to reflect the impact of distributed photovoltaic grid-connection on the voltage drop and power loss of the system. The output voltage and power of photovoltaic power generation are observed by simulating the grid-connected model of photovoltaic power generation, and the voltage fluctuation and harmonic influence of photovoltaic power generation connected to the power grid are analyzed through waveforms.

2. A brief analysis of distributed photovoltaic power generation

2.1. photovoltaic power generation

Composition of photovoltaic power generation system: photovoltaic array, battery, inverter, controller and other parts.The core of the photovoltaic power generation system is the photovoltaic module, which receives sunlight and generates a photocurrent (direct current), which is converted into usable alternating current (ac) by an inverter. On the other hand, energy can also be stored in batteries and other energy storage devices, such as when the need to release[7].

2.2. Grid-connection principle and characteristics of distributed photovoltaic power generation

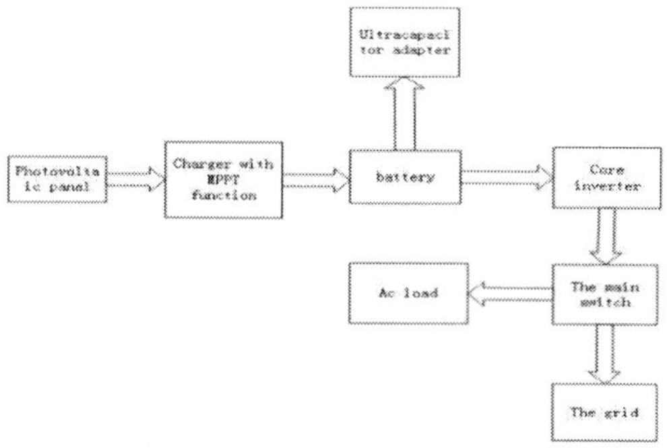

Figure.1 Structure diagram of photovoltaic power generation system

Figure 1 is the structure diagram of a typical photovoltaic power generation system. The photovoltaic panel directly converts light radiation into electric energy. The stronger the radiation, the greater the current.Figure 2 shows the equivalent simplified circuit of the photovoltaic power generation system[8],

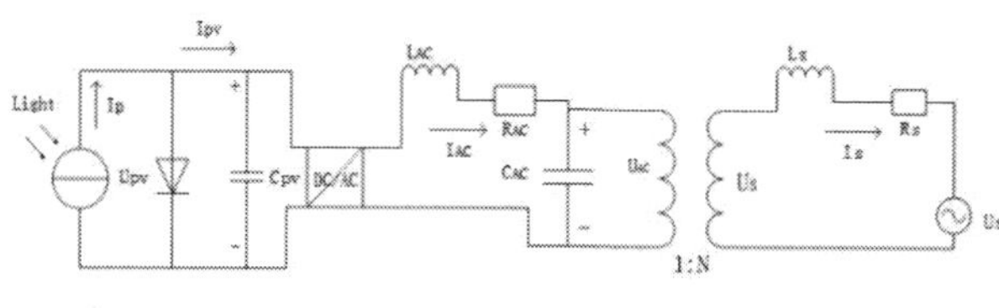

Figure.2 The equivalent circuit of a photovoltaic power system

I_P is the photocurrent generated by the solar panel, I_{PV} is the output current of the photovoltaic array,C_{PV} is the filter capacitance at the outlet side of photovoltaic array,U_{PV} is the dc voltage on both sides of the capacitor,L_{PV} is the filter inductance on the ac side of the inverter,R_{AC} is the equivalent resistance of the filter inductance,C_{AC} is the filter capacitance of the ac side,U_{AC} is the voltage of the filter after passing through the transformer (transformer ratio N),L_S and R_S are the equivalent inductance and equivalent resistance from the transformer to the grid node, respectively.I_S is the current integrated into the grid by the photovoltaic power station,U_S is the bus voltage of the union point.

The current relationship of photovoltaic panels:

$$I_P - I_{PV} = I_0(\exp\frac{qu}{AkT} - 1) \tag{1}$$

Where: I_0 is the reverse saturation current of the PN junction, k is the boltzmann constant 0.86×10^{-4}eV/K,q is the electron charge 1.6×10^{-19}C, A is the curve constant of the PN junction, and T is the absolute temperature.

2.3. Analysis of influence of photovoltaic grid-connection on power flow distribution

The user side of the traditional distribution network is not connected to any power supply. After the access of the distributed photovoltaic power generation system, the power grid becomes a multi-source network, so there may be counter-current and complex voltage distribution[9].

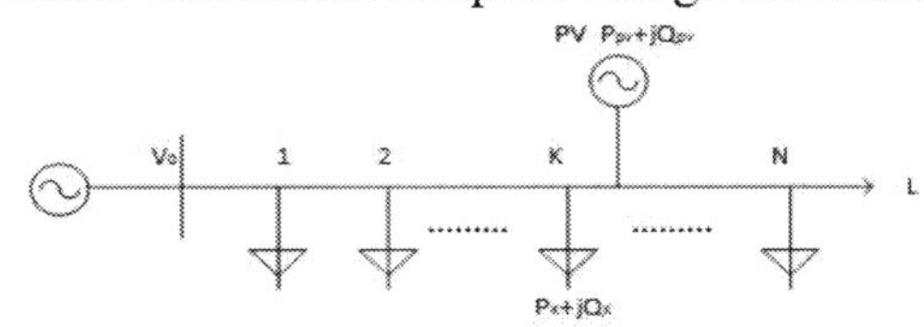

Figure.3 Schematic diagram of access node of distributed photovoltaic power generation system to distribution network

As shown in figure 3 N nodes distribution feeders L, the NTH node voltage is V_N,the initial voltage is V_0,the impedance between nodes is set to $Z_i = R_i + jX_i$ (i=1,2,3...N),distribution network load uniformly release feeder point end,the pv system capacity at node i is $P_{pv}+jQ_{pv}$.Rated voltage is U_N,calculation of power loss and voltage loss at i node:

$$\Delta S_i = (\frac{S_i}{U_N})^2 (R_i + jX_i) \quad \text{(i=1,2,3...N)} \tag{2}$$

ΔS_i is the power loss between $i-1$ and i, S_i is the power of i node, R_i is the resistance between $i-1$ and i circuit, X_i is the impedance between $i-1$ and i.

$$\Delta U_i = \frac{P_i R_i + Q_i X_i}{U_N} \quad \text{(i=1,2,3...N)} \tag{3}$$

ΔU_i is the voltage loss between $i-1$ and i, P_i is the node active power, Q_i is the node reactive power.For ΔU_i and S_i, we have:

$$\begin{gathered} S_1 = (P_1+Q_1)+(P_2+Q_2)+(P_3+Q_3)+......+(P_K+Q_K)+......+(P_N+Q_N) \\ S_2 = (P_2+Q_2)+(P_3+Q_3)+......+(P_K+Q_K)+......+(P_N+Q_N) \\ S_3 = (P_3+Q_3)+......+(P_K+Q_K)+......+(P_N+Q_N) \\ ....S_K = (P_K+Q_K)+......+(P_N+Q_N).........S_N = (P_N+Q_N) \end{gathered} \tag{4}$$

$$\Delta U_1 = U_{0\_1} = \frac{P_1R_1+Q_1X_1}{U_N} \quad \Delta U_2 = U_{1\_2} = \frac{P_2R_2+Q_2X_2}{U_N} \quad \Delta U_3 = U_{2\_3} = \frac{P_3R_3+Q_3X_3}{U_N}$$
$$...........\Delta U_K = U_{K-1\_K} = \frac{P_KR_K+Q_KX_K}{U_N}............ \quad \Delta U_N = U_{N-1\_N} = \frac{P_NR_N+Q_NX_N}{U_N} \tag{5}$$

$$U_1 = U_0 - \Delta U_1 \qquad U_2 = U_1 - \Delta U_2.........U_N = U_{N-1} - \Delta U_N \tag{6}$$

When a distributed photovoltaic system is connected to the power grid, reactive power output is usually not adjusted.When distributed pv is connected, the voltage drop at node i:

$$\Delta U_i = \frac{(P_i - P_{pv})R_i + (Q_i - Q_{pv})X_i}{U_N} \tag{7}$$

As can be seen from formula (7), the apparent power of the origin of $S_i = P_i + Q_i$ is partially offset by the power of photovoltaic power generation, so that the power flow of the line decreases, and then the line voltage decreases and the node voltage increases, playing the role of node "support".However, if the photovoltaic power is too large and exceeds the load power, the "over-offset" will be formed, and the node voltage will be too high. If the terminal voltage is exceeded, the voltage drop will become negative. Then the current will appear countercurrent phenomenon.

3. Influence analysis of distributed photovoltaic power generation grid connection based on MATLAB simulation

3.1. MATLAB simulation analysis of grid connection of photovoltaic power generation system

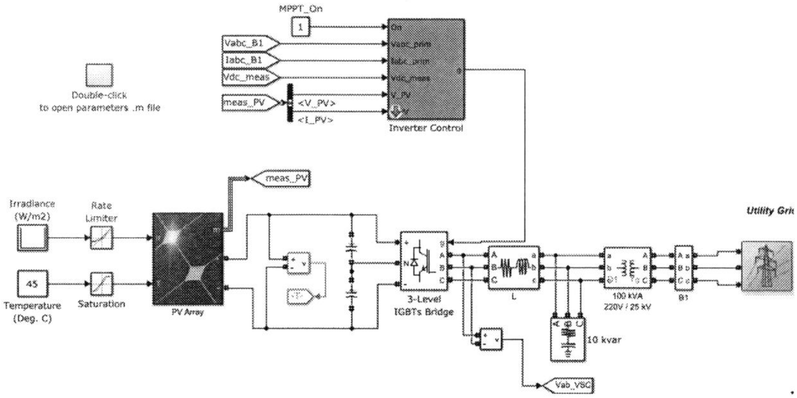

Figure.4 Photovoltaic power system grid-connected MATLAB simulation circuit

In figure 4, the model of solar panel adopts SPR-45E-WHT-D, and the PV curve model:ref S=1000W/m^2, sunshine intensity and battery temperature 45℃, temperature fluctuation limit between 55℃~5℃.The change of radiation intensity is1000W/m^2-200W/m^2,Effective working time accounts for 40% of the cycle.The three-level inverter is used in the inverter. The switching frequency of IGBT is controlled by PWM modulation technology, and the discharge signal is generated to IGBT according to the required reference voltage.In the simulation in this paper, the carrier frequency is set to 1650 Hz(33*50). The MPPT algorithm adopts the perturbation observation method. The MPPT system automatically changes the VDC reference signal of the voltage stabilizer in the inverter to obtain the dc voltage with the maximum power extracted from the PV array.PLL and measurement: synchronous and voltage/current measurements are required. The direct current of the inverter is converted into alternating current, which is then absorbed into the large power grid after LC filtering and transformer boosting. It includes two 35-kv feeders, load, grounding transformer and an equivalent 110-kv transmission system.

3.2. Analysis of influence on power quality

The generation of power quality problem lies in the asymmetry and nonlinearity of power transmission equipment, which leads to the distortion of waveform.The influence of power quality is mainly manifested in two aspects: voltage fluctuation and harmonic wave.

3.2.1. Voltage fluctuations and deviations. Generally, the power curve of photovoltaic power generation is normally distributed according to the change of day and night. Meanwhile, the power curve will fluctuate with the change of cloud cover and temperature, leading to the fluctuation of power grid voltage.

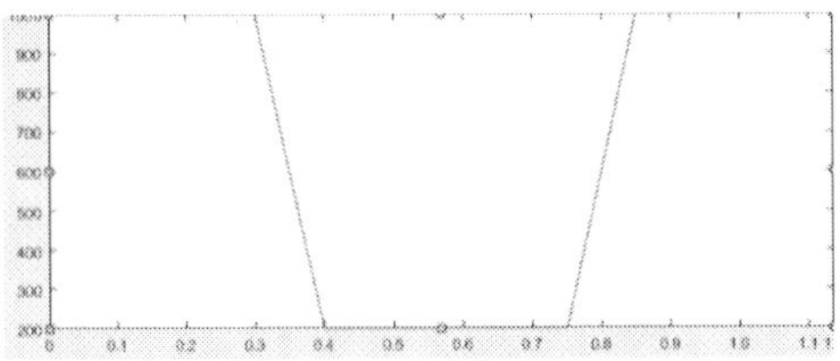

Figure.5 Waveform diagram of solar radiation intensity variation

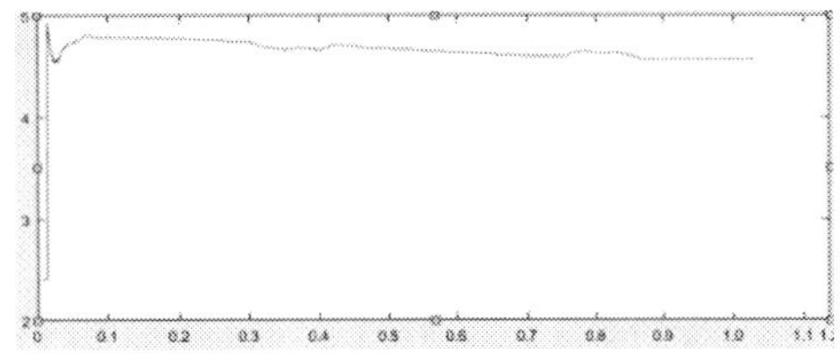

Figure.6 Solar panel output voltage

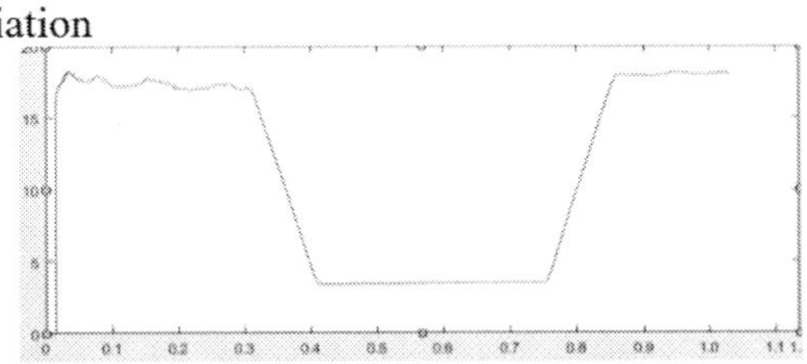

Figure.7 PV side output power

The waveform curve in figure 5 is the irradiation intensity, which starts to change at 0.3 sampling cycles. The irradiation intensity decreases from 1000W/m2 to 200W/m2, and rises to the original value at the end of a sampling cycle.The red waveform curve in figure 6 shows the PV dc side voltage, the reference value of the voltage VB is 100V, and the set dc voltage is 4.8. The actual curve is stable around 4.6, with a certain voltage drop.The waveform curve in figure 7 shows the actual output power of PV18kV, and the reference power is SB=10kV. Because the irradiation intensity causes changes in the current of photovoltaic power generation, the minimum power affected is 4kV.From the figure, we can clearly see the power change of the solar panel caused by irradiation. At the same time, you will find that the small fluctuation of voltage at the beginning is also obviously reflected in the power.

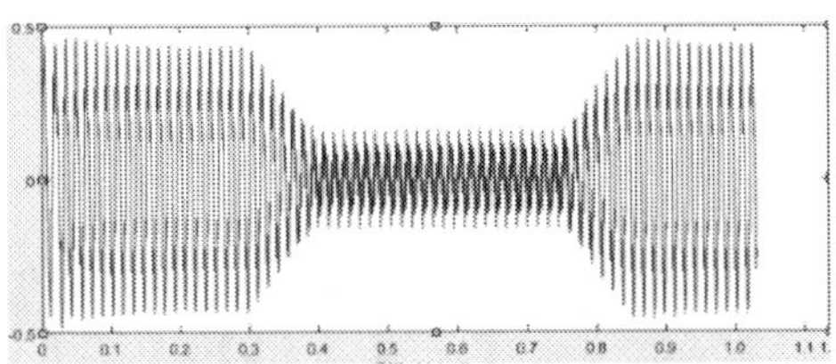

Figure.8 Transformer primary side grid current

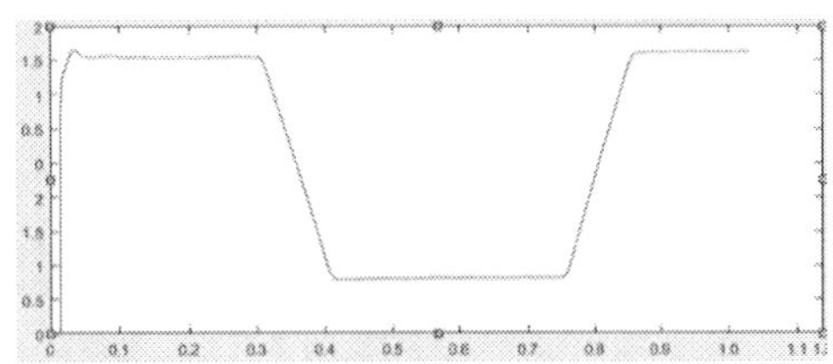

Figure.9 Photovoltaic grid-connected power

Figure 8 and figure 9 for the transformer primary side current and power grid, can also see that current is affected by the irradiation intensity and change.It is normal for the actual power input into the power grid by photovoltaic generation to fluctuate due to the change of current. However, as shown in figure 9, the power fluctuates due to the slight fluctuation of photovoltaic power generation voltage at the beginning of sampling time. If a large number of photovoltaic power generation is connected, it will bring adverse effects on the operation of the power grid. The reference value of current IB=10A, and the reference value of grid-connected power on the primary side of transformer is PB=100kV·A.

3.2.2. Harmonic. Nowadays, with the continuous development of photovoltaic power generation systems, more photovoltaic power generation systems will be connected to the power grid, and the problem of harmonic pollution is inevitable.The main reason of harmonic generation in photovoltaic power generation comes from inverter. Inverter has a large number of electronic components, which will produce high harmonic when making dc to ac. Although there are corresponding filtering devices

IOP Conf. Series: Materials Science and Engineering **685** (2019) 012019 doi:10.1088/1757-899X/685/1/012019

before being connected to the power grid, with the continuous development of photovoltaic grid-connection, the number of harmonic sources connected to the power grid will continue to increase. The superposition of multiple harmonics will cause certain security risks[10].I observed the influence of harmonics in this simulation and compared it with the waveform after filtering to observe its changes.

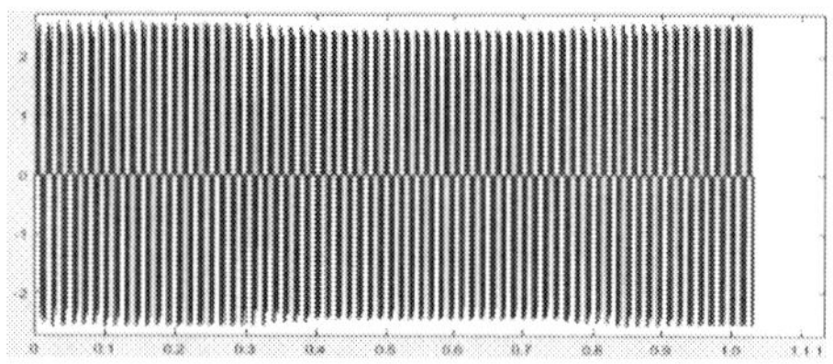

Figure.10 Ac voltage output by the inverter

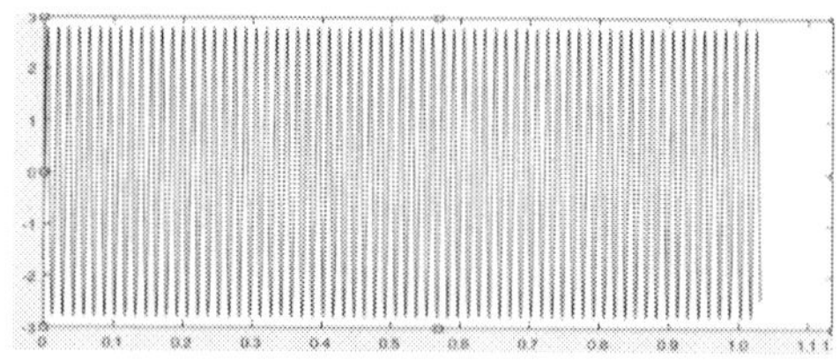

Figure.11 The output voltage on the primary side of the transformer

Figure 10 shows the voltage waveform of the output of the inverter with the reference value of V2=220V. It can be seen from the waveform that there are many clutter signals. These high-frequency harmonics seriously pollute the voltage output signal and it is difficult to distinguish the three-phase ac voltage signal of the sinusoidal waveform.In figure 11, the reference voltage is V3=10kV, and when compared with it, it can be more obvious that the ac voltage connected to the grid after filtering is smoother, so the filtering effect of three-phase voltage is still very good. The FFT Analysis toolbox is used to analyze the voltage harmonic content below.

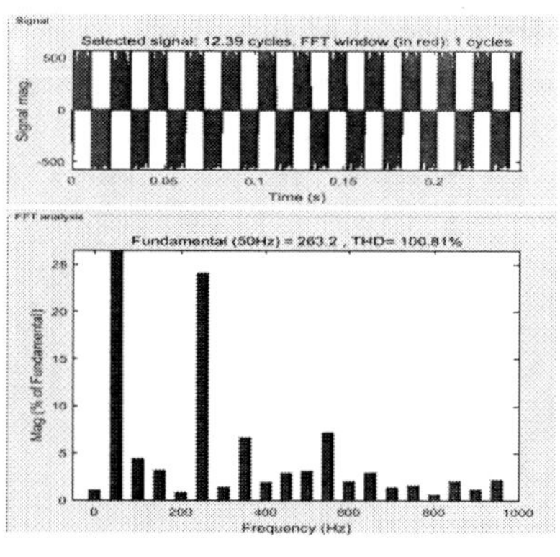

Figure.12 Unfiltered three-phase voltage harmonic analysis

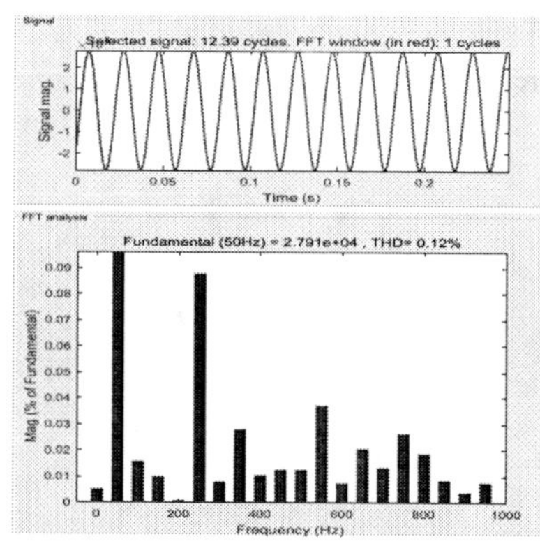

Figure.13 Three-phase voltage analysis of the filtered power grid

From figures 12 and 13, we can see more intuitively that the harmonic content of the ac voltage converted by the dc power through the inverter is THD=100.81%, and the main high harmonic frequency is around 250Hz. After filtering, the harmonic content THD=0.12%, the high-frequency harmonics of 250Hz and above are significantly reduced, and the waveform becomes smooth and can be directly connected to the distribution network.

However, it is found that the waveform of grid-connected output three-phase current is not smooth, as shown in figure.14 FFT analysis. This indicates that there are harmonics in the three-phase current. This is because a large number of high-order harmonics are generated when the inverter switch device IGBT is on and off. These harmonics are still difficult to be filtered out and always exist in the power grid. Although the voltage and current are of the same frequency and phase, the filtering effect of the current is much worse than that of the voltage. Therefore, if a large number of distributed photovoltaic power generation should be connected to the power grid in the future, the filtering system of the current should be further improved.

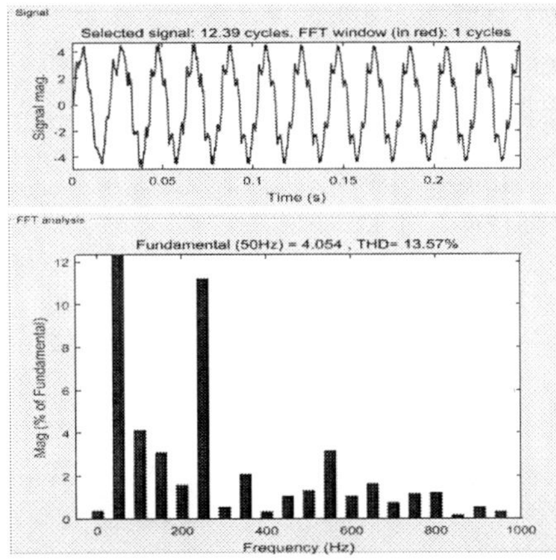

Figure.14 Three phase current FFT analysis of grid-connected output

4. Conclusion

Distributed photovoltaic power generation has a broad prospect of development. However, when distributed photovoltaic is connected to the grid, it will bring changes in power quality, power flow direction and other aspects, which will affect the safe operation of large power grid. This paper elaborates on this problem from two aspects, namely, theoretical formula derivation and numerical simulation experiment. The analysis shows that the distributed photovoltaic power should not be too large, otherwise it is easy to generate counter-current phenomenon, and the node voltage will increase due to the access of photovoltaic power generation. During the grid-connected process, a large number of high-order harmonics are difficult to be completely filtered, especially those carried by the current, and the voltage fluctuation is more obvious when the radiation intensity changes.

In the future, the author will further analyze other influencing factors of distributed photovoltaic, and based on the analysis results, propose control strategies and solutions conducive to photovoltaic grid-connection, so as to promote the safer and more stable development of distributed photovoltaic.

Acknowledgments

Thanks for the financial support provided by the national undergraduate innovation project (201810429202,201810429220).

References

[1] YAN Shizhi 2009 The research on solar energy photovoltaic generation and grid-connected system *D.* (CHENG DU: Southwest Jiaotong University)

[2] LI Bijun,FANG Yongjie,YANG Weidong and XUN Taishan 2010 Problems and countermeasures for large power grids in connection with photovoltaic power *J. Power System and Clean Energy* .26(04):52-59

[3] JIN Chu,LI Jiaming,XU Shenzhi and WEN Jinyu 2013 Probabilistic Load Flow Calculation and Influence Analysis for Power Grid Connecte with Large Scale Photovoltaic Generation System *J. Electric Power Engineering Technology* .30(06):32-37

[4] MA Jie and ZHANG Jiancheng 2010 Harmonics measurement and analysis of grid-connected PV system *J. Power System and Clean Energy*.26(11):112-115

[5] LIU Jin 2009 Analysis of main factors affecting the harmonic level of photovoltaic grid-connected power generation *J. Electric Engineering*.(12):66-67

[6] NIU Qian 2016 Study on grid connected distributed photovoltaic effect on distribution network voltage *D.*(NAN JING:Southeast University)

[7] CHEN Wei, AI Xin, WU Tao and LIU Hui 2013 Study on the influence of photovoltaic grid-connected power system on power grid *J. Electric Power Automation Equipment.* 33(02):26-32+39.

[8] VARMA R K, SALAMA M, SEETHAPATHY R, et al. 2009 Large-scale photovoltaic solar power integration in transmission and distribution networks C. IEEE Power & Energy Society General Meeting (Calgary, AB, Canada) IEEE, 2009:1

[9] ZHAO Weiran,WANG Haijiao,LI Guanghui,HE Guoqing and SUN Jian 2017 Emergency Power Support Control Strategy of VSC-HVDC and LCC-HVDC Coordination *J. Electric Power Engineering Technology*. 36(06):20-26.

[10] LIU Wenxia, XIE Jiang, XU Pengcheng, XU Huiting, YANG Xiaoming and XIAO Fengjun 2017 Simulation method and system development for voltage control of distribution network with large-scale photovoltaics *J. Modern Electric Power*. 34(06):45-52.

Study on the Effect of Excitation Circuit of Spectral Lamp on the Temperature Coefficient of Physics Package of Rubidium Atomic Clock

Yonggang Guo[1], Jun Zhang[1], Shiwei Wang[1], Jianxiang Wang[1] , Lili Zhu[1*], Jianhui Tu[1] and Zhidong Liu[1]

[1]Lanzhou Space Technology Institute of Physics, Science and Technology on Vacuum Technology and Physics Laboratory, No.100 Feiyan Roud, Chengguan District, Lanzhou, Gansu Province, 730000, China

[*]lilizhu2015@foxmail.com

Abstract. As one of the important factors affecting the long-term frequency stability of Rubidium atomic frequency standards (RAFS), the temperature coefficient has a great influence on the performance of RAFS. The temperature coefficient of physics package has an important influence on the temperature coefficient of rubidium atomic clock. The instability of temperature is often the most important error term of RAFS. The variation of temperature to several degrees means a month's aging. As an important part of rubidium atomic clock physics package, the Lamp Excitation Circuit directly affects the voltage adaptation area and temperature coefficient of Rubidium gas cell physics package. In order to satisfy the long-term frequency stability and adapt extraterrestrial environment, we propose the improvements of Lamp Excitation Circuit in physics package of Rubidium Atomic Clock, since minute current change of Lamp Excitation Circuit are mapped onto the clock's long-term frequency stability, and the stabilization of the lamp excitation circuit has the potential to significantly improve the RAFS' long-term frequency stability and temperature coefficient. We demonstrate that the temperature coefficient of physics package can be reduced obviously by improved Lamp Excitation Circuit. Suggesting that performance of the RAFS could be improved greatly further.

1. Introduction

The lamp-pumped Rubidium Atomic Frequency Standard (RAFS) is the most widely used atomic clock in the world[1]. In the past decades, with the continuous development, RAFS is arguably the "workhorse" of atomic timekeeping in space; it has a very good long and short-term stability[2],excellent performance and low radiation sensitivity. Its longevity in space extends to decade[3,4].and Rb atomic clocks are attractive candidates for space programs due to their small size, low weight, and low power consumption[5]. It uses the transition frequency between the ground state $5^2S1/2$ of ^{87}Rb atom and two hyperfine Zeeman sub-levels F=1, mF=0 and F=2, and mF=0 to discriminate the frequency. The crystal excitation frequency is locked in this transition frequency through a frequency-locked loop.

The principle block diagram of passive atomic frequency standard is shown in figure 1. RAFS is essentially a frequency-locked loop, in which the physics package(PP) is equivalent to quantum frequency discriminator[1]. It includes lamp excitation circuit, ^{87}Rb lamp, filter cell, absorption cell,

IOP Conf. Series: Materials Science and Engineering **685** (2019) 012020 doi:10.1088/1757-899X/685/1/012020

photo detector, magnetic shield, C-field coil and temperature control system. The physics package generates transition spectral lines and converts the difference between the excitation signal frequency and the center frequency of the spectral line into a discriminator signal. The light emitted by the spectral lamp filters the ^{87}Rb atom in the absorber cell at the ground state superfine level F=1 to the F=2 level, which causes the number of particles to be reversed. The crystal excitation frequency is synthesized by frequency doubling to obtain the microwave excitation signal similar to the frequency of the above transition spectral line, which excites the atom from the ground state superfine level F=1 to the level F=2 level. F = 2, mF = 0 to F = 1, mF = 0 level transition[6], when the transition occurs, the light intensity through the absorption cell detected by the photocell becomes weaker, which is the detection signal of light. The photometric signal is input into the circuit system to produce the correction signal of crystal excitation, which can lock the crystal excitation frequency to the atomic transition frequency. From the working principle of the rubidium atomic frequency standard, it can be seen that the stability of the rubidium spectral lamp directly affects the stability of the atomic transition spectral line frequency. The stability of lamp excitation circuit is one of the important factors affecting the stability of rubidium spectral lamp. Therefore, how to improve the stability of lamp excitation circuit, then improve the stability of spectral lamp, and ultimately reduce the temperature coefficient of physics package becomes a key issue. Finally, by analyzing and improving the lamp excitation circuit, and testing the performance under atmospheric and vacuum conditions, the testing results is helpful to optimize the temperature coefficient and improve the long-term stability of rubidium-loaded clocks in the application of space.

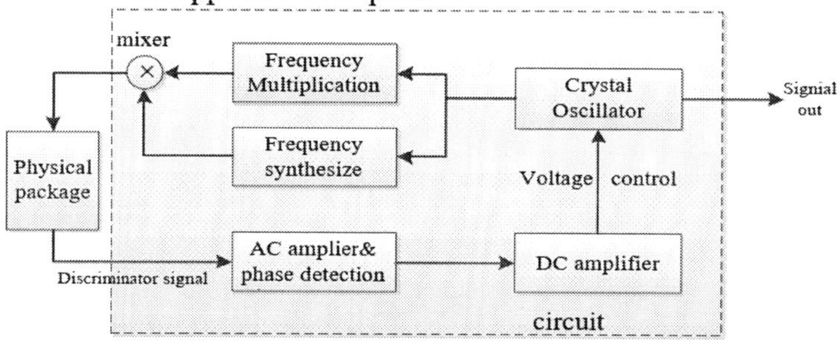

Figure 1. Basic Block Diagram of a Passive atomic frequency standard.

2. Rubidium gas cell physics package

As the core component of rubidium gas cell atomic frequency standards, the physics package mainly consists of lamp excitation circuit, ^{87}Rb lamp, filter cell, absorption cell, photo detector, magnetic shield, C field coil, cavity and oven temperature control system. The physics package mainly plays the role of frequency discriminator, providing a stable frequency and narrow line width atomic resonance absorption line. By comparing the frequency of the microwave excitation signal with the atomic transition spectrum, the output discrimination signal of physics package is then conducted by phase-sensitive detecting, so that the output frequency of the crystal excitation is locked to the central frequency of the transition spectrum. As shown in figure 2, the light beam from the lamp finally gets to the rubidium absorption cell, placed in the microwave cavity, as to deplete the population of F=1 level and build a population inversion between the F=1 and F=2 levels of ^{87}Rb ground state[7]. The excitation circuit is used to generate the excitation frequency, and the rubidium lamp is used to emit ultra-fine level spectrum a and b lines, and the filter cell is used to remove the hyperfine component of the a lines and retain b lines[8]. The absorption cell is used to reverse the number of ground-state particles, pumping ^{87}Rb F1 ground-state atoms to F2 level (F1-> F2). The photo detector converted the optical signal through the absorbed cell into electrical signals and supplied to the servo system; the C-field is used to fine-tune the transition frequency of ^{87}Rb and exclude them from the phase-locked loop capture band by sensitivity of the non-clock transition signals to the C-field. the magnetic shielding is mainly used to shield the influence of the geomagnetic field and the spatial stray magnetic field on the transition frequency of rubidium clock atoms.in the PP, in order to obtain the best SNR (signal to noise ratio), The filter cell and absorption cell are separately temperature controlled[9,10].

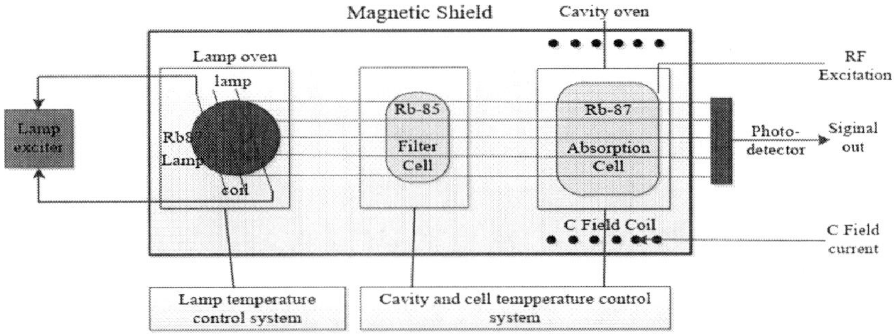

Figure 2. Rubidium gas cell physics package.

3. Lamp excitation circuit

3.1. Traditional Lamp excitation circuit

One of the important components of the Physics Package is Rb lamp.The Rb lamp consists of electrode less Rb bulb and a lamp exciter[11]. Rubidium spectrum lamp is a kind of electrodeless discharge lamp by using high frequency excitation applying to the rubidium atomic gas. The lamp excitation circuit provides a radio frequency between 100MHz and 150MHz for the rubidium spectrum lamp. The rubidium lamp emits pumping light under the excitation.

Considering the stability of lamp excitation circuit, we ultimately choose the Clapp oscillator which is an improved Three-point capacitance oscillator. the feedback coefficient and frequency stability of the excitation will be affected slightly when the external environment changes. The circuit principle is shown in figure. 3(a) and (b). L3 is the coupling coil around the rubidium lamp, and $\mathbf{C_{ce}}$ is the interelectrode capacitance between the emitter and collector of transistor.

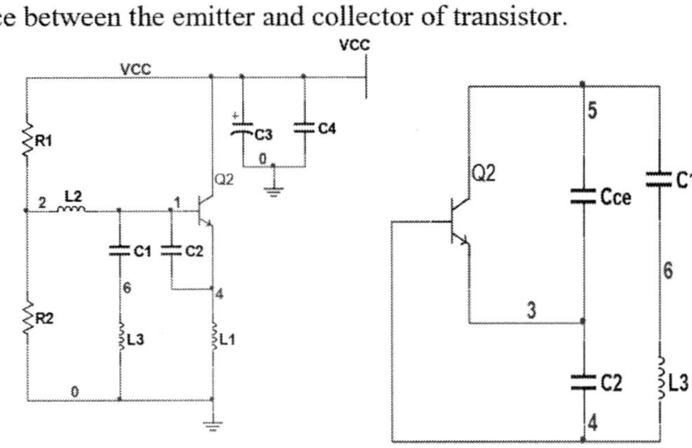

(a) basic schematic circuit diagram. (b) alternating current path.

Figure 3. The principle chart of the spectral lamp excitation circuit.

From the alternating current path shown in figure 3(b), it can be seen that the excitation frequency of the circuit is as follows:

$$f_0 = \frac{1}{2\pi\sqrt{L_0 C_\varepsilon}} \tag{1}$$

Considering the transistor equivalent output capacitance $\mathbf{C_{ce}}$, the total capacitance $\mathbf{C_\varepsilon}$ of the oscillating circuit is:

$$\frac{1}{C_\varepsilon} = \frac{1}{C_{ce}} + \frac{1}{C_1} + \frac{1}{C_2} \tag{2}$$

Reaction coefficient F is:

IOP Conf. Series: Materials Science and Engineering **685** (2019) 012020 doi:10.1088/1757-899X/685/1/012020

$$F = \frac{C_{ce}}{C_2 + C_{ce}} \tag{3}$$

Because the junction capacitance of the transistor and $\mathbf{C_2}$ are parallel, when capacitance $\mathbf{C_2}$ values is larger, the frequency of the exciter circuit will not affected by the junction capacitance. In addition, $\mathbf{C_{ce}}$ and $\mathbf{C_2}$ are both larger than $\mathbf{C_1}$, so the oscillating frequency mainly depends on $\mathbf{C_1}$ and $\mathbf{L_0}$.

Equation (3) shows that the feedback coefficient will not be influenced when the excitation frequency is changed by adjusting C1 values. In addition, because f0 is almost independent of the transistor equivalent input capacitance and output capacitance, it improves the frequency stability.

3.2. Improvement of Lamp Excitation Circuit

The main problem of Clapp circuit is that the excite current is affected by the temperature sensitivity of the transistor. Considering that the transistor is an active device, the working state and the output of the lamp will be changed when the temperature of the excite circuit or the surrounding environment changes which is mainly the change of light intensity and spectrum distribution.

Zero drift is usually called as temperature drift, the main cause of zero drift is variation of transistor parameters caused by temperature, Therefore, in order to restrain this temperature drift, we adopt temperature compensation measurement in traditional lamp excite circuit. The improved schematic diagram of the lamp excitation circuit is shown in figure 4.

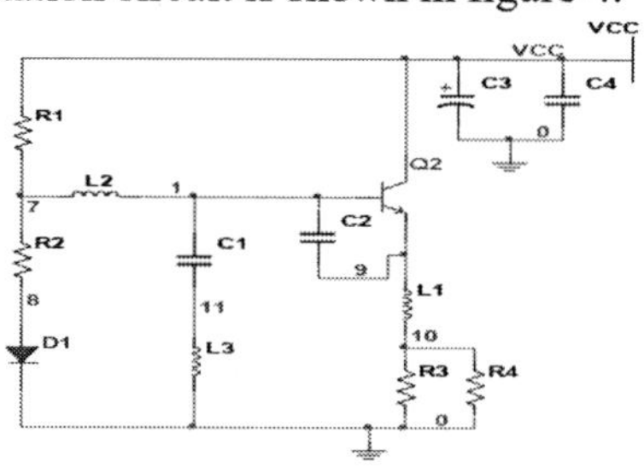

Figure 4. Improved lamp excitation circuit.

There are two main measurements to restrain the temperature drift. One is to introduce the DC negative-response into the circuit as to add the resistance R3 and R4 in series at the emitter of the transistor. When the temperature increases, the collector current of transistor (I_C) increases, and then the emitter current of transistor (I_E) increases correspondingly. Therefore, the resistance R3 and R4 increase accordingly. Because U_{BQ} basically remains constant and $U_{BE} = U_B - U_E$, the U_B will decrease which leads to a dropping of the base current I_B and I_C. As a result, the increase of collector current with temperature almost equal to the decrease of base current. therefore, I_C and U_{CE} are basically unchanged. Similarly, when the temperature decreases, each variable changes in the opposite direction, I_C and U_{CE} are basically unchanged.

The other method is to adopt the temperature compensation by using thermal elements to offset the temperature drift, in which a diode is connected in the bottom of the base circuit, and the temperature compensation is carried out by using the forward characteristics of the diode. When the temperature increases, collector current increases, and then emitter voltage U_E increases. On the other hand, the dropped voltage of diode inevitably decreases, and then U_B decreases. As a result, U_{BE} decreases and collector current I_C decreases.

4. Expeiment

4.1. Current Test

The traditional lamp excitation circuit and the improved lamp excitation circuit are tested separately at different temperatures. The lamp voltage is set to be 17V and 19V respectively. The test results are shown in Figure 5(a) and (b).

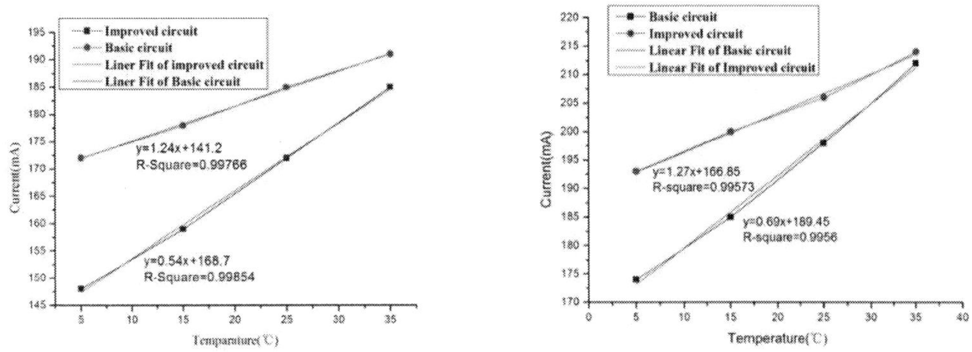

Figure 5. (a) Lamp Voltage 17V (b) Lamp Voltage 19V

As shown in figure 5(a) and (b), the black and red line respectively show the trend of the current changing with temperature in traditional circuit and the improvement of the lamp excitation circuit. It can be seen that the slope of the red curves is obviously smaller than that of black. The current temperature sensitivity of the improved lamp excitation circuit has been significantly improved.

4.2. the lamp voltage test

Test frequency accuracy is conducted by changing the lamp voltage starting from 19V with an decreased step of 0.5V. The black and red line in figure 6 shows the trend of voltage adaptation zone in the original lamp excitation circuit and the improved lamp excitation circuit. The lamp voltage adaptive zone test data are shown in table 1.

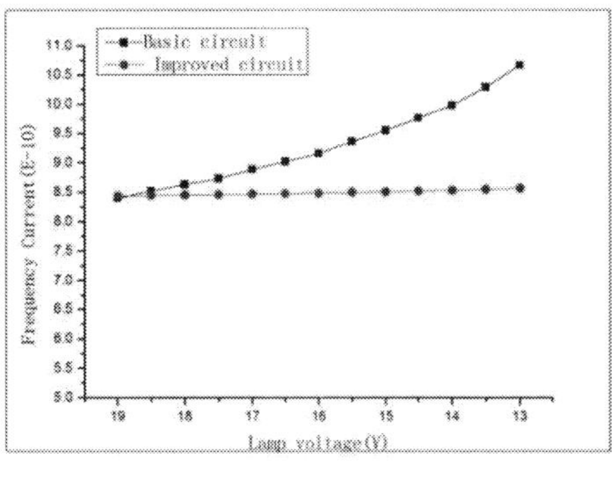

Figure 6. lamp voltage adaptive zone

Table 1. Debugging record of lamp voltage adaptive zone.

Voltage change	19V-18V	18V-17V	17V-16V	16V-15V	15V-14V	14V-13V
Original circuit	2.272E-11	2.599E-11	2.926E-11	3.630E-11	4.409E-11	6.824E-11
Improved circuit	1.540E-11	1.300E-11	1.740E-11	2.190E-11	2.930E-11	3.500E-11
promote	7.32E-12	1.30E-11	1.19E-11	1.44E-11	1.48E-11	3.32E-11

It can be seen from figure 6 that the test curve of lamp voltage adaptation zone becomes flat by adding temperature compensation measurement to the original lamp excitation circuit. It can be seen from Table 1 that the change of frequency stability corresponding to each 1V lamp voltage decreases between 7.32E-12/V and 3.32E-11/V. It can be concluded from the test that the lamp voltage adaptation zone improves obviously with the improvement of lamp excitation circuit. The test results show that the lamp voltage adaptation zone has a direct impact on the temperature coefficient of the physics package. When the test results of lamp voltage adaptation zone become smaller, the temperature coefficient of physics package decreases obviously, which is helpful to improve the temperature coefficient of rubidium clock and further enhance the long-term frequency stability of rubidium clock.

4.3. temperature coefficient

The temperature coefficient[12] of PP has an important influence on the temperature coefficient of and long-term stability of rubidium atomic frequency standards (RAFS).

The temperature coefficient of PP was tested by changing the temperature of vacuum tank. Test temperature coefficient of physics package is conducted by changing the temperature of vacuum tank, which starts from 32℃ with each step of 2℃.

The temperature coefficient test curve of physics package of original lamp excitation circuit and improved lamp excitation circuit is shown in figure 7 and figure8 respectively. The experimental results of the improved temperature coefficient of physics package in original and the improved circuit are shown in Table 2.

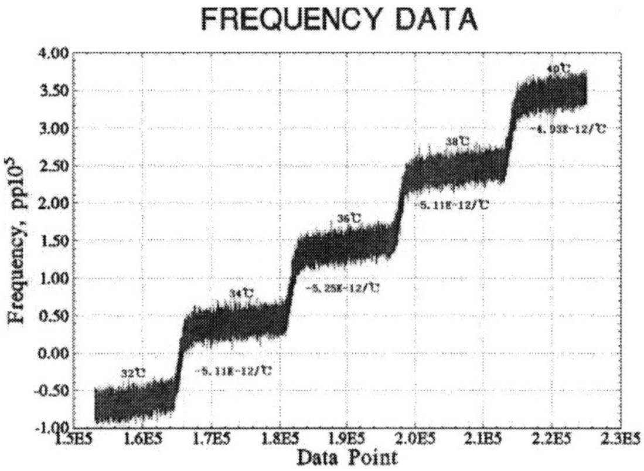

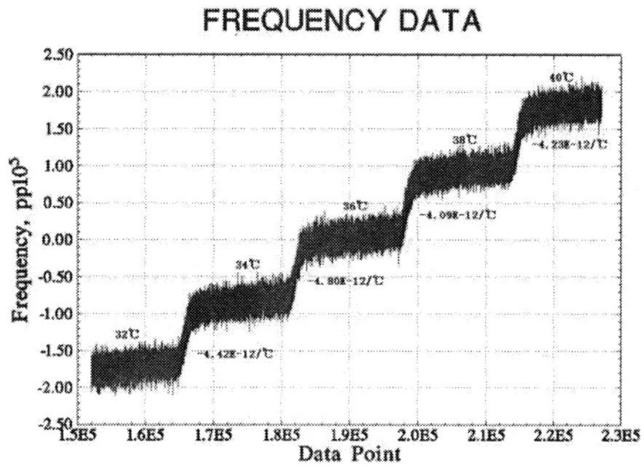

Figure 7. temperature coefficient of PP . Figure 8. temperature coefficient of PP.

Table 2. debugging record of temperature coefficient of physics package.

Temperature change	32℃-34℃	34℃-36℃	36℃-38℃	38℃-40℃
Original	-5.11E-12/℃	-5.25E-12/℃	-5.11E-12/℃	-4.93E-12/℃
improved	-4.42E-12/℃	-4.80E-12/℃	-4.09E-12/℃	-4.23E-12/℃
promote	6.90E-13/℃	4.50E-13/℃	1.02E-12/℃	7.00E-12/℃

As can be seen from the above table, the temperature coefficient of the physics package in each temperature section decreases. The temperature coefficient decreases by 6.90E-13/ ℃ when temperature of the vacuum tank rises from 32℃ to 34℃. The temperature coefficient decreases by 4.50E-13/C when the temperature of vacuum tank rises from 34 ℃ to 36℃. The temperature coefficient decreases by 1.02E-12/℃ when temperature of the vacuum tank rises from 36℃ to 38℃. The temperature coefficient decreases by 7.00E-13/℃ when temperature of the vacuum tank rises from 38℃ to 40℃.

5. Conclusion

The improved lamp excitation circuit can obviously improve the temperature coefficient of the physics package of the rubidium clock, and the influence of temperature on the frequency stability of the physics package is reduced, which is helpful to improve the temperature coefficient and long-term frequency stability of the whole rubidium clock. The lamp oscillation circuit has a great influence on the temperature coefficient of the physics package, to reduce the temperature coefficient of the physics package and then improve the frequency stability of rubidium clock, A new improved lamp excitation circuit has been proposed by analyzing the principle of the lamp excitation circuit. The rationality and feasibility of the improved circuit are proved by the experiment of lamp voltage adaptive zone and temperature coefficient of the physics package. Through the specific experimental, it is shown that the lamp excitation circuit has a great influence on the voltage adaptation zone and temperature coefficient of the physics package. The improved circuit in this paper can help to improve the performance of the physics package, and can be applied for improving the temperature coefficient and long-term frequency stability of the rubidium atomic clock.

References

[1] Hao Q , Li W , He S , et al. A physics package for rubidium atomic frequency standard with a short-term stability of 2.4 × 10-13 τ-1/2[J]. Review of Scientific Instruments, 2016, 87(12):123111.
[2] Dupuis R T , Lynch T J , Vaccaro J R . Rubidium Frequency Standard for the GPS IIF program and modifications for the RAFSMOD Program[C]// IEEE International Frequency Control Symposium. IEEE, 2008.
[3] Camparo J C , Hagerman J O , Mcclelland T A . Long-term behavior of rubidium clocks in space[C]// European Frequency & Time Forum. IEEE, 2012.
[4] V.Formichella, J. Camparo, and P. Tavella, On-orbit GPS RAFS lamplight variations: Statistics of lamplight jump. in Proc. 2017 Precise Time and Time Interval Meeting, ION PTTI 2017 (The Institute of Navigation, Manassas, VA, 2017)pp.291-298
[5] Camparo J C , Moss S C , Lalumondiere S D . Space-system timekeeping in the presence of solar flares[J]. IEEE Aerospace and Electronic Systems Magazine, 2004, 19(5):3-8.
[6] Camparo J C . The Rubidium Atomic Clock and Basic Research[J]. Physics Today, 2007, 60(11):12.
[7] Dingjan J , B. Darquié, Beugnon J , et al. A frequency-doubled laser system producing ns pulses for rubidium manipulation[J]. Applied Physics B (Lasers and Optics), 2006, 82(1):47-51.
[8] Camparo J C, Mackay R. Spectral mode changes in an alkali rf discharge[J]. Journal of Applied Physics, 2007, 101(5):27-1022.
[9] W J Riley,"A Rubidium clock for GPS",in proceedings of the 13th Precise Time and Time Interval (PTTI) Applications and Planning Meeting(1982)
[10] Li W , Kang S , Ming G , et al. Demonstration of a Physics Package with High SNR for Rubidium Atomic Frequency Standards[C]// China Satellite Navigation Conference (CSNC). 2013.
[11] Singh S , Ghosal B , Saxena G M . A New Rb Lamp Exciter Circuit for Rb atomic clocks and Studies on, Transition from Ring to Red mode[J]. Physics, 2010.
[12] Feng H , Cui J Z , Tu J H . An Enhanced Physics Package Used in Rubidium Atomic Frequency Standards[J]. Applied Mechanics and Materials, 2014, 687-691:3179-3182.

IOP Conf. Series: Materials Science and Engineering **685** (2019) 012021 doi:10.1088/1757-899X/685/1/012021

Conceptual Multidisciplinary Design Optimization of Commercial Aero-engine System

Fushui Guo[1], Shiqi Zhao[1], Xiaodong Zhou[2,*], Jun Fan[3] and Dong Mi[4]

[1] AECC Commercial Aircraft Engine Co., Ltd., Shanghai, China
[2] State Grid General Aviation Company Limited, Beijing, China
[3] Army Aviation Institute, Beijing, China
[4] AECC Hunan Aviation Powerplant Research Institute, Zhuzhou, China

[*]819331507@qq.com

Abstract. Commercial aero-engine design is a complicated task, which involves many disciplines, such as thermodynamics, aerodynamics, solid mechanics, heat transfer and materials science. Besides, for the sake of market requirements, designing a commercial aero-engine should not only consider the performance, fuel consumption, thrust and weight, but also pay special attention to noise, emissions and other environmental indices. In order to balance conflicts among various discipline indicators, and design an aero-engine with high comprehensive performance, we apply the multidisciplinary design optimization (MDO) technology to the conceptual design of an aero-engine system in this paper. Particularly, the system conceptual design of a high-bypass-ratio turbofan aero-engine is selected as a case study. A total of five disciplines, including thermal cycle, aerodynamics, structural mechanics, emission and noise, are considered. Meanwhile, six modules are integrated and analysed carefully. These modules include performance evaluation at the design point, performance evaluation at fifteen different off-design points, engine size assessment, engine weight assessment, engine emission assessment as well as engine noise assessment. On the basis of these modules, a MDO platform is set up. In addition, one optimization study with a single target of fuel burn is carried out by utilizing the adaptive simulated annealing (ASA) algorithm. Moreover, another optimization with multiple targets of fuel burn as well as nitrogen oxide emission is also studied by employing the non-dominated sorting genetic (NSGA-II) algorithm. The results shows that the MDO method for the commercial aero-engine can obtain better solutions for the system by fully taking into account the mutual effects among different disciplines.

1. Introduction

Green, environmental friendliness, low cost as well as high comprehensive performance are the future development trends of the advanced commercial aircraft engine. Compared with the traditional design method, the multidisciplinary design optimization (MDO) method has a great advantage in meeting demands of modern aircraft engines. First proposed by Sobieski [1], MDO has drawn much attention to many researchers and has been widely applied in the field of aero-engine.

In academia, there have been a great number of research achievements on the MDO of aero-engine components, such as: the fan MDO [2], the compressor MDO [3], and the turbine MDO [4]. Additionally, some researchers explored the MDO of aero-engine system. Considering the disciplines such as aerodynamics, structural mechanics and noise, Leborgne [5] studied the optimization for low-noise open-rotor engine. The results show that the MDO method can not only satisfy all constraints of aerodynamics and structural mechanics, but also improve the comprehensive performance of aerodynamics and noise. Jonathan [6] carried out the research of design and optimization for contra-rotating open rotors. In the paper, he presented a novel multidisciplinary design methodology which involved CFD, FEM and acoustics as well. These studies show that the MDO method can not only improve the efficiency of design, but also design products with high comprehensive performance by fully considering the interactions among various disciplines [7].

In industry, aero-engine manufacturers are also actively developing their multidisciplinary design systems. In 1999, NASA [8] published the Numerical Propulsion Systems Simulation (NPSS), which integrates the multi-fidelity simulations for different components or disciplines of the aero-engine. NPSS is aimed at reducing the development cost of the aero-engine by utilizing high fidelity models in the early design stage. In 2003, Jones [9] introduced the preliminary multidisciplinary design system of Rolls-Royce named Genesis, which integrates the thermodynamic cycle analysis, flow path design, structural mechanics analysis, cost assessment, and so on. In 2011, Donus [10] introduced the MTU's preliminary design system Modular Performance and Engine Design System (MOPEDS). MOPEDS involves many disciplines such as thermodynamics, aerodynamics, structural mechanics, weight, noise and cost. Furthermore, Brophy [11] described the Whitney & Pratt Canada's preliminary multidisciplinary design optimization system PMDO. These multidisciplinary design systems can effectively improve product performance, increase productivity as well as reduce design cost by balancing conflicts among various indicators in the early design stage.

It can be seen that the MDO method has obvious advantages over the traditional design method. In order to balance conflicts among various discipline indicators, and design an aero-engine with high comprehensive performance, we will apply the MDO technology to the system conceptual design of a commercial aero-engine in this paper.

2. Physical models

2.1. Performance evaluation at the design point

Performance evaluation at the design point is the core content of the aero-engine system design. One of its purposes is to select aerodynamics performance parameters by taking into account the technical levels of engine components. Another main purpose is to select the appropriate cycle parameters by employing thermal cycle analysis. According to the previous design experience, we choose the cruise installation state as the design point in this paper. Besides, the software GasTurb11 is re-developed to achieve the background operation. As a result, the software can be utilized by the Isight optimization platform to automatically assess the performance at the design point. Particularly, reasonable iterations are set in the software. For example, in order to obtain the desired thrust F, the air mass flow W_2 is chosen as an iterative parameter.

2.2. Performance evaluation at off-design points

On the basis of the engine working envelope, fifteen typical operating states are selected as off-design points. Similarly, the software GasTurb11 is re-developed to correlate the component design points with the component maps. Besides, the maps are scaled by utilizing standard maps and standard design point setting. Therefore, the performance at each off-design point could be evaluated. As a result, some calculated parameters at the off-design points will be used as input parameters of the noise assessment module or the emission assessment module, while others will be used as part of the objectives or constraints.

2.3. Engine size assessment

In 1970s, NASA released the WATE-series methods for evaluating the aero-engine size and the aero-engine weight. On the basis of WATE-series methods, the NPSS evaluation method was developed subsequently. Referring to these methods, the self-developed software is selected to assess the engine size in this paper. In detail, the preliminary geometry structure can be obtained with the help of Smith diagrams of the engine components. Particularly, we will take the compressor size estimation as an example to explain the method. First, the inlet and outlet thermal parameters of the components can be obtained through the performance evaluation at the design point. Then, the flow coefficient and work factor can be estimated with the help of Smith diagram of the components.

2.4. Engine weight assessment

Having obtained the main geometry structure according to the engine size assessment, we can easily estimate the weight of the main components, which include fan, booster compressor, high pressure compressor, combustion chamber, high pressure turbine, low pressure turbine, nozzles, shafts, supporting structure and accessories. Particularly, we will select a single stage compressor as an example to illuminate the process roughly. In details, the single stage compressor is mainly composed of four parts, namely the blade, disc, connecting piece and casing. Through reasonable simplification, we can calculate the volumes at first. Then, the weight can be estimated by combining the characteristics of the material. It should be noted that the material characteristics are default in the process. Accordingly, the whole engine weight can also be obtained.

2.5. Engine emission assessment

ICAO simplified a complete flight process into a LTO (Land and take off) cycle. That is to say, the complete flight process is divided into four off-design point states, namely LTO taking off state, LTO climbing state, LTO approaching state and LTO idling state. On the basis of the LTO cycle, various emission assessment methods have been proposed. These methods include P3T3 method, BFFM2 method, DLR method and ICAO method. In this paper, the BFFM2 method is chosen to evaluate the engine emission. In detail, the BFFM2 method utilizes the ICAO emission database to get the empirical correlation between the emission and the fuel flow.

2.6. Engine noise assessment

Generally speaking, for the high-bypass-ratio turbofan engine, there are four main noise sources, namely the fan noise, core noise, turbine noise and jet noise. Nowadays, various noise forecasting methods have been proposed to evaluate the four different noise sources by many researchers. For example, the H-W model and Rawls-Yeager model have been used to estimate the fan noise, the Mathews-Rekos model and ICAO model have been used to assess the core noise, the KV model and Matta-Sandusky-Doyle model have been used to evaluate the turbo noise, and the Stone model has been used to forecast the jet noise. In this paper, we employ the self-developed software to assess the major noise sources. And the whole aero-engine noise can therefore be estimated. Considering requirements of CCAR36 for noise, we should note that only noise in the overflying state, approaching state and boundary state is evaluated in this paper.

3. Optimization models

3.1. Single-objective optimization

3.1.1. Objective. The thermal cycle parameters of the aero-engine can affect not only the engine thrust and fuel consumption, but also the engine weight and nacelle size. Furthermore, these influences can cause changes of the aircraft's weight as well as the aircraft's flight performance. Therefore, in order to obtain the optimal performance for the aero-engine, it is very necessary and important to conduct the aircraft\engine integrated design. So the fuel burn FB can be calculated. The detailed procedure needs to discretize the entire mission into very small segments at first and then calculate FB step-by-step. However, on one hand, it will take a long time to conduct the detailed procedure. On the other hand, it is difficult to ensure that every calculation step converges when the design variables vary in large ranges. Therefore, in this paper we introduce the weighted fuel consumption sfc , which can be expressed as Eq. (1). In Eq. (1), $sfc_{(i)}$ is a parameter which denotes the fuel consumption at each typical operating state, and w_i is a weighted coefficient which can determine the contribution of the fuel consumption at each typical operating state. In addition, by conducting the flight mission analysis, we can obtain the approximation functional relationship among δFB, δsfc and δm_E. The relationship can be represented as Eq. (2), where δFB denotes the changing rate of FB, δsfc denotes the changing rate of sfc, and δm_E denotes the changing rate of m_E. At last, δFB is chosen as the optimization objective in the case of single-objective optimization (SOO).

$$sfc = \sum w_i * sfc_{(i)} \tag{1}$$

$$\delta FB = f_F\left(\delta m_E, \delta sfc\right) \tag{2}$$

3.1.2. Constraints. Considering emission requirements, we choose NO_x as a constraint.

- Considering noise requirements, we choose $dB_{(9)}$, $dB_{(10)}$ and $dB_{(11)}$ as constraints.
- Considering size requirements, we choose Dt as a constraint.
- Considering the technical levels of components, we choose $\pi_{TH(0)}$ and $\pi_{TL(0)}$ as constraints.
- Considering material characteristics, we choose $T_{3(1)}, T_{3(2)}, T_{3(4)}, T_{4(1)}, T_{4(2)}, T_{4(3)}, T_{4(4)}, T_{45(2)}$ and AN^2 as constraints.

In summary, a total of 16 constraints are selected in the case of SOO.

3.1.3. Design variables. In order to meet various indicators discussed above, the cycle parameters should be chosen carefully. According to the previous design experience, we choose five design variables in this paper. They are B, π_F, π_{CL}, π_{CH} and $T_{4(0)}$.

3.2. Multi-objective optimization

In order to meet demands of the green and environmental friendliness, we should pay more attention to the performance and emission. Therefore, δFB and δNO_x are selected as optimization objectives in the case of multi-objective optimization (MOO). Compared to the SOO problem, the design variables of the MOO problem remain unchanged, but NO_x is no longer treated as a constraint.

4. Optimization algorithms

With the help of the Isight software, a conceptual multidisciplinary design optimization platform for the system design of the commercial aero-engine is set up. The computing platform is an HP Z820 desktop workstation, which is configured with a 64GB memory, and a 20 Intel (R) Xeno (R) CPU E5-2680e clocked at 2.8GHz.

The Adaptive Simulated Annealing (ASA) is chosen to deal with the SOO problem. First proposed by Lester on the basis of traditional simulated annealing method and simulated quenching process, the ASA algorithm has become more and more popular in both science research and engineering tasks.

The Non-dominated Sorting Genetic Algorithm (NSGA-II) is chosen to deal with the MOO problem. The method was improved by Deb on the basis of the NSGA algorithm. It has the advantage of selecting excellent individuals by using the non-dominated sorting method.

5. Results and discussion

5.1. Single-objective optimization

The optimization converges after 2188 iterations, which takes 63 hours 36 minutes and 5 seconds in total. Figure 1 shows the convergence process of the objective in the case of SOO, where the red points represent infeasible solutions, the green ones represent feasible solutions, and the blue circle represents the optimal solution. From it we can see: 1) In the optimization, 566 feasible solutions are obtained. 2) The optimal solution is obtained at the 2066th iteration. 3) Particularly, the result from the vicinity of the 500th iteration is close to the optimal solution. It shows that the ASA algorithm can converge quickly to the optimal solution.

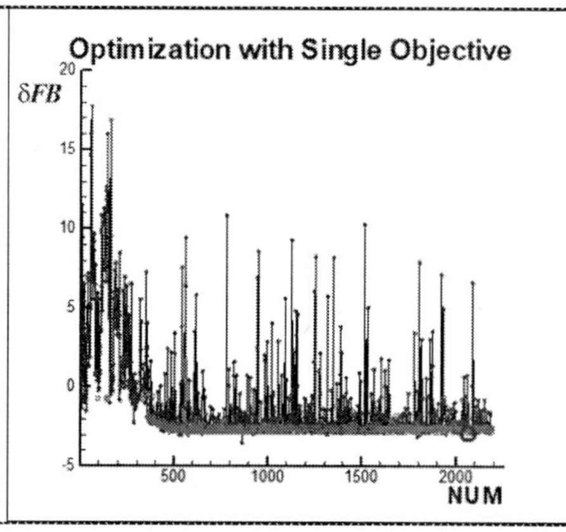

Figure 1. Convergence process of the objective function in the case of single-objective optimization.

Comparing the optimal scheme with the initial scheme, we can calculate the changing rate of the optimal solution. The detailed information is showed in Table 1, from which we can see: the fuel burn FB is reduced by 2.85%; the weighted fuel consumption sfc is reduced by 1.93%; the total weight of the aero-engine m_E is reduced by 3.23%; the emission of nitrogen oxides NO_x is reduced by 7.21%; the noise index $dB_{(9)}$ and $dB_{(10)}$ are reduced by 0.19% and 0.26% respectively. Meanwhile, the bypass ratio B is increased by 16.46%; the fan pressure ratio π_F is reduced by 0.66%; the pressure ratio of the low pressure compressor π_{CL} is reduced by 2.42%; the pressure ratio of the high pressure compressor π_{CH} is reduced by 7.40%; the total temperature at the combustor exit $T_{4(0)}$ is reduced by 1.04%.

Table 1. Optimal solution in the case of single-objective optimization.

	FB	sfc	m_E	NO_x	$dB_{(9)}$	$dB_{(10)}$	$dB_{(11)}$	Dt
changing rate	-2.85%	-1.93%	-3.23%	-7.21%	-0.19%	-0.26%	0.12%	2.35%
	$\pi_{TH(0)}$	$\pi_{TL(0)}$	$T_{3(1)}$	$T_{3(2)}$	$T_{3(4)}$	$T_{4(1)}$	$T_{4(2)}$	$T_{4(3)}$
changing rate	-15.08%	13.26%	-3.22%	-3.22%	-2.85%	-1.50%	-1.47%	-1.48%
	$T_{4(4)}$	$T_{45(2)}$	AN^2	B	π_F	π_{CL}	π_{CH}	$T_{4(0)}$
changing rate	-1.01%	4.82%	-11.82%	16.46%	-0.66%	-2.42%	-7.40%	-1.04%

As expected, the results indicate that the MDO method of aero-engine system based on ASA algorithm can obtain a better solution of the system with all the constraints being satisfied. It just shows that the proposed method has an obvious advantage in considering the mutual effects among all the five disciplines discussed in this paper.

5.2. Multi-objective optimization

The optimization converges after 2241 iterations, which takes 62 hours 3 minutes and 26 seconds in total. Specifically, 941 feasible solutions and 2 Pareto solutions are obtained in the process of optimization. Figure 2(a) and figure 2(b) show the convergence process of the objectives, namely FB and NO_x, in the case of MOO. Figure 3(a) shows the Pareto front in the case of MOO. Besides, the local information near the optimal solutions is amplified in figure 3(b). From these figures, we can see: 1) In the optimization, two Pareto solutions are obtained respectively at the 2239th iteration and the 2241th iteration. 2) Particularly, the objectives reduce significantly in the vicinity of 1300th iteration, which shows that the NSGA-II algorithm has the ability of avoiding trapping in the local optimum.

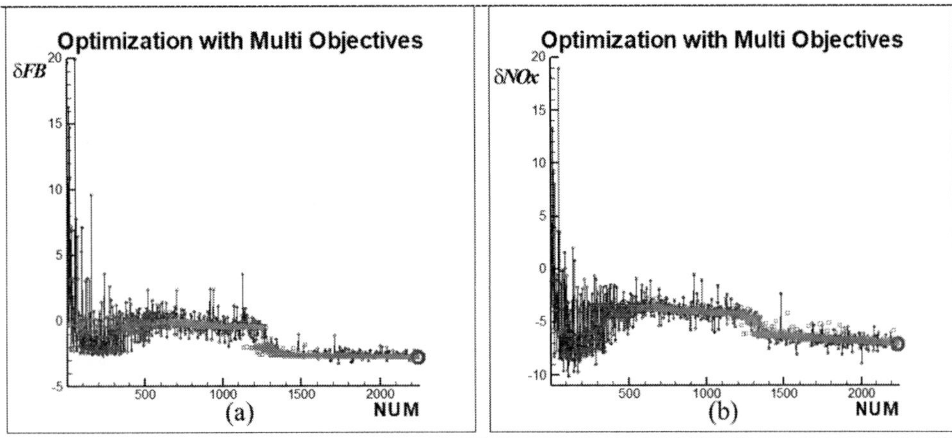

Figure 2. Convergence process of the objective function in the case of multi-objective optimization.

The solution with the minimal FB is selected as the optimization scheme 1. And the solution with the minimal NO_x is selected as the optimization scheme 2. The detailed information is listed in Table 2, from which we can see: 1) When comparing optimization scheme 1 with the initial scheme, the fuel burn FB is reduced by 2.82%, meanwhile the emission of nitrogen oxides NO_x is reduced by 7.06%. 2) When comparing the optimization scheme 2 with the initial scheme, the fuel burn FB is reduced by 2.80%, meanwhile the emission of nitrogen oxides NO_x is reduced by 7.08%. 3) It's obvious that the two optimization schemes are similar. In detail, the difference for the fuel burn FB is only 0.02%, and the difference for the emission of nitrogen oxides NO_x is still 0.02%.

Table 2. Pareto solutions in the case of multi-objective optimization.

	FB	NO_x	sfc	m_E	B	π_F	π_{CL}	π_{CH}	$T_{4(0)}$
1	-2.82%	-7.06%	-1.98%	-2.20%	18.34%	-0.57%	-10.16%	-0.75%	-1.02%
2	-2.80%	-7.08%	-1.95%	-2.33%	18.42%	-0.37%	-10.07%	-0.70%	-1.05%

As expected, the results discussed above indicate that the MDO method of aero-engine system based on NSGA-II algorithm has the advantage of "Optimize at first, and then make decisions". That

IOP Conf. Series: Materials Science and Engineering **685** (2019) 012021 doi:10.1088/1757-899X/685/1/012021

is to say, it doesn't need to consider the preference of the designer at first. On the contrary, it can obtain a set of Pareto solutions for the designer to make final decision.

5.3. Comparison and discussion

We find some differences in results between the SOO case and MOO case. Therefore, it's necessary to discuss the advantages and disadvantages of the two cases.

From Figure 1 and Figure 2, we can see the result at the 500th iteration is close to the optimal solution in the SOO case, yet in the MOO case the results at the 1200th iteration is still far away from the Pareto solutions. Obviously, the SOO case performs better than the MOO case in efficiency. But what leads to such a result? We think the reason is that the MOO case traps in the local optimum before the 1200th iteration.

When comparing the optimization results, we can find that the optimal scheme in the SOO case is better than both the Pareto solutions in the MOO case. Similarly, what leads to such a result? Figure 3(a) shows that there is a positive correlation between FB and NO_x. That is to say, there is no obvious conflict between the two indicators. Therefore, in the SOO case, NO_x decreases with the reduction of FB.

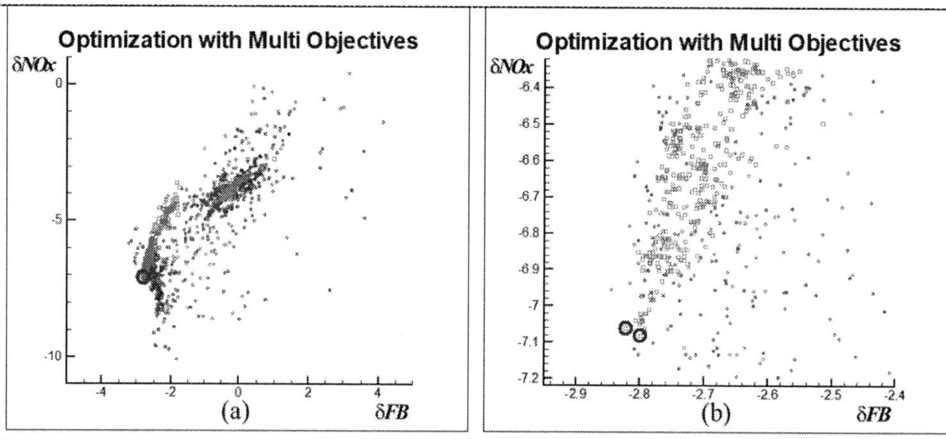

Figure 3. Pareto front in the case of multi-objective optimization.

In summary, for the problem in this paper, the SOO case performs better than the MOO case both in results and efficiency.

6. Conclusion

In this paper, a high-bypass-ratio turbofan aero-engine was selected as a case study. A total of five disciplines were considered. Six modules were integrated and analysed carefully. Besides, a MDO platform for the system conceptual design of the commercial aero-engine was set up. Through the research, we can get the following conclusions.

- The conceptual MDO method for the commercial aero-engine system is an advanced design method. It has a great advantage in considering the mutual effects among all the five disciplines discussed in this paper. It can obtain better solutions of the system.
- The MDO method of aero-engine system based on ASA algorithm can converge quickly to the optimal solution of the system with all the constraints being satisfied. The optimal fuel burn FB is reduced by 2.85%; the optimal emission of nitrogen oxides NO_x is reduced by 7.21%.
- For the problem in this paper, the SOO case performs better than the MOO case in both results and efficiency.

IOP Conf. Series: Materials Science and Engineering **685** (2019) 012021 doi:10.1088/1757-899X/685/1/012021

References

[1] Sobieszczanski-Sobieski J. A linear decomposition method for large optimization problems. Blueprint for development [J]. 1982.
[2] Deng X, Guo F, Liu Y, et al. Aero-Mechanical Optimization Design of a Transonic Fan Blade[C]//ASME Turbo Expo 2013: Turbine Technical Conference and Exposition. American Society of Mechanical Engineers, 2013: V06BT43A014-V06BT43A014.
[3] Demeulenaere A, Bonaccorsi J C, Gutzwiller D, et al. Multi-disciplinary multi-point optimization of a turbocharger compressor wheel[C]//ASME Turbo Expo 2015: Turbine Technical Conference and Exposition. American Society of Mechanical Engineers, 2015: V02CT45A020-V02CT45A020.
[4] Lu S, Chi Z, Wang S, et al. Full Three-Dimensional Optimization Platform of Turbine Blades Considering the Film Cooling[C]//ASME Turbo Expo 2013: Turbine Technical Conference and Exposition. American Society of Mechanical Engineers, 2013: V06BT43A002-V06BT43A002.
[5] Leborgne M, Lonfils T, Lepot I. Development and Application of a Multi-Disciplinary Multi-Regime Design Methodology of a Low-Noise Contra-Rotating Open-Rotor[C]//ASME Turbo Expo 2015: Turbine Technical Conference and Exposition. American Society of Mechanical Engineers, 2015: V02CT45A019-V02CT45A019.
[6] Vlastuin J, Dejeu C, Louet A, et al. Open Rotor Design Strategy: From Wind Tunnel Tests to Full Scale Multi-Disciplinary Design[J]. Power, 2015, 12: 12.
[7] Luo L, Wang S, Wang Z. Aerodynamic and heat transfer coupling optimization considering the blade profile and cooling structure in gas turbine stage[J]. 2014.
[8] Lytle, J. K., 1999. The Numerical Propulsion System Simulation A multidisciplinary design system for aerospace vehicles. ISABE 99-7111, Florence, Italy.
[9] Jones M J, Bradbrook S J, Nurney K. A preliminary engine design process for an affordable capability[R]. ROLLS-ROYCE PLC BRISTOL (UNITED KINGDOM) NEW PROJECTS ENGINEERING, 2003.
[10] Donus F, Bretschneider S, Schaber R, et al. The architecture and application of preliminary design systems[C]//ASME 2011 Turbo Expo: Turbine Technical Conference and Exposition. American Society of Mechanical Engineers, 2011: 801-810.
[11] Brophy F, Mah S, Turcotte J. Preliminary Multi-Disciplinary Optimization (PMDO) an example at engine level.[J]. 2009.

IOP Conf. Series: Materials Science and Engineering **685** (2019) 012022 doi:10.1088/1757-899X/685/1/012022

A Transmission System Planning Method Considering Demand-side Response and Capability for Accommodating Wind Power

Bo Sun[1], Bo Zou[2], Jubin Yan[1], Xiaoming Wang[1], Yuxuan Yang[3] andYanHuang[1]

[1] Chengdu Chengdian Electric Power Engineering Design Co. Ltd. Chengdu, China
[2]Economic Research Institute of State Grid Zhejiang Electric Power Co. Ltd. Hangzhou,China
[3]Economic Research Institute of State Grid Sichuan Electric Power Co. Ltd.Chengdu, China

1321600622@qq.com

Abstract. The large-scale development of grid-connected wind power, which is intermittent, random and uncontrollable, has introduced great challenges to power system planning and operation. Power system planning should fully consider the capability for accommodating wind power, as well as system's regulation capability and spare capacity. Given this background, a bi-level transmission system planning model for this purpose is proposed in this paper. In the proposed model, demand-side response resources, such as incentive load and interruptible load, are used for peak-load shifting so as to optimize power flow distribution, reduce transmission investment and improve the utilizaiton level of wind power. The two levels are implemented interactively and iteratively, and finally converge to an optimized planning scheme considering both security and economy. The essential features of the developed model and adopted algorithms are demonstrated by a modified 18-bus test system.

1.Introduction

These With the increasing exhaustion of fossil energy, many countries have developed clean energy such as wind power and photovoltaics in recent years[1]. The intermittent, random and uncontrollable nature of wind power has resulted in new problems and challenges to the secure and economic operation of the power system concerned. On the one hand, the capability for accommodating wind power should be considered for the outstanding phenomenon of wind curtailment. On the other hand, higher requirement is imposed on system's regulation capability and spare capacity, and thus additional spare capacity are needed to cope with the random fluctuation of wind power and maintain power balance and power system stability.

As for the planning of transmission systems with wind farms, some research has been conducted at home and abroad. In [2], a transmission planning model of the wind-storage joint system is established with the goal of maximizing the total benefit of the system. The authors in [3] combine two planning problems into a single planning problem and numerically show the benefits of their combination. In [4], a transmission planning approach considering correlation of wind power is proposed. The spatial dependence of neighboring wind farm outputs is modeled by Copula theory. A BGTEP model is proposed in [5] and the optimal planning for along-term period is obtained such that the cost of installation and operation would be minimized. In [6], a novel generation and transmission expansion planning model considering positive and negative climbing ability of conventional units and N-1 security constraints is proposed. In [7], a multi-objective integrated generation and transmission expansion planning taking into account the uncertain load and wind power production with one objective function is proposed. In [8], a transmission network planning model which considers the influence of the wind-storage system is established. Generally speaking, existing research on transmission system planning with wind power is relatively preliminary. The development of demand-side response resources, however, provides a new measures to cope with the uncertain output of wind power and improve the capability for accommodating wind power.

Given this background, a bi-level transmission system planning model considering demand-side response and capability for accommodating wind power is proposed. Demand-side management and wind power abandonment are regarded as reserve resources. Finally, the essential features of the developed model and adopted algorithms are demonstrated by a modified 18-bus test system.

2.Transmission System Planning with Wind Farms

Traditionally, transmission system planning is to seek a planning scheme that can minimize investment costs and operating costs while ensuring the reliability of the power system concerned.

As the penetration of wind power in the power system reaches a certain level, it is necessary to properly account for its impact on the planning of power transmission system in order to solve or avoid related problems. The main factors considered in this paper are as follow: 1) Investment budget constraints and investment costs minimization requirements; 2) Impact of wind power output fluctuation on power system security; 3) System adjustment capability and spare capacity requirements; 4) The capability for accommodating wind power.

2.1. Demand-side Response

Traditionally, demand-side load is regarded as a passive and unresponsive one. With the development of power market, the concept of relying on compulsory means to schedule power cuts has been gradually abandoned. The regulation effect of demand-side resources in electricity markets is becoming more and more important. Demand-side response (DR) is a market participation behavior that power users respond to market price signals or incentives, which can change normal power consumption patterns.

Power users are supposed to take actions when electricity price is high or power system reliability is in emergency, so as to alleviate the severe power supply situations. From the perspective of resources, DR is a kind of resource for the effect of peak load or installed capacity reduction; from the perspective of capability, DR can improve the operational reliability of power transmission system and enhance the capability of emergency response; from the perspective of user behavior, DR means the participate in load management and adjustment of power consumption mode.

Demand-side response should be taken into consideration in transmission system planning stage when DR is widely used in the operation of power system. There have been some research work reported in this area. In [9], penalty costs for N-1 expected accidents is calculated as expected value of load shedding amount multiplied by the unit penalty value. Due to the limited peaking capacity and reserve level, the measure of load shedding penalty is not enough to guarantee the safe and stable operation of the power system concerned. In power market environment, it is necessary to introduce DR mechanism for peak-load shifting and change the normal power consumption mode.

Price-based DR is the main form of DR in electricity market, where power users make arrangement and adjustment of power consumption time and mode according to the time or space-based price signals. For another, incentive-based DR is a market-oriented measure where power users are motivated by direct rewards to participate in load adjustment.

Power system is a unified whole composed of power generation, transmission, distribution and consumption. DR can be regarded as a kind of virtual power generation resource, which can provide spare capacity for wind farms and is an effective way to improve the utilization level of wind power

[7]. In this paper, DR mechanism is introduced in the planning of transmission system with wind farms for the advantages as follows. On the one hand, power load is reduced in case of peak load and shortage of wind power output after a interruptible contract with power users. On the other hand, wind curtailment has to be done in case of valley load and sufficient wind power output, but the introduction of DR can help increase the demand for electricity through electricity price incentive during the valley period, which can improve the utilization level of wind power. Hence, the introduction of DR into transmission system planning problem can help shift peak load, and thus reduce transmission investment cost and improve the utilization level of wind power.

2.2. Capability for Accommodating Wind Power

The adoption of wind power can save the generation cost of thermal power units and reduce sulfur dioxide emissions. Existing research on transmission system planning with wind farms is insufficient in measuring the accomodation capability for wind power and the corresponding social benefits. In the proposed model, power generation of wind farms is maximized during a given period while satisfying safety and reliability constraints of power system. Thus, the proposed model can comprehensively measure the advantages of wind farms in various aspects such as economy and environmental protection.

A bi-level transmission system planning model is proposed in this paper, where demand-side response, wind power accomodation capability are embeded in the transmission system planning optimization problem.In the upper level, the objective is to maximize the generating quantity from wind farms by per unit transmission investment within a given time period. In the lower level, the objective is to minimize the cost for DR and penalty for abandoned wind power in the given time period. These two levels are implemented interactively and iteratively, and the optimal transmission system planning scheme will ultimately be attained by the upper level. The bi-level model has been reported in power system planning area[9], [10].

3.Transmission System Planning Model with Wind Farms

3.1. Upper Level Model

(1) Objective function

The objective function of the upper level model considering DR and capability for accommodating wind power is:

$$f_1: \quad \max R = W/(C_L + C_W) \tag{1}$$

$$W = \sum_{t\in T}\sum_{j\in N_W}\left(p_{wj}(t) - p_{fj}(t)\right)\Delta t \tag{2}$$

$$C_L = \frac{r(1+r)^m}{(1+r)^m - 1}\sum_{i\in N_B} C_{Li} Z_i \tag{3}$$

$$C_W = \sum_{t\in T}\sum_{i\in N_L}[\beta_{ai}(t)p_{ai}(\mathrm{t}) + \beta_{bi}(t)p_{bi}(\mathrm{t})]\Delta t + \sum_{t\in T}\sum_{j\in N_W} K_j p_{fj}(t)\Delta t \tag{4}$$

Where R is generation quantity from wind farms per unit transmission investment during a given time period, W is generation quantity from wind farms transferred from the lower level model, $p_{Wj}(t)$ is active power output of wind farm j during time period t, $p_{fj}(t)$ is the amount of abandoned power for wind farm j during time period t, Δt is the duration of period t, which is taken as 1h in case study, C_L is line investment cost, r is the discount rate of funds, m is the number of apportionment years, C_{Li} is the cost of candidate line i (￥10000), Z_i is the number of new lines in the corresponding corridor, C_W is the sum of DR cost and abandoned wind power cost, which is transferred from the lower level model, $\beta_{ai}(t)$ is the unit incentive cost of load i for increasing power demand as needed, $p_{ai}(t)$ is the power comsumption increased by load i during time period t, $\beta_{bi}(t)$ is unit compensation cost of interruptible load i for load shedding; $p_{bi}(t)$ is the power interrupted during time period t, K_j is the unit

IOP Conf. Series: Materials Science and Engineering **685** (2019) 012022 doi:10.1088/1757-899X/685/1/012022

cost of abandoned wind power from wind farm j, N_B、N_L、N_W and T are candidate line sets, load bus sets, wind farm sets, and scenario sets, respectively.

(2) Constraints

Newly-constructed lines constraint:

$$0 < Z_i \le Z_i^{\max} \tag{5}$$

Where $Z_i^{\max}$ is the maximum number of new lines that can be constructed in the corresponding corridor.

3.2. Lower Level Model

The lower level model is used to minimize DR cost and abandoned wind power cost.

(1) Objective function

$$f_2 = \min \sum_{t\in T}\sum_{i\in N_L}[\beta_{ai}(t)p_{ai}(t)+\beta_{bi}(t)p_{bi}(t)]\Delta t + \sum_{t\in T}\sum_{j\in N_W} K_j p_{fj}(t)\Delta t \tag{6}$$

(2) Constraints

1) Power flow constraints

The DC power flow model is used in this paper:

$$\boldsymbol{P_G} + \boldsymbol{P_W} + \boldsymbol{p_b} - \boldsymbol{p_f} - \boldsymbol{P_D} - \boldsymbol{p_a} = \boldsymbol{B\theta} \tag{7}$$

$$f_{ij}(t) \le \left|\bar{f}_{ij}\right| \tag{8}$$

Where $\boldsymbol{P_G}$, $\boldsymbol{P_W}$, $\boldsymbol{p_b}$, $\boldsymbol{p_f}$ $\boldsymbol{P_D}$ and $\boldsymbol{p_a}$ are power output vector of conventional power plants, wind farm power output vector, interruptible load vector, abandoned wind power vector, load vector and incentive load vector respectively, $\boldsymbol{B}$ is bus admittance matrix, $\boldsymbol{\theta}$ is node voltage phase angle vector, $f_{ij}(t)$ and $\bar{f}_{ij}$ are power flow of branch ij at time t and the upper limit.

2) Power output constraints of conventional power plants

$$\underline{p}_{Gk} \le p_{Gk}(t) \le \bar{p}_{Gk} \tag{9}$$

Where $p_{Gk}(t)$, $\bar{p}_{Gk}$ and $\underline{p}_{Gk}$ are active power output of conventional generator at time t and its upper and lower limits.

3) Reserve capacity constraints

$$\sum_{i\in N_L} p_{bi}(t) + \sum_{j\in N_W} p_{wj}(t) + \sum_{k\in N_G} p_{Gk}^{\max}(t) \ge \sum_{m\in N_L} p_{Dm}(t) + R_u(t) \tag{10}$$

$$p_{Gk}^{\max}(t) = \min(\bar{p}_{Gk}, \omega_{Uk}\times\Delta t + p_{Gk}(t)) \tag{11}$$

$$\sum_{i\in N_L} p_{ai}(t) + \sum_{j\in N_W} p_{wj}(t) + \sum_{k\in N_G} p_{Gk}^{\min}(t) \le \sum_{m\in N_L} p_{Dm}(t) - R_d(t) \tag{12}$$

$$p_{Gk}^{\min}(t) = \max(\underline{p}_{Gk}, p_{Gk}(t) - \omega_{Dk}\times\Delta t) \tag{13}$$

Where N_G is conventional power plant sets, $p_{Gk}^{\max}(t)$ is the upper limit of conventional power plant k at time t, $p_{Dm}(t)$ is the power demand of load m at time t, $R_u(t)$ is the lower limit for the positive rotation reserve capacity required at time t, ω_{Uk} is the climbing rate of conventional power plant k, $p_{Gk}^{\min}(t)$ is the lower limit of conventional power plant k at time t, $R_d(t)$ is the lower limit of the negative rotation reserve capacity required at time t, ω_{Dk} is the landslide rate of conventional power plant k.

4) Demand-side response cost and wind power abandonment constraints

$$p_{ait}^{\min} \le p_{ai}(t) \le p_{ait}^{\max},\ i\in N_L \tag{14}$$

$$p_{bit}^{\min} \le p_{bi}(t) \le p_{bit}^{\max},\ i\in N_L \tag{15}$$

$$0 \le p_{fj}(t) \le p_{wj}(t) \tag{16}$$

Where $p_{ait}^{\min}$ and $p_{ait}^{\max}$ are respectively the lower and upper limits of the increasable load of load i at time t, $p_{bit}^{\min}$ and $p_{bit}^{\max}$ are respectively the lower and upper limits of the interruptible load of load i at time t.

4. Case Study

The modified 18-bus system[10], [11] is employed to demonstrate the performance of the proposed model and algorithm. It is assumed that load rate of each bus is the same at the same time. Detail information about the case is available from the authors upon request.

Take the data of a real wind farm for example, and given the following assumptions: 1) The system needs 9% positive and negative spinning reserve provided by conventional power plants; 2) The failure rate of conventional generators and power lines during each time period is 0.01; 3) The fund discount rate is 10%; 4) The climbing and landslide rates of conventional power plants are 1% of the rated capacity per minute; 5) The funding period for transmission system planning is 20 years; 6) The abandoned wind power penalty for a given wind farm is ￥6,100 / MWh; 7) The number of particle swarms is 50; 8) The maximum number of allowed iterations is 300.

The 18-bus system currently includes 10 nodes and 9 lines. At the end of planning horizon, the system will increase to 18 buses, including 7 power buses and 17 load buses, with a total load of 35,870 MW[10], [11]. The wind farm is located at bus 2, and the wind farm has a rated capacity of 3,600 MW, which accounts for about 10% of the total power load. Each line corridor can be expanded by up to three lines. The unit investment cost for a given line is ￥2 million/km.

Three preferred schemes A, B and C can be obtained by solving the proposed model, as shown in Table 1.

As can be seen from Table 1, all three schemes have wind power abandonment and demand-side response. Compared with scheme B, the investment cost of scheme A is 6% lower, and the demand-side response cost and the abandoned wind power cost is 11.8% lower. However, the sum of line investment cost and operation cost of scheme A is 6.4% lower than that of scheme B, and the annual power generation of wind farm per unit of investment is 6.5% higher than that of scheme B. From this aspect, scheme A is much better than scheme B.

The investment cost of scheme C is the largest, and the demand-side response cost and the abandoned wind power cost are the smallest, which means the highest reliability level. However, the annual power generation of wind farms per unit of investment is the smallest, which indicates that the high investment of scheme C does not bring an expected comprehensive benefit. In contrast, scheme A has the largest target value and is the best among the three.

Table 1. Three network planning schemes for the modified 18-bus system

Scheme	*New line*	*Number of new lines*	*W (MW·h)*	*C_L (￥10,000)*	*Demand-side response cost(￥10,000)*	*Abandoned wind cost(￥10,000)*	*C_W per unit of investment (%)*	*R (MW·h/￥10,000)*
A	2(1), 5(2), 7(1), 9(1), 10(2), 11(1), 12(2), 13(3), 16(2), 18(1), 19(2), 21(2), 23(3), 25(1), 26(2), 27(3)	29	14321648	69161	912	2997	5.3%	196
B	1(1), 2(1), 3(1), 4(1), 5(1), 7(3), 10(2), 11(2), 12(1), 13(2), 16(2), 18(2), 19(1), 21(3), 22(1), 23(2), 24(1), 25(1), 26(1), 27(2)	31	14357914	73602	846	3584	5.7%	184
C	1(2), 2(1), 5(1), 7(1), 8(1),	34	14400197	76070	962	3868	6%	178

	10(1), 11(2), 12(1), 13(2), 15(2), 16(2), 18(1), 19(2), 21(1), 22(2), 23(2), 24(2), 25(3), 26(1), 27(2)							

5. Conclusion

With the gradual depletion of fossil energy and rapid development of renewable energy, demand-side response and capability for accommodating wind power are introduced into the transmission system planning with wind farms. Demand-side response mechanism is applied to cope with the uncertainty of wind power and optimize the planning shceme, for the effect of peak-load shifting, system flow optimization, transmission system investment reduction, and wind power utilization level improvement. Under the premise of ensuring the reliability of power supply, the utilization level of wind power is improved, transmission line investment plan is optimized, and the economics and reliability of transmission investment is increased. Finally, the essential features of the developed model and adopted algorithms are demonstrated by the modified 18-bus test system.Study results show that demand-side response mechanism can cope with wind power uncertainty and improve the economics of transmission investment and the level of wind power utilization.

References

[1] Renewable Energy Policy Network for the 21st Century[Z]. Renewable global status report, 2010.
[2] A. Wang, et al, "Study on transmission planning of combined wind and storage system."*2018 2nd IEEE Conference on Energy Internet and Energy System Integration (EI2)*. IEEE, 2018.
[3] Ziaee, Omid, Omid Alizadeh-Mousavi, and F. Fred Choobineh, "Co-optimization of transmission expansion planning and TCSC placement considering the correlation between wind and demand scenarios." *IEEE Transactions on Power Systems* 33.1 (2018): 206-215.
[4] Y. Zhang, et al, "Transmission planning considering spatial correlation of wind power." *Transmission & Distribution Construction, Operation & Live-Line Maintenance (ESMO), 2016 IEEE PES 13th International Conference on*. IEEE, 2016.
[5] Baharvandi, Arash, et al, "Bundled generation and transmission planning under demand and wind generation uncertainty based on a combination of robust and stochastic optimization."*IEEE Transactions on Sustainable Energy*(2018).
[6] H. Zhang, H. Z. Cheng, and S. X. Zhang, "Research on generation and transmission expansion planning with large-scale wind farms integration."*2018 International Conference on Power System Technology (POWERCON)*. IEEE, 2018.
[7] Praveen, P., et al, "Multi-Objective power system expansion planning with renewable intermittency and considering reliability." *2018 International Conference on Computation of Power, Energy, Information and Communication (ICCPEIC) ional conference on computation of power, energy, Information and Communication (ICCPEIC)*. IEEE, 2018.
[8] Y. Wang, et al, "Transmission planning considering the influence of wind-storage system and technical-economic analysis."*2018 2nd IEEE Conference on Energy Internet and Energy System Integration (EI2)*. IEEE, 2018.
[9] H. Fan, et al, "Transmission network bi-level programming model considering economy and reliability and hybrid algorithm." *Proceedings of the CSEE* 16 (2008): 1-7.
[10] J. Zheng,, F. S. Wen, and L. Li, "Two-level planning of transmission system with optimal placement of efficient power plant." *Electric Power Automation Equipment* 33 (2013): 13-33.
[11] X. F. Wang, "Power system optimization planning." China Water & Power Press, 1990.

IOP Conf. Series: Materials Science and Engineering **685** (2019) 012023 doi:10.1088/1757-899X/685/1/012023

Analysis and Application of Distributed Navigation Spoofers

Chao Ma[1], Jun Yang[1*] and Chao Zhou[1]

[1] College of Intelligence Science and Technology, National University of Defense Technology, Changsha, 410073, China

*mcmtwjyy@gmail.com

Abstract. As a typical aerospace technology application, Global Satellite Navigation System (GNSS) affects the national Positioning Navigation Timing (PNT) service. However, not all users are making good use of GNSS. In recent years, the most prominent phenomenon is the Unmanned Aerial Vehicle (UAV) "black flight", which seriously affects the normal operation of civil aviation. To solve such problem, navigation spoofing can be used as an effective method to disrupt the flight of malicious UAVs. Currently, a centralized antenna is the most common way to transmit false navigation signals, but when the target receiver contains two receiving antennas or spatial filtering technology, this spoofing method is ineffective. Therefore, we aim at the distributed spoofers that each antenna emits a false navigation signal, and point out that this sophisticated attack will become the development trend of navigation spoofing in the future.

1. Introduction

The low power and openness of civil navigation signals make them vulnerable to interference or spoofing. In 2002, Warner and Johnston conducted the first GPS spoofing experiment by using a navigation signal simulator [1]. This test has shown that a hand-held GPS civilian receiver inside a truck can be kept locked on false navigation signals at a distance of no more than 30 feet. In 2008, Prof. Todd modified the GRID receiver to design a real navigation spoofer [2], and conducted spoofing experiment [3-5]. In 2006, Mark and Todd systematically concluded the types and methods of navigation spoofing [6].

Navigation spoofer can be divided into three categories: simplistic, intermediate and sophisticated. Among them, intermediate spoofer is the most widely used. The characteristic is the "Receiver-Spoofer" structure. The function of Receiver is to transmit the space-time system of real navigation satellites to the spoofer by a commercial navigation chip, and the spoofer uses an antenna to transmit the simulated constellation to the target receiver. However, this type of spoofer can be effectively detected by spatial filtering methods such as antenna array [7]. So it is necessary to design sophisticated spoofer to solve this problem. The sophisticated spoofer is based on distributed antennas, so we regard this spoofer as distributed navigation spoofer.

The next section covers navigation basics and three types of distributed navigation spoofer. Section 3 discusses key technologies and considerations. Section 4 simulates distributed navigation spoofers and gives some conclusions.

2. Navigation, spoofing and sophisticated spoofer design

2.1. Satellite navigation basics

The navigation signal belongs to the spread spectrum signal, which is divided into three signal levels: carrier, pseudo-code and date code. The data code and pseudo-code perform an exclusive or operation to realize spread spectrum. The combined code then modulates the carrier through Binary Phase Shift Keying (BPSK). Taking Global Positioning System (GPS) as an example [8], without considering P(Y) code, the navigation signal $s^{(i)}(t)$ emitted by satellite i can be expressed as

$$s^{(i)}(t) = A\left(x^{(i)}(t)D^{(i)}(t)\right)\sin\left(2\pi ft + \theta\right) \tag{1}$$

Where A is the signal amplitude. $x^{(i)}(t)$ is the C/A code of satellite i. $D^{(i)}(t)$ is the data code. f and θ are carrier frequency and initial phase respectively.

The target receiver calculates the pseudorange and satellite position from the navigation signal, and obtains the self-position and time results by solving the following four-element nonlinear equations [8]:

$$\begin{cases} \sqrt{\left(x^{(1)}-x\right)^2+\left(y^{(1)}-y\right)^2+\left(z^{(1)}-z\right)^2}+\delta t_u=\rho_c^{(1)} \\ \sqrt{\left(x^{(2)}-x\right)^2+\left(y^{(2)}-y\right)^2+\left(z^{(2)}-z\right)^2}+\delta t_u=\rho_c^{(2)} \\ \cdots \\ \sqrt{\left(x^{(N)}-x\right)^2+\left(y^{(N)}-y\right)^2+\left(z^{(N)}-z\right)^2}+\delta t_u=\rho_c^{(N)} \end{cases} \tag{2}$$

Where $[x,y,z]^{\mathrm{T}}$ is the coordinate vector of the target receiver. $[x^{(i)},y^{(i)},z^{(i)}]^{\mathrm{T}}$ is the coordinate vector of the satellite i. $\rho_c^{(i)}$ the pseudorange after correction. δt_u is the clock error. Changing user positioning results can be realized by changing satellite positions or pseudoranges, and the common part of pseudoranges affects timing results.

2.2. Definition and Goal of navigation spoofing

Navigation spoofing refers to the process that makes errors in the position or time information of the target receiver by false navigation signal of replaying or self-generating. From the view of application, the goal of navigation spoofing can be divided into two types:

- $G1$: The accuracy of the receiver positioning (or timing) is reduced by navigation spoofing. Standard Positioning Service (SPS) cannot be satisfied.
- $G2$: The receiver positioning (or timing) is mistakenly locked to the designated point. When the receiver is installed in a vehicle such as UAV, the goal of navigation spoofing can also be defined that the vehicle is mistakenly lured to the designated area.

2.3. Design of distributed navigation spoofer

The distributed navigation spoofer takes "Receiver-spoofer" structure as the basic unit. The platform such as UAV, hot air balloon, base station or low-orbit satellite can be used as carrier to realize the false constellation design in figure 1. Different platform equipped with basic unit is called as a node. Three distributed navigation spoofers are proposed based on this structure.

- Independent: node i uses the C/A code of real satellite j, and the navigation messages are filled according to the own simulated satellite motion. It can be seen that this kind of spoofer is treated as an independent navigation constellation for spoofing.
- Hybrid: node i uses the C/A code and the navigation messages of real satellite j.
- Duplicated: node i uses the C/A code and the navigation messages of real satellite j, and compensates pseudorange. Compared with Hybrid spoofer, the duplicated spoofer only treats the node

as a relay station. For the target receiver, the node is basically transparent. The characteristics of different spoofers are illustrated in table 1.

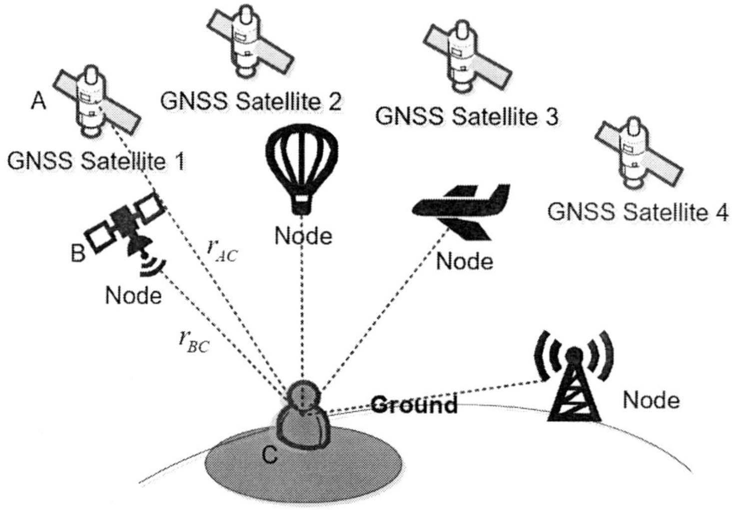

Figure 1. Sophisticated spoofer structure.

Table 1. Comparison of three distributed navigation spoofers.

	Independent	Hybrid	Duplicated
Implementation complexity	Intermediate	Low	High
Spoofing goal	$G1, G2$	$G1$	$G1, G2$
Detection probability	Easy	Hard	Intermediate
Dilution of precision (DOP)	Decided by nodes	Decided by real satellites	Decided by real satellites
Recommended carrier	Low-orbit satellites or high-speed aircraft	none	Low-speed stabilized platform

3. Key technologies and considerations of distributed navigation spoofer

3.1. Compensation for pseudorange and doppler frequency

Depending on the different spoofer types, the node needs to decide whether to do pseudorange or Doppler frequency compensation. We arrange them as shown in table 2.

Table 2. Compensation for pseudorange and doppler frequency.

	Independent	Hybrid	Duplicated
Pseudorange compensation	×	×	√
Doppler compensation	×	√	√

Pseudorange offset is compensated by delay filter [9] or ephemeris modification [10]. The offset value ΔD can be expressed as

$$\Delta D = r_{AC} - r_{BC} - c \cdot \Delta T_B \tag{3}$$

Where r_{AC} and r_{BC} are the geometric distances between AC and BC respectively in figure 1. c is the speed of light. ΔT_B is the transmission delay required for each node to transmit, receive and process.

Doppler frequency offset is compensated by Numerically Controlled Oscillator (NCO). The offset value Δf can be expressed as

$$\Delta f = \frac{f_{L1}}{c}\left(\left(v_A - v_C\right)\cdot\vec{l}_{AC} - \left(v_A - v_B\right)\cdot\vec{l}_{AB} - \left(v_B - v_C\right)\cdot\vec{l}_{BC}\right) \quad (4)$$

Where Δf_{L1} is the carrier frequency. $\left(v_i - v_j\right)\cdot\vec{l}_{ij}$ is the vector projection of the velocities of i and j point along the line of ij. c is the speed of light.

3.2. Influence factors on the spoofing accuracy of the target receiver

For the target receiver, the positioning/timing accuracy can be expressed as the product of DOP factor and User-Equivalent Range Error (UERE). DOP is determined by the geometric relationship between satellites and the receiver. UERE is the equivalent error of different errors in space. Under the condition of the distributed navigation spoofer, UERE will be greatly affected by time synchronization between nodes.

3.2.1. The effect of time synchronization between nodes. Compared with the real navigation constellation, each node of the distributed navigation spoofer does not have a unified time system because there is no highly stable atomic clock. And this factor will seriously affect the spoofing accuracy.

The typical UERE budget for a single frequency receiver is shown in table 3 [8]. Different types of distributed navigation spoofer will introduce new errors. For example, each node of the independent spoofer uses navigation chip to complete positioning and timing, so the positioning result affects the broadcast ephemeris, and the timing result affects the broadcast clock in table 3. The positioning/timing errors of the hybrid/duplicated spoofer can be expressed by the equivalent range error to join in table 3. It is worth mentioning that the timing accuracy of commercial receiver is about 20 ns at present, and it can reach 10 ns with timing receivers. When we adopt Real-Time Kinematic (RTK) or Precise Point Positioning (PPP) technology, the timing accuracy can reach 2-5 ns.

Table 3. GPS Standard Positioning Service typical UERE budget.

Segment Source	Error Source	1σ Error (m)
	Broadcast clock	1.1 (Independent)
Space/control	L1 P(Y) - L1 C/A group delay	0.3
	Broadcast ephemeris	0.8 (Independent)
	Ionospheric delay	7.0
	Tropospheric delay	0.2
User	Receiver noise and resolution	0.1
	Multipath	0.2
	Equivalent range error	Hybrid, Duplicated
System UERE	Total (RSS)	7.1

3.2.2. DOP selection. The independent spoofer as an standalone system, DOP can be determined by itself. The nodes' constellation can be arranged by numerical analysis or intelligent algorithm [11-13]. For the hybrid and duplicated spoofer, DOP is determined by the real navigation constellation. When the number of nodes is not consistent with the number of visible satellites, an exhaustive method or k-means clustering algorithm [14] can be adopted to select visible satellites to ensure the DOP does not decrease obviously.

3.2.3. Isolation of reception and transmission. A serious problem faced by the "Receiver-Spoofer" structure in the practical application is the isolation of receiving and transmitting navigation signals. Each node must ensure that it does not affect the reception of real navigation signals while emitting spoofing signals. The isolation technologies mainly include:

- Time division of reception and transmission: the receiving and transmitting time of each node is completely independent. Because the receiving time of real satellite signals is generally within a few minutes, it is feasible to apply this method.
- Reducing receiving sensitivity and transmitting power control: since the distributed navigation spoofer is usually in the air, reducing the sensitivity of the receiving of each node will not improve the missed detection rate obviously, and the transmitting power can be appropriately reduced according to the specific scene.
- Space isolation technology: Spiral or horn antenna is used to improve antenna directivity, or physical isolation medium is added to reduce antenna coupling.

4. Simulation analysis and conclusion

Using the Satellite Tool Kit (STK) and MATLAB, three distributed navigation spoofer are simulated in typical scenarios. A typical scenario:

- Time: 8 a.m. on July 3, 2019.
- The effective coverage area for target receivers: 1×1 km^2. The longitude and latitude coordinate of the center: 112.997620, 28.217295.
- The antenna elevation angle of target receivers: 30°. The number of visible satellites is 8.

4.1 The independent spoofer

The orbital elements of real satellite are modified to make it conform the low-orbit satellite motion. Using dual-frequency commercial or timing receiver applies on each node [15-16], UERE budget is obtained as shown in table 4. The positioning errors in the coverage area are shown in figure 2. If the spoofer can make the target coincide with the real position, it would be feasible to get the target anywhere by reasonably varying pseudoranges. So the positioning errors are the differences between the spoofing positioning results and the real target positions. This type of spoofer can achieve navigation spoofing for multiple targets. However, the spoofing accuracy is lower than SPS because the positioning/timing accuracy of each node is limited.

Table 4. UERE budget of the independent spoofer.

Segment Source	Error Source	1σ Error (m) commercial	1σ Error (m) timing
Space/control	Broadcast clock	6.1	3.2
	L1 P(Y) - L1 C/A group delay	0.3	0.3
	Broadcast ephemeris	3.1	2.1
User	Ionospheric delay	0.2	0.2
	Tropospheric delay	0.2	0.2
	Receiver noise and resolution	0.1	0.1
	Multipath	0.2	0.2
System UERE	Total (RSS)	6.860	3.856

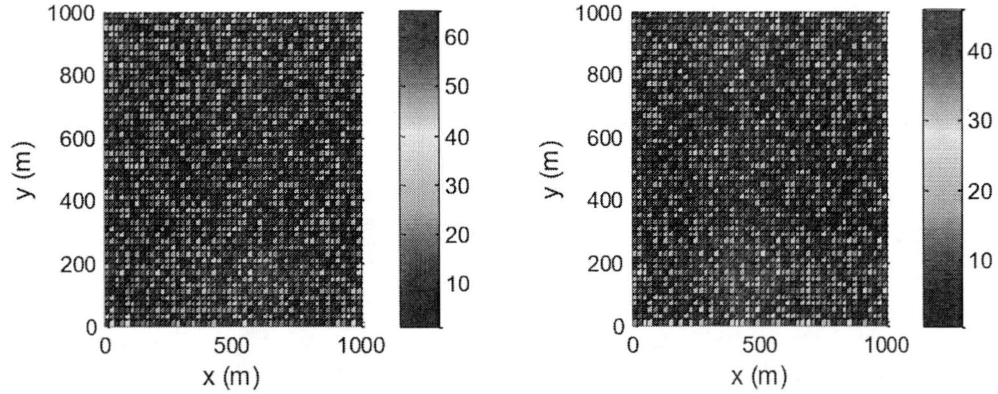

Figure 2. The spoofing accuracy under different UERE budget (independent).

4.2 The hybrid spoofer

UERE budget is expressed as shown in table 5, and the positioning errors in the coverage area are shown in figure 3. It can be seen that hybrid spoofer can only implement spoofing goal $G1$, and the spoofing accuracy generates overall deviation within the coverage area. This is mainly determined by the relative position of each node, each satellite and the target receiver. The effect of UERE on positioning results is basically negligible.

Table 5. UERE budget of the hybrid spoofer.

Segment Source	Error Source	1σ Error (m) commercial	1σ Error (m) timing
Space/control	Broadcast clock	1.1	1.1
	L1 P(Y) - L1 C/A group delay	0.3	0.3
	Broadcast ephemeris	0.8	0.8
User	Ionospheric delay	0.2	0.2
	Tropospheric delay	0.2	0.2
	Receiver noise and resolution	0.1	0.1
	Multipath	0.2	0.2
	Equivalent range error	6.7	3.6
System UERE	Total (RSS)	6.853	3.877

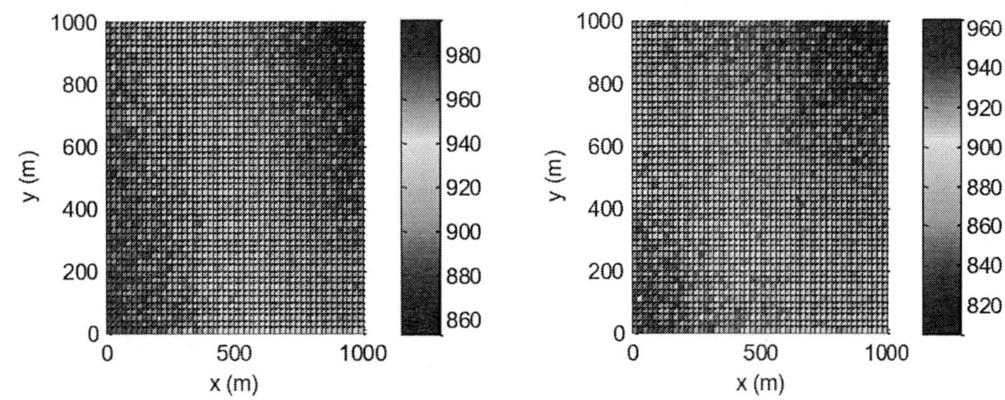

Figure 3. The spoofing accuracy under different UERE budget (hybrid).

4.3 The duplicated spoofer

UERE budget is the same as table 5. The positioning errors in the coverage area are shown in figure 4. The pseudorange compensation can improve the spoofing accuracy, but at the same time, this type spoofer can only keep a target receiver with high spoofing accuracy. With the increasing of the relative offset to the specified target, the spoofing accuracy decreases continuously. We should be flexible in selecting distributed navigation spoofer based on specific scenarios.

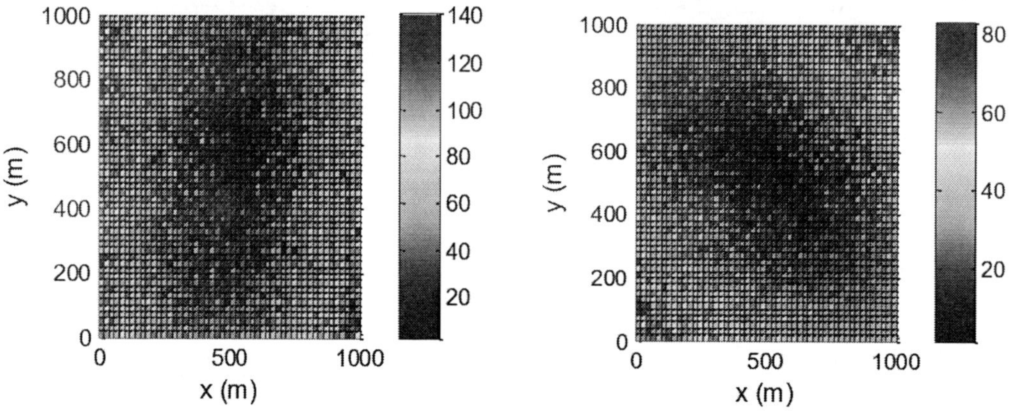

Figure 4. The spoofing accuracy under different UERE budget (duplicated).

References

[1] J. S. Warner and R. G. Johnston. (2002) *A Simple Demonstration that the Global Positioning System (GPS) is Vulnerable to Spoofing.* The Journal of Security Administration

[2] Humphreys TE, Ledvina BM, Tech V, et al. (2009) *Assessing the Spoofing Threat : Development of a Portable GPS Civilian Spoofer.*

[3] Shepard DP, Bhatti JA, Humphreys TE, Fansler AA. (2012) *Evaluation of Smart Grid and Civilian UAV Vulnerability to GPS Spoofing Attacks.*

[4] Bhatti J, Humphreys T E. (2017) *Hostile Control of Ships via False GPS Signals: Demonstration and Detection.* Navigation, 64(1):51-66.

[5] Shepard DP, Bhatti JA, Humphreys TE, Fansler AA. (2012) *Evaluation of Smart Grid and* Civilian UAV Vulnerability to GPS Spoofing Attacks.

[6] Psiaki, M. L., & Humphreys, T. E. (2016). GNSS Spoofing and Detection. Proceedings of the IEEE, 104(6), 1258-1270.

[7] Jafarnia-Jahromi A., Broumandan A., Nielsen J. and Lachapelle Gerard. (2012) *GPS vulnerability to spoofing threats and a review of antispoofing techniques.* International Journal of Navigation & Observation.

[8] Kaplan, D. (1996) *Understanding GPS: Principles and Applications.* Artech House Publishers, Norwood.

[9] Yang, J., Chen, J. Y., Zhong, X. P., & Ming, D. X.. (2010) *High Precision Delay Satellite Navigation Signal Generation Theory and Application In Navigation Satellite Simulation Test //* CSNC 2010.

[10] Chao M., Jun Y., Jianyun C. and Yinyin T.. (2018) *Indoor and outdoor positioning system based on navigation signal simulator and pseudolites.* Advances in Space Research.

[11] BlancoDelgado, Nunes and SecoGranados. (2011) *Relation between GDOP and the geometry of the satellite constellation.* International Conference on Localization & Gnss. IEEE.

[12] Nielsen R. O.. (1997) *Relationship between dilution of precision for point positioning and for relative positioning with GPS.* IEEE Transactions on Aerospace & Electronic Systems, 33(1), 333-338.

[13] Chien-Sheng Chen. (2015) *Weighted geometric dilution of precision caliculations with matrix multiplication.* Sensors, 15, 803-817.

[14] Kanungo T., Mount D. M., Netanyahu N. S., Piatko C. D., Silverman R. and Wu A. Y.. (2002) *An efficient k-means clustering algorithm: analysis and implementation.* IEEE Transations on Pattern Analysis and Machine Intelligence, 24(7), 0-892.

[15] http://www.unicorecomm.com/product/content_972.html

[16] http://www.septentrio.com/products

Aerodynamic Thermodynamic Modeling and Simulation of Turbofan Engine

Renjun Qian[1,*], Benwei Li[1], Hanqiang Song[2], Qing Dong[1] and Yonghua Wang[1]

[1]Aviation Foundation College, Naval Aviation University, Yantai, Shandong, China
[2]Naval Research Academy, Shanghai, China

[*]18902016296@163.com

Abstract. The mathematical model of turbofan engine is established and simulation calculation is carried out in this paper. An engine component model is established by means of the gas flow path of the engine. Then, according to the common working conditions, common working equations are established and the N+1 residual method is used to solve the nonlinear equations. The general model of an engine is established based on the design point parameters, and the specific performance parameters of the intermediate state are calculated under standard atmospheric conditions. The simulation results are compared with the measured performance data of the engine, which shows that: there are errors between the calculated value of the model and the measured value of the engine. The maximum error is 2.85%, which meets the requirement of model error less than 3% in general engineering.

1. Introduction

There are many kinds of aeroengine models, and the classification criteria are also different. From a mathematical point of view, engine models can be divided into non-linear aerothermodynamic model and linear small deviation model [1]. The former is an accurate description of the working process of the engine. It is based on the main components along the flow path of the engine, the characteristic data of engine components and aerothermodynamic laws [2]. It is applicable to both steady and dynamic processes. The latter assumes that the input parameters vary in a very small range. Based on the overall engine, the model obtained by linearizing the engine non-linear model with given input parameters, output parameters and control parameters is usually applicable to the dynamic characteristics of the engine near a given working state [3]. In this paper, the steady-state mathematical model of the engine is studied, and the non-linear aerothermodynamic model is adopted.

2. Establishment of Engine Component Level Model

The basic idea of engine component level model is to model the relatively independent components separately according to the sequence of gas flow based on the known component characteristic data [4]. Then, according to the conditions of continuous flow rate and power balance, the common working equations are established. Finally, the non-linear equations are solved simultaneously to determine the common working point of the engine. The section diagram and definition of the engine studied in this paper are shown in Figure 1 and Table 1.

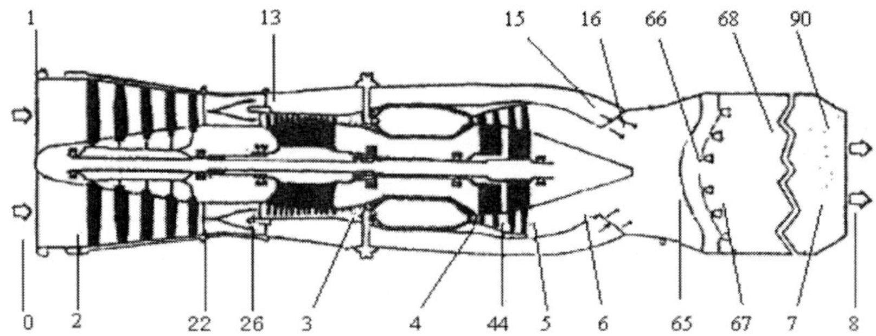

Figure 1. The sketch map of engine section

Table 1. Definition of Engine Sections

Section Number	Actual definition	Section Number	Actual definition
0	Far ahead of engine inlet	1	Entrance inlet
2	Fan inlet	22	Fan outlet
26	Connotation inlet of high pressure compressor	13	Outer culvert entrance
3	Combustor inlet	4	High pressure turbine inlet
44	Low pressure turbine inlet	5	Low pressure turbine outlet
15	Outlet of outer culvert	6	Heat flow inlet of mixing chamber
16	Cold flow inlet of mixing chamber	65	Mixing chamber outlet
67	Afterburner inlet	68	Afterburner outlet
7	Nozzle inlet	8	Nozzle outlet

When the engine components work together, the working conditions of aerothermodynamics and rotor dynamics must be followed. That is the basic equations of flow balance, power balance and pressure balance [5]. The common working equations of engine being made up of these basic equations are specifically as follows:

(1) Flow balance of low pressure turbine/fan

The inlet flow rate of low pressure turbine can be obtained by linear interpolation of low pressure turbine characteristics. The equilibrium equation is:

$$(M_{44} - M_{44}^{'}) / M_{44} = 0 \tag{1}$$

(2) Flow balance equation of high pressure turbine/compressor

The flow rate M_{40} can be obtained by linear interpolation of high pressure turbine characteristics. The equilibrium equation is:

$$(M_{40} - M_{40}^{'}) / M_{40} = 0 \tag{2}$$

(3) Power balance equation of low pressure turbine/fan is:

$$(N_{TL} - N_{cL}) / N_{TL} = 0 \tag{3}$$

(4) Power balance equation of high pressure turbine/compressor

$$(N_{TH} - N_{cH}) / N_{TH} = 0 \tag{4}$$

(5) Static pressure balance at mixing chamber inlet:

$$(P_5 - P_{15}) / P_5 = 0 \tag{5}$$

(6) Air flow balance of afterburner outlet/tail nozzle:

$$(M_8 - M_7) / M_8 = 0 \tag{6}$$

Two parameters are needed to determine the working state of fan, compressor, high and low pressure turbines. n_L, π^*_{cL}, n_H and π^*_{cH} are selected respectively as the parameters to characterize the working state of fan and compressor. Because of the mechanical connection between compressor and turbine, n_L and n_H have been determined [6]. So enthalpy drop DH_H and DH_L are selected as parameters to characterize the working state of high and low pressure turbines. Since the four components work together and are subject to the constraints of common working conditions, the values of the six parameters ($n_L, n_H, \pi^*_{cL}, \pi^*_{cH}, DH_H, DH_L$) determining the four components' states must meet the common working equations, which is:

$$\begin{cases} G_1(n_H, n_L, \pi^*_{cH}, \pi^*_{cL}, DH_H, DH_L) = (N_{TL} - N_{cL}) / N_{TL} = 0 \\ G_2(n_H, n_L, \pi^*_{cH}, \pi^*_{cL}, DH_H, DH_L) = (N_{TH} - N_{cH}) / N_{TL} = 0 \\ G_3(n_H, n_L, \pi^*_{cH}, \pi^*_{cL}, DH_H, DH_L) = (M_{44} - M'_{44}) / M_{44} = 0 \\ G_4(n_H, n_L, \pi^*_{cH}, \pi^*_{cL}, DH_H, DH_L) = (M_{40} - M'_{40}) / M_{40} = 0 \\ G_5(n_H, n_L, \pi^*_{cH}, \pi^*_{cL}, DH_H, DH_L) = (P_5 - P_{15}) / P_5 = 0 \\ G_6(n_H, n_L, \pi^*_{cH}, \pi^*_{cL}, DH_H, DH_L) = (M_8 - M_7) / M_8 = 0 \end{cases} \tag{7}$$

3. Algorithms for Solving Engine Model

When the engine is in steady state, the model solving problem is transformed into solving a set of nonlinear equations composed of six common working equations. N-R method and N+1 residual method are commonly used to solve the common working equations of engines. N-R method has high accuracy, but its operation speed is slow [7][8]. N+1 residual method has the advantages of high accuracy and fast convergence in solving nonlinear equations. The principle and solving process of N+1 residual method are briefly introduced below.

The systems of non-linear equations are:

$$\begin{cases} G_1(x_1, x_2, \ldots, x_N) = 0 \\ G_2(x_1, x_2, \ldots, x_N) = 0 \\ G_3(x_1, x_2, \ldots, x_N) = 0 \\ G_4(x_1, x_2, \ldots, x_N) = 0 \\ G_5(x_1, x_2, \ldots, x_N) = 0 \\ G_6(x_1, x_2, \ldots, x_N) = 0 \end{cases} \tag{8}$$

There is an approximate solution $X = (x_1, x_2, \ldots, x_N)$. Substituting it into equations (8) will obtain:

$$\begin{cases} G_1(x_1,x_2,...,x_N)=e_1 \\ G_2(x_1,x_2,...,x_N)=e_2 \\ G_3(x_1,x_2,...,x_N)=e_3 \\ G_4(x_1,x_2,...,x_N)=e_4 \\ G_5(x_1,x_2,...,x_N)=e_5 \\ G_6(x_1,x_2,...,x_N)=e_6 \end{cases} \tag{9}$$

Noting $E=(e_1,e_2,...,e_6)^T$, which is a residue of the approximate solution X of the system of equation (9).

Assume that equations (8) have N+1 approximate solutions: $X_0, X_1,..., X_N$. Their corresponding residues are $E_0, E_1,..., E_N$. Let $Y=(y_0, y_1,..., y_N)$, then a system of linear equations of order N+1 can be constructed:

$$\begin{bmatrix} 1 & 1 ... 1 \\ E_0 & E_1 ... E_N \end{bmatrix} \cdot \begin{bmatrix} y_0 \\ y_1 \\ ... \\ y_N \end{bmatrix} = \begin{bmatrix} 1 \\ 0 \\ ... \\ 0 \end{bmatrix} \tag{10}$$

From this linear equations we can obtain $y_0, y_1,..., y_N$. A new approximate solution X_{N+1} of the equations (8) can be formed by using them.

$$X_{N+1} = \sum_{i=1}^{N} y_i \cdot X_i \tag{11}$$

Check whether the corresponding residual satisfies the following inequalities:

$$\| E_{N+1} \| = \sqrt{\sum_{i=1}^{N} e_i^2} \le \varepsilon \tag{12}$$

If the inequality is satisfied, the solution vector corresponding to the norm is the solution. Otherwise, compare the norm of N+1 residue and select the maximum value:

$$\| E_{\max} \| = \max(\| E_0 \|, \| E_1 \|, ..., \| E_{N+1} \|) \tag{13}$$

In that way, the corresponding solution vector X_{max} is the worst solution. Then latest solution X_{N+1} and its norm are used to replace the worst solution, and the calculation is restarted in the order of formula (10) ~ (13) until the inequality (12) is satisfied. The new one is the solution that satisfies the accuracy requirement [9].

How to determine the N+1 approximate solutions are the first problem in the application of N+1 residual method. Generally, an approximate solution is given and its corresponding residual is calculated.

$$X_0 = (x_1^0, x_2^0, ..., x_N^0) \tag{14}$$

$$E_0 = (e_1^0, e_2^0, ..., e_N^0) \tag{15}$$

Then each component of X_0 is given a random increment Δ_i in turn, which forms N approximate solutions and their corresponding residual vectors.

$$X_i = (x_1^0, ..., x_i^0 + \Delta_i, ..., x_N^0) \quad (16)$$

$$E_i = (e_1^i, e_2^i, ..., e_N^i) \quad (17)$$

Since N approximate solutions are generated randomly on the basis of X_0, the initial solution vectors generated by each running simulation are different. Therefore, the key to the application of N+1 residual method is to give a set of better initial solutions. In this paper, the experience value in the process of engine design is taken as the initial guess value, and the calculation results are fine.

4. Simulation Results and Analysis

Taking an engine for an example, a general model of the engine is established based on its design point parameters, and the specific performance parameters of the engine are calculated in the intermediate state under standard atmospheric conditions. The simulation results are compared with the measured performance data of the engine. The results are shown in Table 2. It can be seen that there are some deviations between the calculated value of the model and the measured value of the engine. The maximum error is 2.85%, which meets the requirement of the model error within 3% in general engineering. The error is mainly due to the simplification of many conditions such as heat exchange between air flow and engine components in the thermodynamic modeling process.

Table 2. Comparisons between calculated and measured values of the model

performance parameter	relative error
P22	-0.50%
P3	-1.08%
T3	1.66%
T55	-1.91%
P55	1.85%
F	2.85%
WFT	-2.05%

5. Conclusions

A component model is established according to the gas flow path of the engine. A set of common working equations is established according to the common working conditions of aerothermodynamics and rotor dynamics. And the N+1 residual method is used to solve the nonlinear equations. Taking an engine for an example, the performance of intermediate state in standard atmosphere is calculated. Comparing with the measured data, the maximum error between the calculated and measured values is 2.85%, which meets the requirement of model error less than 3% in general engineering.

Acknowledgement

Thanks for the financial support from special funds of Taishan Scholar Project.

References

[1] Sun Jianguo. *Advanced Multivariable Control Systems of Aeroengines*. Beijing: Beijing University of Aeronautics and Astronautics Press, 2005

[2] F. Wei, *Study on Thermodynamic Calculation and System Level Simulation of Specific Heat Capacity of Aeroengine*. Shanghai: Shanghai Jiaotong University, 2006.

[3] D. W. YIN, *Study on Parameter Estimation in Aeroengine Model Solving Algorithm and Performance Optimization Control*. Changsha, Hunan: National University of Defense Technology, 2011.

[4] G. Q. Luo, *Numerical Simulation of Aero Gas Turbine Engine*. Beijing: National Defense Industry Press, 2007.

[5] S. Q. Fan, *Aeroengine Control*. Xian, Shanxi: Northwestern Polytechnical University Press, 2008.
[6] S. Garg, J. Peter, "Integrated Flight/Propulsion Control Design for a STOVL Aircraft Using H-infinity Control Design Techniques", *Proceedings of 1991 American Control Conference*.
[7] J. W. Denham. "STOVL Integrated Flight and Propulsion Control: Current Successes and Remaining Challenges", *AIAA-2002-6021*.
[8] M. C. Turner, D. G. Bates, P. Ian, "Input/Output Conditioning of Robust Integrated Flight/Propulsion Control", *Journal of Guidance, Control and Dynamics*, 2001,24(6).
[9] X. Wang, *Study on Mathematical Model and Control Law of Aeroengine*. Xian, Shanxi: Northwestern Polytechnical University, 2007.

IOP Conf. Series: Materials Science and Engineering **685** (2019) 012025 doi:10.1088/1757-899X/685/1/012025

Study on the Forecasting of the Hot Corrosion Resistance of Typical Superalloys for Aeroengines

Bingxiu Guo, Xiaohui Wang*, **Yanyan Wang and Yue Shao**

School of Reliability and Systems Engineering, Beihang University, Beijing, P. R.China

*Email: xiaohuiw@buaa.edu.cn

Abstract. The high-temperature components of aeroengines are in contact with gas flow for a long time, making them susceptible to hot corrosion, which can affect the reliability and lifespan of aeroengines. In this study, five types of superalloys commonly used in the high-temperature components of Aeroengines are selected for gas-based hot corrosion tests, and the corrosion rates are calculated using the weight loss method. Gradient Boosting Regression Tree (GBRT) machine learning algorithm is utilized to establish a corrosion rate forecasting model. The evaluation results show the predictability of this method. The effect of input parameters, including main alloy chemical composition and corrosion time, on the corrosion rate was discussed using GBRT and critical factors are obtained. These results provide a reference for the protection of aeroengines from hot corrosion.

1. Introduction

The hot-end components of aeroengines, including combustion chambers, gas turbines, power turbines and exhaust systems, are directly exposed to high-temperature gas for a long time. Sulfates accumulated on the surface after combustion will cause accelerated oxidation, resulting in serious thermal corrosion[1]. As materials with multiphase structure and complex components, superalloys have good impact resistance and oxidation resistance, making them widely used in the hot-end parts of aeroengines[2-4].

Scholars have studied the hot corrosion resistance of different types of superalloys under various preparation processes in complex environments. The literature has established that the research methods for the hot corrosion resistance of superalloys to mostly focus on the performance test and corrosion products analysis after the test. However, the hot corrosion test has certain limitations in studying many corrosion influencing factors and corrosion laws of alloy, due to the complex reacting process, long cycle and high cost of test data sets. Corrosion prediction technology, which analyzes, models, verifies and analyzes corrosion test data, can obtain information from the existing experimental data through the establishment of dynamic model and statistical model. Further, it provides a new framework for further research and real-world application. At present, the corrosion prediction statistical model of poor information system, including Cellular Automata, Principal Component Analysis, and Grey System Theory, has been applied in aviation materials, reinforced concrete structures, and other fields.

A rapidly growing field, machine learning has been largely employed by many scholars in recent years to solve corrosion problems. For example, Fang et al.[5] proposed a genetic algorithm (GA) and support vector regression (SVR) combined method and can successfully predict atmospheric corrosion of metallic materials such as zinc and steel. Chou et al.[6] applied a single model and a set model for machine learning prediction to predict the pitting corrosion risk of reinforced concrete and the rate of marine corrosion of carbon steel. Mousavifard et al.[7] used the artificial neural network (ANN) and adaptive neuro-fuzzy inference systems (ANFIS) to model the corrosion rate as a function of pH, zirconium concentration, temperature, and immersion time for a hot dip galvanized steel.

In this paper, five typical superalloys used in aeroengines were selected as test objects for the gas hot-corrosion test, and the corrosion rate was calculated by the weight loss method. Gradient Boosting Regression Tree (GBRT) was applied to model the corrosion rate with corrosion time and main chemical composition for five types of superalloys after test. The influence of the main chemical composition in the superalloy on the corrosion rate was explored using GBRT. The effects of important alloy constituents such as Ni, Al, Ti, Mn, Mo and Co on the corrosion rate were obtained. The results contributed to better understand hot corrosion protection and provided a reference for the selection of aeroengines.

2. Materials and Methods

2.1. Test materials

Table 1 describes the five test materials selected in this paper. Table 2 gives the chemical composition of the materials.

Table 1. Test materials

Name	Attribute	Use
GH3625	Cr-Mo-Nb solid solution strengthened nickel-base superalloy	Engine combustion chamber, exhaust system, tool
GH2132	Fe-25Ni-15Cr base superalloy	Engine turbine disk, compressor disk, etc.
GH605	20Cr-15W solid solution strengthened cobalt-based superalloy	Engine blades, combustion chambers, etc.
GH3536	Cr-Mo solid solution strengthened nickel-base superalloy	Engine combustion chamber
GH4738	Ni-Cr-Co-based superalloy	Engine gas turbine

Table 2. superalloy chemical compositions

Name	(wt,%)											
GH3625	Ni	Cr	Mo	Nb	Fe	Co	Si	Mn	Ti	Al	C	Cu	P, S
	61.85	21.50	9	3.65	2.50	0.50	0.25	0.25	0.20	0.20	0.05	0.04	0.01
GH2132	Fe	Ni	Cr	Ti	Mo	Mn	Si	Al	C	V	P, S		
	54.16	25.50	15	2.025	1.25	1	0.50	0.20	0.04	0.03	0.025		
GH605	Co	Cr	W	Ni	Mn	Fe	Si	C	P, S				
	51.67	20	15	10	1.50	1.50	0.20	0.10	0.04				
GH3536	Ni	Cr	Fe	Mo	Co	W	Mn	Si	Cu	C	Ti	Al	P, S
	46.96	21.75	18.50	9	1.50	0.60	0.50	0.50	0.25	0.10	0.25	0.08	0.02
GH4738	Ni	Cr	Co	Mo	Ti	Al	Fe	Si	C	Cu	Mn	P,S,B	
	57.04	19.50	13.50	4.25	3	1.40	1	0.08	0.07	0.05	0.05	0.07	

2.2. Experimental

The gas hot-corrosion test in this study is a cold-heat alternating cycle test of the sample in a gas formed under the conditions of a specified temperature, a fuel flow rate, an oil-gas ratio and sea salt content. This method can simulate the working environment of an aeroengine and evaluate the hot corrosion resistance of materials and has been widely used[8,9]. The conditions under which the gas is formed in this study are as follows:

Test Temperature (℃): 900;

Gas-oil ratio: 1/45;

Fuel flow (L/h): 0.2;

Seawater discharge: 0.2;

Air flow rate (L/h): 9;

Seawater concentration (×10-6): 20.

Each material consists of three samples. All samples have no surface treatment. The samples were placed in the prescribed gas environment for 100h, heat preservation 55 min and cooling 5 min per hour, every 25 hours take out and weigh the samples.

2.3. Calculation of the corrosion rate

In this study, the corrosion rate was selected to evaluate the corrosion resistance of the materials. The weight loss method is a method for determining the corrosion rate of a sample before and after corrosion. Generally, the metal is made into the shape and size of the test piece and placed in a corrosive environment. After a certain period of time, the weight change is measured, and the corrosion rate v can be calculated.

$$v = \frac{m_0 - m_t}{St} \tag{1}$$

where v represents the corrosion rate, m_0 represents the precorrosion weight, m_t represents the weight after the test, t represents the corrosion time, S represents the surface area of a sample.

2.4. Gradient boosting regression tree (GBRT)

Gradient boosting regression tree (GBRT) is a combinatorial algorithm that uses the decision tree (CART) (as the basic learning device) and gradient boosting (to train several times)[10]. Each decision tree is used to evaluate the residuals of all previous trees. The results of each tree are added up to obtain the final prediction value. GBRT is regarded as one of the best algorithms for prediction. Its good reputation for its capacity to sort and analyze the importance of the input of the model,as well as to express complex laws.

3. Result and Discussion

3.1. Description of the test results

Table 3 shows the corrosion rate of 15 items. Negative values in Table 3 indicates that the weight of the corrosion product exceeds the corrosion loss. Combined with the corrosion rate data, it obtained that GH605 exhibits the most serious corrosion 100h after the test, while GH3625 and GH2132 have some slight corrosion. GH3536 and GH4738 have almost no corrosion.

Table 3. corrosion rate data set

Materals	Number	Corrosion rate(g/(m^2*h))			
		Time 25/h	Time 50/h	Time 75/h	Time 100/h
GH3625	1	0.5775	0.3108	0.2355	0.1834
	2	0.5911	0.3227	0.2582	0.1987
	3	0.5741	0.3125	0.2514	0.1546
GH2132	4	0.1732	0.0934	0.1268	0.2217
	5	0.1529	0.0798	0.1008	0.2285
	6	0.2955	0.1189	0.1042	0.1512
GH605	7	0.9262	5.3206	6.751	8.0735

	8	0.6849	5.6546	6.8339	7.7944
	9	1.6882	3.8241	6.4715	8.0446
GH3536	10	-0.0815	-0.0815	-0.0566	-0.0391
	11	-0.0951	-0.073	-0.06	-0.0433
	12	-0.1461	-0.09	-0.0566	-0.0467
GH4738	13	-0.1393	-0.0883	-0.0396	-0.028
	14	-0.1155	-0.0764	-0.0328	-0.0238
	15	-0.1393	-0.0985	-0.0442	-0.0314

3.2. Corrosion rate forecasting model

3.2.1. Model establishment. Scikit-learn is a machine learning library based on python, which can facilitate the implementation of machine learning algorithms. In this research, the 11 components of the superalloys and the corrosion time were selected as features in GBRT. The corrosion rate was used as objective value. 90% and 10% of total experimental data (60 data) were used as training and testing data set randomly. The process was repeated 10 times. Ten different training processes and test sets were used to get the average performance of GBRT. A 8-fold cross-validation was used to train data to improve the generalization of the model, which means the training data set was randomly divided into eight parts and each of parts taken turns as the validation data set.

3.2.2. Model parameter selection. Combining the characteristics of the experimental data, this study selected the main parameters of each model by using GridsearchCV in Scikit-learn to adjust the parameters of the models in the Python environment. The specific description of the main parameters is as follows:

The maximum depth of the decision tree depends on the interaction between the input variables. The number of decision trees represents the number of gradient enhancements, which has strong robustness to overfitting of the model. Figure 1 shows the relationship between the number of decision trees (Boosting Iterations) and the mean square error (MSE) in GBRT. This study determined that the number of decision trees is 500 and the maximum depth is 3. Other parameters were set as the default parameters in sklearn.

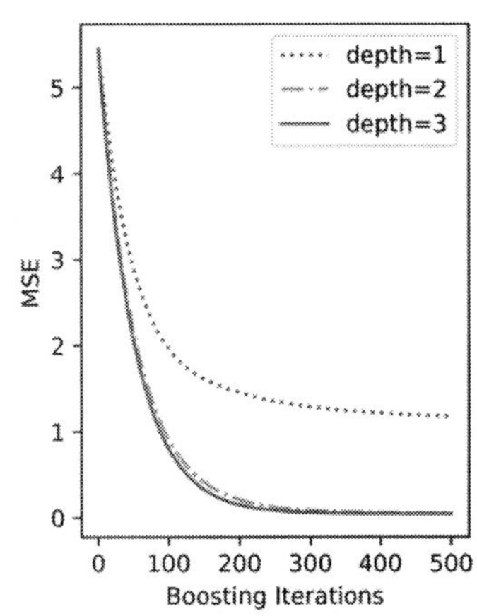

Figure 1. Relationship between Boosting Iterations and MSE in GBRT

3.2.3. Evaluation of predictive effects. The mean square error (MSE) and the determination coefficient (R2 score) are selected as the evaluation indicators of the model forecasting effect. It was calculated MSE=0.0451,R2=0.9922 in the training set, MSE=0.0477,R2=0.9696 in the testing set.It can be seen that the GBRT could predict the corrosion rate with a good reliability for the training set and the testing set.

3.3. GBRT model

3.3.1. Evaluation of feature importance. Feature_importance in Scikit-learn is used to output the importance ranking and score for each feature. Figure 2 shows the importance of 11 features. In GBRT, time is the factor that affects the maximum corrosion rate. Among chemical composition of superalloys, Mn, Co, Al, Ni, Ti, and Mo have the greatest influence on the corrosion rate, and the remaining components have less influence.

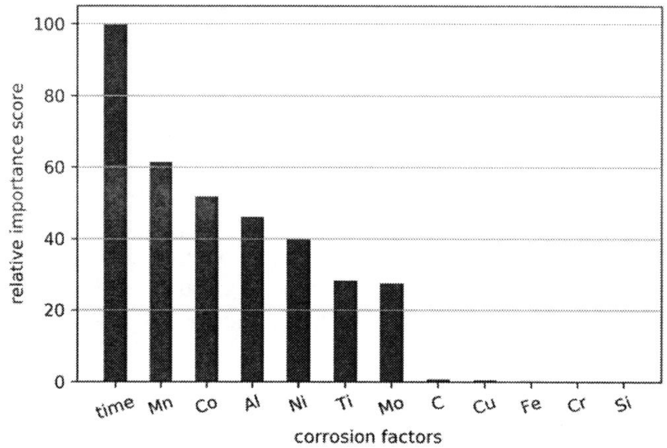

Figure 2. Importance order of corrosion factors

3.3.2. Effect of influencing factors on the corrosion rate. The "partial_dependence" function in GBRT shows the independence between the target correspondence and a set of features, excluding all other features. Partial_dependence can be interpreted as expected target responses, and function of target characteristics.

A one-dimensional partially dependent as a function to study the relationship between corrosion rate and the chemical compositions for Mn, Co, Ti, Ni, Mo and Al, is calculated by partial_dependence. Figure 3 shows the following:

- Increasing the content of Ti, Al and Mo with a certain specific gravity could reduce the corrosion rate very effectively. After a certain proportion, the increase of Al and Mo would not affect the corrosion rate, the effect of Ti would also be greatly reduced.
- When the content of Ni fell within the range of 10% ~ 25%, the corrosion of superalloys could be inhibited, but the corrosion rate was not affected when the content was more than 30%.
- Corrosion would be aggravated when the content of Co exceeds 20%, so the content of Co should not be large in the design. In combination with Figure 2 and Figure 3, a small amount of Mn would also aggravate the corrosion to a great extent. Therefore, the content of Mn should be strictly controlled in superalloy design not to exceed 1%.

A two-dimensional partial dependence function to calculate the relationship between corrosion rate and four sets of feature combinations. Figure 4 shows the effect of time-Al, time-Ni, time-Co, Ni-Co on corrosion rate. The color in the figure represents an increase in corrosion rate from dark to light. In Figure 4(a), 4(b), inclined trend lines indicate the correlation between the Al, Ni content in a certain range and the corrosion time. With the increase of corrosion time, the inclined degree of trend line is greater, which indicates that the suppression effect of Al, Ni element on corrosion was more obvious with the process of corrosion. The corrosion rate line in Figure 4(c) becomes more dense with the increase of Co content. The increasing trend of corrosion rate is also faster. Figure 4(d) shows the combined effect of Ni and Co on corrosion rate of superalloys. The composition combination of Ni-Co in purple region could effectively inhibit the corrosion.

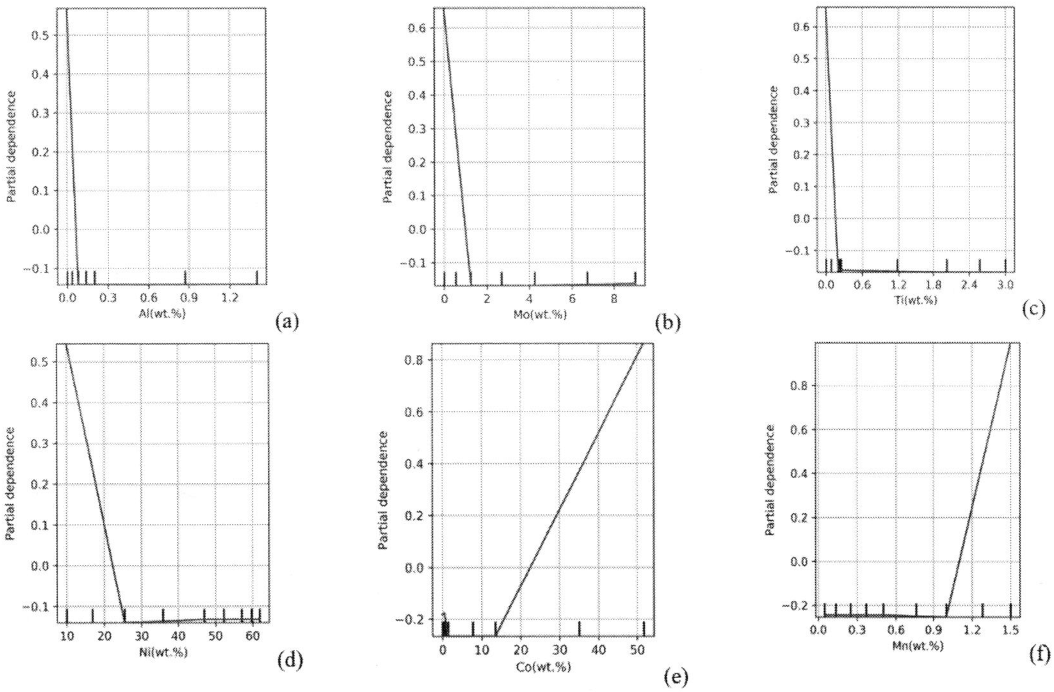

Figure 3. One-dimensional partial dependence function of corrosion factors and corrosion rate (a)Al (b)Mo (c)Ti (d)Ni (e)Co (f)Mn

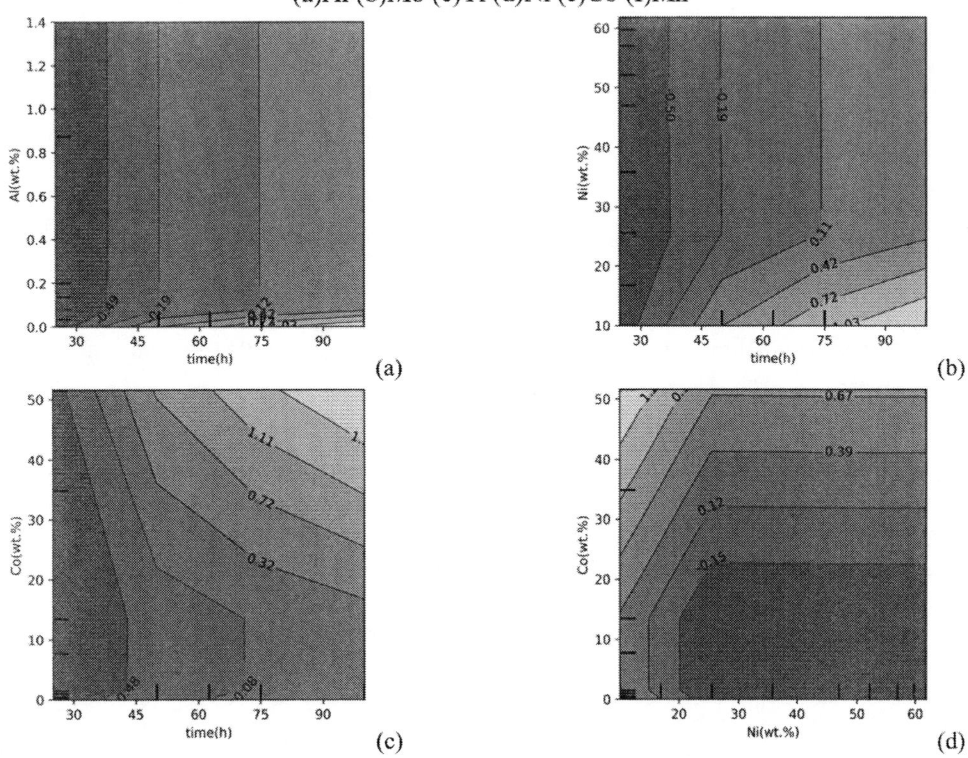

Figure 4. Two-dimensional partial dependence function of corrosion factors and corrosion rate(a)time and Al (b)time and Ni (c)time and Co (d)Ni and Co

IOP Conf. Series: Materials Science and Engineering **685** (2019) 012025 doi:10.1088/1757-899X/685/1/012025

4. Conclusion

In this study, gas hot-corrosion tests of GH3625, GH2132, GH605, GH3536, and GH4738 were carried out, and the corrosion rates at 25h, 50h, 75h, and 100h were calculated by the weight loss method.

Gradient boosting regression tree was used to analyze the relationship between the chemical compositions and corrosion rates. From the analysis of the predicted value and the real value, the evaluation of predictive effects was obtained. GBRT results led to a good understanding of the effect of each factor on the corrosion rate. The significance of superalloy elements was evaluated using feature importance function. The separate effects and co-effects on the important factors was calculated by partial dependent function. The elements of Mn, Co had a great negative influence on the corrosion rate of superalloys in this study while Al, Ni, Ti and Mo were shown to have a positive effect.

This study provides a reference for the selection and protection of aeroengine superalloys. However, it should be noted that this part of the analysis only for the five superalloys under the experimental conditions in this paper, different test conditions could affect the alloy composition of the thermal corrosion resistance. Therefore, more empirical data are needed for future research to explain. Due to the lack of corrosion process data in the test, further research is needed. The conclusions of this paper need to be proved through verification experiments in the future research.

Acknowledgments

Thanks for the technical guidance provided by School of Reliability and Systems Engineering in Beihang University and experimental guidance by Engineer ShaoYue.

References

[1] [Dong J X, Li L H, Li H Y,Zhang M C and Yao Z H 2015 Effect of extent of homogenization on the hot deformation recrystallization of superalloy ingot in cogging process. *Mater. Lett.* **51** 1207-1218

[2] Zhou C, Zhang Q and Yao L I 2013 Thermal shock behavior of nanostructured and microstructured thermal barrier coatings on a Fe-based alloy *Acta Metall. Sinica.* **217** 70 – 75

[3] Yang H, Dong L, Zhu X, Yang Y, Wang J and Wang H 2018 Effect of rolling passes on thermal parameters and microstructure evolution via ring-rolling process of GH4738 superalloy *Surf. Coat. Technol.* **1** 1-10

[4] Deb D, Iyer S R and Radhakrishnan V M 1996 A comparative study of oxidation and hot corrosion of a cast nickel base superalloy in different corrosive environments *Mater. Lett.* **29** 19-23

[5] Fang S F, Wang M P, Qi W H and Zheng F 2009 Hybrid genetic algorithms and support vector regression in forecasting atmospheric corrosion of metallic materials *Comput. Mater. Sci.* **44** 647-655

[6] Chou J S, Ngo N T and Chong W K 2017 The use of artificial intelligence combiners for modeling steel pitting risk and corrosion rate *Eng. Appl. Artif. Intell.* **65** 471-483

[7] Vafakhah M 2012 Application of artificial neural networks and adaptive neuro-fuzzy inference system models to short-term streamflow forecasting *Can. J. Civ. Eng.* **39** 402-414

[8] Warnes B M, Pettit F S and Meier G H 2002 Hot-corrosion resistance of Ni-Cr-Al-Y and Ni-18% Si alloys in sulfate eutectic and sulfate plus vanadate melts at 973K *Oxid. Met.* **58** 487-498.

[9] Raffaitin A, Crabos F, Andrieu E and Monceau D 2006 Advanced burner-rig test for oxidation-corrosion resistance\udevaluation of mcraly/superalloys systems *Surf. Coat. Technol.* **201** 3829-3835.

[10] Friedman J H 2001 Greedy function approximation: a gradient boosting machine *Ann. Stat.* **29** 1189-1232.

IOP Conf. Series: Materials Science and Engineering **685** (2019) 012026 doi:10.1088/1757-899X/685/1/012026

Single-Station TDOA Passive Location Under Multipath Propagation Conditions

Luxi Zhang[1, a], Yuanyuan Song[1, b], Yijun Li[1, c] and Qun Wan[1,d]

[1]School of Information and Communication Engineering, University of Electronic Science and Technology of China, Chengdu, China, 611731

[a]1602295537@qq.com,[b]song1797@yahoo.com,[c]1049383157@qq.com,[d]wanqun@uestc.edu.cn

Abstract. Aiming at the problem of limited information obtained by single-station location, a single-station time difference of arrival (TDOA) location algorithm based on scatterer-assisted single-station location under the condition of multipath propagation is proposed in this paper. The location of scatterers are known, and Chan algorithm of direct wave is applied to geolocation in non-direct wave environment. The simulation results show that the algorithm can achieve the goal of location with high accuracy and low complexity. It has a certain reference value for the engineering realization of target location in complex signal propagation environment.

1. Introduction

Passive location technology [1] is a set of technical means by which the observing station can estimate the space coordinate position of the target by acquiring the location parameters of the target signal under test. According to the number of observation stations, passive positioning technology can be divided into single station positioning technology [2] and multi-station positioning technology [3]. Single station positioning technology has many advantages, such as small equipment, low cost, flexibility and good maneuverability. It has attracted more and more attention, and has become one of the hot issues in current research. However, because the amount of information obtained by a single Observatory is less than that obtained by multiple observatories, it is difficult to realize the single observer positioning technology. Generally speaking, a single passive static observing platform cannot carry out the radiation source. In complex propagation environment, non-line-of-sight (NLOS) propagation [4] between target and Observatory is common, and it is also one of the most important factors affecting positioning accuracy. In this paper, based on scatterer-assisted, the target signal is reflected and transmitted to the observatory through the scatterer to form multipath propagation. In this paper, multi-path transmission of information is used to achieve the target location.

The location method based on scatterer information was put forward later. For the first time, the scatterer information was used in the target location method in document [5]. The location of the target was estimated by using various positioning parameters such as angles of arrival (AOA), time of arrival (TOA) and Doppler frequency shift obtained from the observation station. The basic principle of specular reflection was proposed in document [6], which was measured by AOA. Literature [7] proposes a new single-station multipath passive location technology, which extracts the original location information from both the direct-path signal from the emitter and the multipath signal scattered by the emitter in the environment, and then realizes the estimation of the target's motion state.

However, the location method is based on the existence of the direct-path signal. Therefore, it is impossible to locate in the environment without direct path signals. Document [8] adds clustering technology based on TOA information of scatterers, and uses three base stations to locate targets.

In this paper, the single-station positioning technology in NLOS environment is mainly studied, and the scatterer information is used to realize the single-station positioning of the target. The organization of the remaining papers is as follows: Section 2 introduces the problem formulation; Section 3 describes the location algorithm and discusses the situation; Section 4 simulates the above algorithm; Section 5 is a summary.

2. Problem Formulation

Suppose there is an observation station located in $\mathbf{q}=(x_0,y_0,z_0)$ in three-dimensional space. N_s scatterers with known positions are involved in the location of the target. $\mathbf{s}_i=(x_i,y_i,z_{i_i})(i=1,2,\ldots,N_s)$ represents the position coordinate vector of the i th scatterer, and $\mathbf{p}=(x,y,z)$ represents the position coordinate vector of the target. When the target signal is transmitted to the observatory through direct and non-direct paths, the number of multi-paths of the signal to the observatory is N_s+1. When the target signal is transmitted to the observatory only through non-direct path, the number of paths is N_s. The corresponding location scenario schematic diagram is shown in Figure 1, in which the solid line represents the direct path and the dotted line represents the non-direct path.

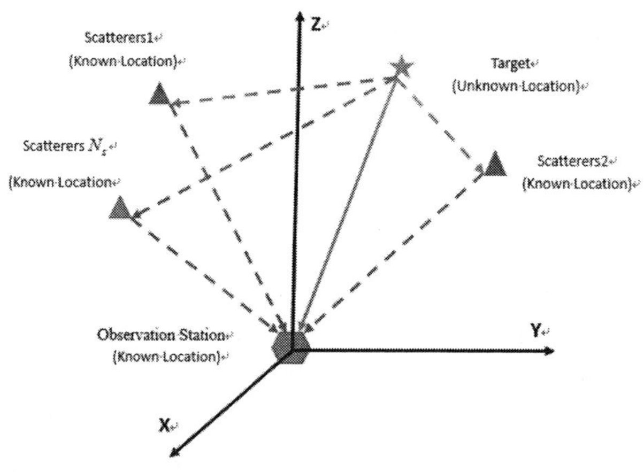

Figure 1 Sketch of 3-D localization scene with known scatterers location.

2.1. Direct and Non-direct Path Propagation Conditions

Where $i=0$ represents the parameters of the direct path. The distance between the target and the observatory is expressed as:

$$r_0=\|\mathbf{p}-\mathbf{q}\| \tag{1}$$

The distance between the scatterer and the target emitter is expressed as:

$$r_i=\|\mathbf{s}_i-\mathbf{p}\| \tag{2}$$

The distance between the scatterer and the observatory is expressed as:

$$h_i=\|\mathbf{s}_i-\mathbf{q}\| \tag{3}$$

The time delays of the direct path and the i th non-direct path are expressed as follows:

$$\tau_0 = \frac{r_0}{c}$$
$$\tau_i = \frac{h_i + r_i}{c} \tag{4}$$

Assuming that the signal and noise are independent of each other and are all zero-mean Gaussian stationary random sequences, and the direct path is taken as the reference path, the TDOA value between the i th non-direct path and the direct path is:

$$\tau_{i,0} = \tau_i - \tau_0 \tag{5}$$

If the propagation speed of the signal is c, then,

$$r_{i,0} = r_i - r_0 = c\tau_{i,0} - h_i \tag{6}$$

In addition,

$$\begin{aligned} r_i^2 &= (x-x_i)^2 + (y-y_i)^2 + (z-z_i)^2 \\ &= K_i - 2x_i x - 2y_i y - 2z_i z + x^2 + y^2 + z^2 \end{aligned} \tag{7}$$

2.2. Non-direct path propagation

Compared with direct path and non-direct path, the information available is less than one direct path. If the first non-direct path is taken as the reference path, the TDOA value between the third non-direct path and the first non-direct path is:

$$\tau_{i,1} = \tau_i - \tau_1 \tag{8}$$

Then,

$$r_{i,1} = r_i - r_1 = c\tau_{i,1} + h_1 - h_i \tag{9}$$

In addition,

$$\begin{aligned} r_i^2 &= (x-x_i)^2 + (y-y_i)^2 + (z-z_i)^2 \\ &= K_i - 2x_i x - 2y_i y - 2z_i z + x^2 + y^2 + z^2 \end{aligned} \tag{10}$$

Among them, $K_i = x_i^2 + y_i^2 + z_i^2$. Formula (6) and formula (9) are a set of non-linear equations about the position of the target. The estimation of the position of the target can be obtained by solving them with appropriate algorithms.

3. Positioning Algorithm

Chan algorithm [9] is a non-recursive hyperbolic equation solution with analytic solution. The feature of the algorithm is that the computation is small and the positioning accuracy is high under the condition that the noise obeys the Gauss distribution. The algorithm uses twice weighted least squares method to get the final target position estimation. Preliminary position estimation can be obtained from the results of one calculation, and the results of one estimation can be improved by the constraints of additional variables. This algorithm is actually an approximation of the maximum likelihood estimation method, so it is also called the approximate maximum likelihood estimation (AMLE) method.

In this paper, Chan algorithm is applied to multipath propagation environment. Scattering body position is regarded as base station position, and it is substituted for the algorithm to estimate target position. For the case of direct wave and non-direct path propagation, Chan algorithm can be divided into two cases: the number of scatterer is two and more than two because of the existence of direct path information. In the case of non-direct path propagation, Chan algorithm can be divided into two cases: the number of scatterer is three and more than three. In this paper, the non-direct path propagation is taken as an example to derive the algorithm.

3.1. Three scatterers

In three-dimensional space, there are three scatterers, the target height is known, and the target is located on the ground or in the air. Then,

$$\begin{bmatrix} x \\ y \end{bmatrix} = \begin{bmatrix} 2x_{2,1} & 2y_{2,1} \\ 2x_{3,1} & 2y_{3,1} \end{bmatrix}^{-1} \times \left\{ \begin{bmatrix} -2r_{2,1} \\ -2r_{3,1} \end{bmatrix} r_1 + \begin{bmatrix} -r_{2,1}^2 - 2z_{2,1}\mathrm{z} + \mathrm{K}_2 - \mathrm{K}_1 \\ -r_{3,1}^2 - 2z_{3,1}\mathrm{z} + \mathrm{K}_3 - \mathrm{K}_1 \end{bmatrix} \right\} \tag{11}$$

By substituting formula (11) for formula (10) and making , a second-order equation about R 1 is obtained. The positive root of equation (10) is substituted for formula (11), and the position estimation of the target is obtained. In some cases, two positive roots may be obtained, and the base station can be limited to a certain area according to prior knowledge to obtain the unique estimation of the target location.

3.2. Four or more scatterers

The results of one estimate are as follows:

$$\begin{aligned} \mathbf{z}_a &= \arg\min\{(\mathbf{h} - \mathbf{G}_a\mathbf{z}_a)^T \boldsymbol{\psi}^{-1}(\mathbf{h} - \mathbf{G}_a\mathbf{z}_a)\} \\ &= (\mathbf{G}_a^T\boldsymbol{\psi}^{-1}\mathbf{G}_a)^{-1}\mathbf{G}_a^T\boldsymbol{\psi}^{-1}\mathbf{h} \end{aligned} \tag{12}$$

Among them, $\mathbf{z}_a = [\mathrm{x}, \mathrm{y}, r_1]^T$, $\boldsymbol{\psi} = \mathrm{c}^2\,\mathbf{BQB}$, $\mathbf{B} = diag\{r_2^0, r_3^0, \ldots, r_{N_s}^0\}$, $\mathbf{Q}$ represents the covariance matrix of the error vector of TDOA measurements. And,

$\mathbf{h} = \frac{1}{2}\begin{bmatrix} r_{2,1}^2 - K_2 + K_1 + 2z_{2,1}z \\ r_{3,1}^2 - K_2 + K_1 + 2z_{3,1}z \\ \vdots \\ r_{N_s,1}^2 - K_{N_s} + K_1 + 2z_{N_s,1}z \end{bmatrix}$, $\mathbf{G}_a = -\begin{bmatrix} x_{2,1} & y_{2,1} & r_{2,1} \\ x_{3,1} & y_{3,1} & r_{3,1} \\ \vdots & \vdots & \vdots \\ x_{N_s,1} & y_{N_s,1} & r_{N_s,1} \end{bmatrix}$. In one estimate, Chan method considers that the variables r_1 are not related to x and y. In fact, r_1 is a function of x and y, and the result of one estimate has a great error.

The results of the second estimation are as follows:

$$\mathbf{z}_a' = (\mathbf{G}_a'^T\mathbf{B}'^{-1}\mathbf{G}_a\mathbf{Q}^{-1}\mathbf{G}_a\mathbf{B}'^{-1}\mathbf{G}_a')^{-1}(\mathbf{G}_a'^T\mathbf{B}'^{-1}\mathbf{G}_a\mathbf{Q}^{-1}\mathbf{G}_a\mathbf{B}'^{-1})\mathbf{h}' \tag{13}$$

Among them, $\mathbf{h}' = \begin{bmatrix} (\mathbf{z}_{a,1} - x_1)^2 \\ (\mathbf{z}_{a,2} - y_1)^2 \\ \mathbf{z}_{a,4}^2 \end{bmatrix}$, $\mathbf{G}_a' = \begin{bmatrix} 1 & 0 \\ 0 & 1 \\ 1 & 1 \end{bmatrix}$, $\mathbf{B}' = diag\{x^0 - x_1, y^0 - y_1, r_1^0\}$, The position coordinates of the target are:

$$\begin{bmatrix} x \\ y \end{bmatrix} = \sqrt{\mathbf{z}_a'} + \begin{bmatrix} x_1 \\ y_1 \end{bmatrix} or \begin{bmatrix} x \\ y \end{bmatrix} = -\sqrt{\mathbf{z}_a'} + \begin{bmatrix} x_1 \\ y_1 \end{bmatrix} \tag{14}$$

IOP Conf. Series: Materials Science and Engineering **685** (2019) 012026 doi:10.1088/1757-899X/685/1/012026

4. Simulation

In order to verify the positioning effect of Chan algorithm under the condition of multipath propagation, simulation analysis is carried out on the platform of MATLAB. Assume that the position of the target is $(50,30,40)$ (km), the location of the observatory is $(-5,-10,10)$ (km), and the number of scatterers is three and their positions are: $(-10,0,10)$ (km), $(10,10,-10)$ (km), $(-10,10,0)$ (km). The plane schematic diagram of the target radiation source, observation station and scatterer is shown in Figure 2.

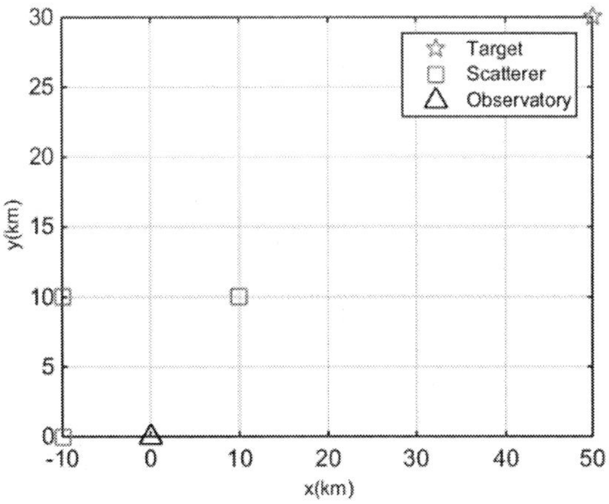

Figure 2 The plane schematic diagram of the target radiation source, observation station and scatterers.

The time difference error of 100ns ~ 1000 ns, the position error of 50 m, 100 m and 200 m observation stations and scatterers are added to the measurement data. The result of positioning error under non-direct path propagation is shown in Figure 3. When the number of scatterers is two under the conditions of direct and non-direct propagation, the algorithm can obtain the target position estimation by solving the second-order equation. The error figure of positioning error results is shown in Figure 4.

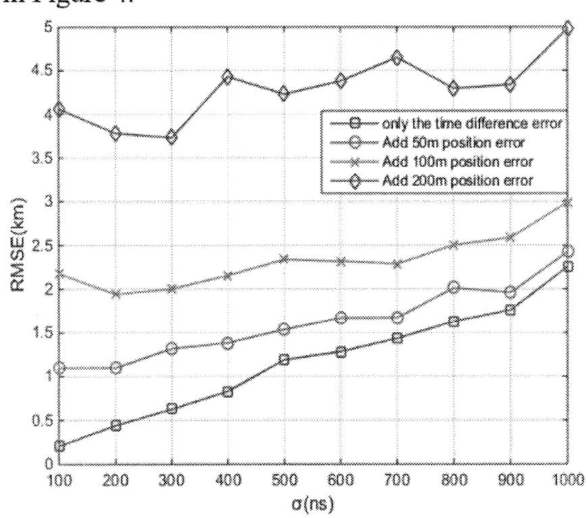

Figure 3 Location Error Curve under Non-direct Path Propagation.

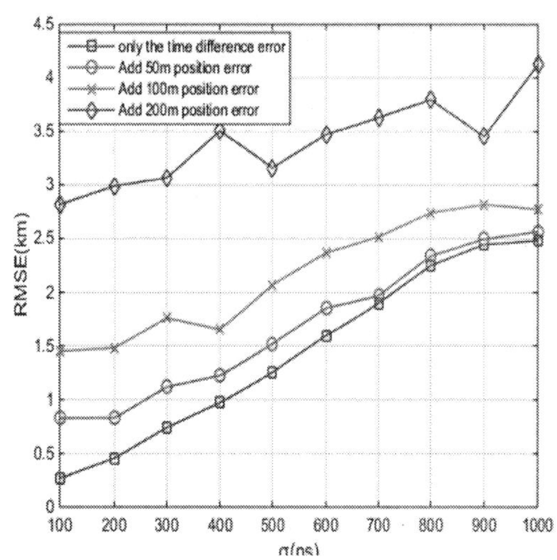

Figure 4 Location Error Curve under Direct and Non-Direct Path Propagation Conditions.

From Figure 3 and Figure 4, it can be seen that the Chan algorithm can be applied to single station TDOA location under multipath propagation conditions to achieve the goal of target location. Moreover, with the increase of time difference error, the positioning error becomes larger and larger. When the position errors of scatterers and observation stations are added to the measured data, the positioning errors become larger and larger.

5. Conclusion

In this paper, single-station passive location under multipath propagation is studied. When both the target and the observation station are stationary, single-station cannot locate. With the help of scatterer, more parameter information can be obtained. Chan algorithm is applied to multipath environment. The simulation results show that this method can locate the target. The algorithm has the characteristics of high positioning accuracy and low computational complexity, but in the non-direct path environment, the positioning accuracy of the algorithm has a downward trend.

Acknowledgments

This work was supported in part by the National Natural Science Foundation of China under Grant U1533125, and Grant 61771108, in part by the National Science and Technology Major Project under Grant 2016ZX03001022, in part by the Fundamental Research Funds for the Central Universities under Grant ZYGX2015Z011 and Sichuan science and technology planning project (key R & D project 18ZDYF0990).

6. References

[1] Shan Y H, Sun Z K and Huangfu K 2001 Research on single observer-passive location technology based on method of phase difference rate of change *Journal of National University of Defense Technology.*

[2] Zhang H, Guo F, Lu T T, et al. 2014 Study on distance and angle measurement for single-base-station UWB positioning system with circular antenna array *International Computer Conference on Wavelet Active Media Technology & Information Processing.*

[3] Tao Y U 2008 Approach Guidance System for Shipboard Plane Based on Multi-station Positioning Technology *Journal of System Simulation.*

[4] Wang X, Wang Z, O'Dea B 2001 A TOA-Based Location Algorithm Reducing the Errors Due to Non-Line-of-Sight (NLOS) Propagation *Journal of China Institute of Communications* **52** 112-116.

[5] Thomas N J,Cruickshark D G M and Laurenson D I 2001 Calculation of Mobile Location Using Scatter Information .*Electronic Letters* **37** 1193-1194.

[6] Honglei miao, Kegen Yu, et al.. 2006 Positioning for NLOS propagation: algorithm derivations and Cramer-Rao bounds *IEEE ICASSP* **2006** 1045-1048.

[7] Coutts S D 2000 Passive localization of moving emitters using out-of-plane multipath *IEEE Transactions on Aerospace & Electronic Systems* **36** 584-595.

[8] Y.T. Huang 2001 Real-Time Passive Source Localization: A Practical Linear-Correction Least-Squares Approach *IEEE Trans. on Speech and Audio Processing* **9** 943-956.

[9] Chan Y.T and Ho K.C. 2002 A simple and efficient estimator for hyperbolic location *IEEE Transactions on Signal Processing* **42** 1905-1915.

IOP Conf. Series: Materials Science and Engineering **685** (2019) 012027 doi:10.1088/1757-899X/685/1/012027

Performance Analysis for MIMO SWIPT Systems with Space-Time Coding

Tao Cui[1,2], Jun Li[1,2], Jun Shen[2] and Shiwei Dong[1,2]

[1]National Key Laboratory of Science and Technology on Space Microwave, CAST, Xi'an, Shaanxi 710100, China

[2]China Academy of Space Technology (Xi'an), Xi'an 710100, China

[*]conan915@163.com

Abstract. Energy supply is a key factor affecting the normal operation of energy-constrained networks. Simultaneous wireless information and power transfer (SWIPT) technology can be used to obtain a stable and continuous energy source. But SWIPT need to take into account data transmission rate and energy harvesting performance, it is necessary to design a complex physical layer method to improve the energy efficiency of SWIPT system. The paper studies a multiple input multiple output (MIMO) SWIPT system based on space-time coding, the transmitter sends the information using the space-time encoder through multi-antenna transmission, the receiver uses the space-time decoder to decode the information and collect the energy. The expressions of channel capacity and energy harvesting under MIMO are given respectively. The energy harvesting optimization target equation under coding conditions is established. The simulation results show that the using of space-time coding can reduce the data transmission error rate of MIMO SWIPT system and improve the energy harvesting performance of the user terminals.

1. Introduction

Energy-limited networks such as wireless sensor networks and micro-nano satellite networks use the power supply method of saving batteries due to the particularity of the working environment. However, due to the limitation of volume and weight, the storage capacity of the carried savings battery is limited, which leads to great limitations in the energy and power of the communication system, so that the service performance and standby time of the system are restricted. In this regard, considering the wireless energy transmission and wireless communication mechanism, the academic community has proposed a new idea that can transmit information and energy synchronously, namely simultaneous wireless information and power transfer (SWIPT) technology.

SWIPT refers to using the same emitted electromagnetic wave to transport both energy that is harvested at the receiver, and information that is decoded by the receiver[1]. But in the wireless communication, due to channel fading and path loss, the bit error rate of data transmission will increase, and the energy transmitted through the wireless channel is also limited, so that the energy harvesting by the SWIPT receiver is also limited, it is necessary to design a complex physical layer method to improve the performance of SWIPT systems[2]. In this situation, the MIMO technology has drawn more attentions as a technology that can effectively increase the speed of wireless communication systems. The MIMO technology can merge the diversity technologies of transmitting and receiving together. It can take full advantage of the signal space resources through the configuration of multiple antennas to improve the system capacity.

The existing researches on MIMO SWIPT system mainly focus on the energy transmission in the MIMO channel. The energy efficiency has been studied in terms of resource allocation, power allocation, interference channel conditions etc. Reference [3] compared massive MIMO with traditional small-cell networks in the perspective of energy efficiency. It is pointed out that the former has a relatively low short circuit power, so that the energy efficiency of massive MIMO is significantly improved. The authors in [4] studied different MIMO channels and proposed an optimal transmission strategy that can maximize information and energy transmission. Reference [5] studied power allocation for energy harvesting in multi-user MIMO networks. Reference [6] investigated resource allocation in multi-user MIMO networks to maximize energy efficiency. Reference [7] studied the energy and rate of wireless energy transmission in a two-stage massive MIMO heterogeneous network. Reference [8] analyzed the energy efficiency of the cross-layer between the physical layer and the MAC layer in a MIMO network. Reference [9] analyzed the throughput in energy harvesting MIMO relay networks under fading channel conditions. The research of space-time coding is emphasized on improving channel capacity of information. Reference [10] proposed a pre-encoding method for SWIPT in a MIMO relay network, and studied the compromise area of information and energy transmission.

It can be seen that there are few researches on the performance of the SWIPT system in the MIMO system from the coding point of view. This paper studies the space-time coding method in the MIMO system and discusses steering the beam direction of the transmitted signal by coding at the transmitter. The tuning of phase and other parameters can maximize the harvested energy at the receiver while maintaining the high information transmission rate. Thereby, the performance of the MIMO enabled communication system can be improved. The energy harvesting optimization target equations are given, and the BER and energy harvesting performance of the MIMO SWIPT system with space-time block coding are analyzed by simulation. By simulation analysis, the use space-time block coding can reduce the data transmission error rate of MIMO SWIPT system and improve the energy harvesting performance of the user terminals.

2. MIMO SWIPT system model based on space-time coding

In this section, we first introduce the space-time coded MIMO SWIPT system model used in this paper, then give a mathematical description of the system, and deduce the expression of system channel capacity and energy harvested by the receiver based on the mathematical model.

The system includes a transmitter and k users, the transmitter has $N_T > 1$ transmit antennas and each user has $N_R > 1$ receive antennas, as illustrated in Figure1. The received signal is split into two independent streams by each receiver, these two streams are used separately for information decoding (ID) and energy harvesting (EH).

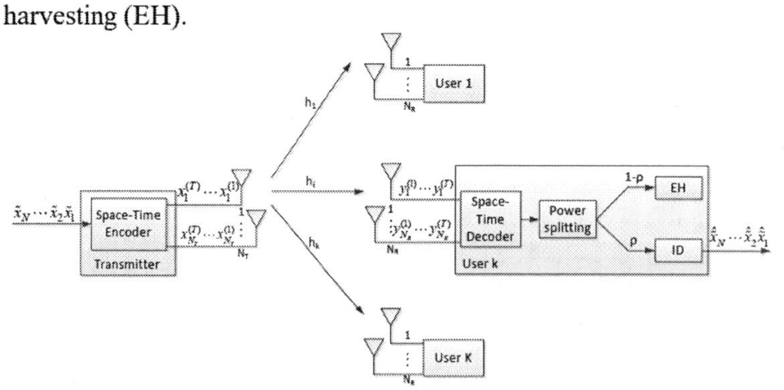

Figure 1. A MIMO SWIPT system with Space-Time coding

In the space-time coded MIMO SWIPT systems, bit stream is mapped into symbol $\{\tilde{x}_i\}_{i=1}^{N}$, As depicted in Figure1, a symbol stream, which size is N, is encoded into $\{\tilde{x}_i^{(t)}\}_{i=1}^{NT}, t=1,2,\cdots,T$ with space-time coding, where **t** is the symbol time index and **i** is the antenna index. Then the equation of the complex baseband transmitted signal at the transmitter is

$$\mathbf{x}=\sum_{k=1}^{K} v_k \tilde{x}_i^{(t)} \tag{1}$$

Where v_k denotes the beamforming vector at the transmitter for user **k** ,. $\mathbf{k}\in\{1,2,\cdots,\mathrm{K}\}$.

In the paper, we use the flat fading channel and define the channel matrix from the transmitter to the **k** th user is $\mathbf{H}_k$, $\mathbf{H}_k \in \mathrm{C}^{N_R \times N_T}$. So the equation of signal received by the **k** th user is:

$$y_k = H_k x + n_k \tag{2}$$

Where $n_k \sim \mathbf{cn}(\mathbf{0},\delta_k^2)$ denotes the antenna noise of the **k** th user.

After power split, the equation of signal used for energy harvesting is:

$$y_k^{EH} = \sqrt{1-\rho_k}\,(H_k \sum_{j=1}^{K} v_k \tilde{x}_j^{(t)} + n_k), \forall k \tag{3}$$

The signal used for information decoding is expressed as:

$$y_k^{ID} = \sqrt{\rho_k}\,(H_k \sum_{j=1}^{K} v_k \tilde{x}_i^{(t)} + n_k), \forall k \tag{4}$$

Correspondingly, the equation of SNR of information decoding and energy harvesting of **k** th user is:

$$SNR_k = \frac{\rho_k \left|H_k x_k\right|^2}{\rho_k \sum_{j\neq k}\left|H_k x_j\right|^2 + \rho_k \sigma_k^2 + \delta_k^2}, \forall k \tag{5}$$

$$E_k = \eta_i (1-\rho_k)(\sum_{j=1}^{K}\left|H_k x_j\right|^2 + \delta_k^2), \forall k \tag{6}$$

Where $\eta_i \in (0,1]$ denotes the **k** th user's energy conversion efficiency.

3. Performance Analysis

In this section, we derive the system capacity and the user collect energy expression based on the mathematical description of the space-time MIMO SWIPT system model, and obtain the energy harvesting optimization formula. The simulation compares the information error rate and energy efficiency performance of the space-time block case for different number of antennas.

3.1. System capacity and energy harvesting expression

In the paper, we assume that the receiver can accurately know the channel state information, and assume that the noise is independent of each other, resulting in a channel capacity and energy harvesting expressions of user terminal.

$$C_k = \log_2(1+\frac{\rho_k |H_k x_k|^2}{\rho_k \sum_{j\neq k} |H_k x_j|^2 + \rho_k \sigma_k^2 + \delta_k^2}), \forall k$$
$$= \log_2(1+\frac{\rho_k Tr(HH^H XX^H)}{\rho_k Tr(HH^H XX^H) + \rho_k \sigma_k^2 + \delta_k^2}), \forall k \quad (7)$$
$$= \log_2(1+\frac{\rho_k Tr(HQ)}{\rho_k Tr(HQ) + \rho_k \sigma_k^2 + \delta_k^2}), \forall k$$

$$E_k = \eta_i (1-\rho_k)(\sum_{j=1}^{K} |H_k x_j|^2 + \delta_k^2)$$
$$= \eta_i (1-\rho_k)(\|HX\|^2 + \delta_k^2)$$
$$= \eta_i (1-\rho_k)(Tr(HH^H XX^H) + \delta_k^2) \quad (8)$$
$$= \eta_i (1-\rho_k)(Tr(HQ) + \delta_k^2)$$

Where $H = H_k H_k^H$, $Q = \sum_{k=1}^{K} E(x_k x_k^H)$, $E(x_k x_k^H)$ denotes the covariance matrix of transmit signals.

3.2. Energy harvesting optimization formula

Here, we assume that the channel is a flat fading channel, the energy harvesting expression from the precious section can be expressed as:

$$E_k = (1-\rho_k)\eta_i Tr(H_k H_k^H \sum_{k=1}^{K} E(x_k x_k^H) + \delta_k^2) \quad (9)$$

Under the condition of guaranteeing the information trans-mission rate of the receiver and the power constraint of the transmitter, the energy value obtained by the user receiver is maximized, and the optimized formula of energy harvesting is obtained. Its optimization formula is as follows:

$$\max_{\{E(x_k x_k^H)\}, k\in K} \quad E_k$$
$$s.t. \log_2(1+\frac{\rho_k Tr(HQ)}{\rho_k Tr(HQ) + \rho_k \sigma_k^2 + \delta_k^2}) \geq r_k \quad (10)$$
$$Tr(\sum_{k=1}^{K} E(x_k x_k^H)) \leq P_T$$
$$0 \leq \rho_k \leq 1, \forall k \in K$$

Where r_k denotes the threshold of the information transmission rate of the receiver; P_T r_k denotes the maximum transmission power constraint value of transmitter.

3.3. Simulation and analysis

This section simulates the BER performance of the space-time coding and the energy values harvested at the receiver for different transmit antenna (N_T) and receive antenna (N_R) numbers. Simulation parameters are shown in Table 1.

Table 1. Simulation parameters

Parameters	Value
Transmit antenna number	2/4/8
Receive antenna number	2/4/8
Modulation	16QAM
Channel model	Gaussian
Antenna spacing	λ/2
Noise of receiver(dB)	5
Path loss model/(l in km)	$128.1+37.6\lg(l)$
Carrier frequency(GHz)	2.4
Transmission power constraint value(W)	1

We compare the BER of space-time coding in three cases: $N_T=4, N_R=4$, $N_T=2, N_R=4$, $N_T=2, N_R=2$, and the Figure 2 shows the BER performance curve. It can be seen that with the same number of transmit antennas, the increase in the number of receive antennas can reduce the BER of space-time coding, at the same time, as the transmit antennas number and the transmit antennas number and the receive antennas number increases, the bit error rate of space-time coding also decreased.

According to the energy harvesting formula of the receiver, based on the known channel parameter information, the value of the harvested energy is related to the trace of the covariance of the transmitted signal without exceeding the maximum power constraint if the transmitter, and the trace of covariance is related to the sum of the diagonal elements, that is the number of transmit antennas. Here we take $\rho_k=0.5, k=1,2,\cdots,K$, that is 50% of the power received by the receiving antenna is used to collect energy; take $\eta_k=0.3, k=1,2,\cdots,K$, that is the energy conversion efficiency of the receiver is 30%. From this we can get the curve of the energy value harvested by the receiver as shown in Figure3. It can be seen that the energy value harvested by the receiver increases with the number if transmitting antennas. The SISO system and the MIMO SWIPT system using the number of transmitting antenna $N_T=8$ is compared. The curve of the relationship between the energy value harvested by the receiver and the information rate is shown in Figure4. It can be seen that the energy value at the receiver is declining with the increases of the information rate, but the MIMO SWIPT system's drop speed is lower than that of the SISO system.

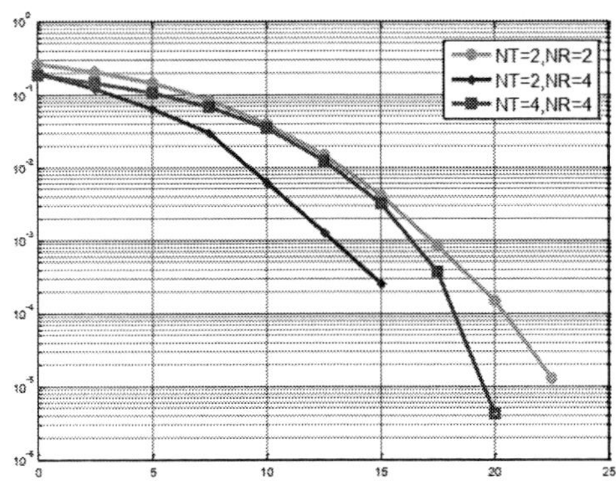

Figure 2. The BER curve of Space-time coding for different number of transmit/receive antennas

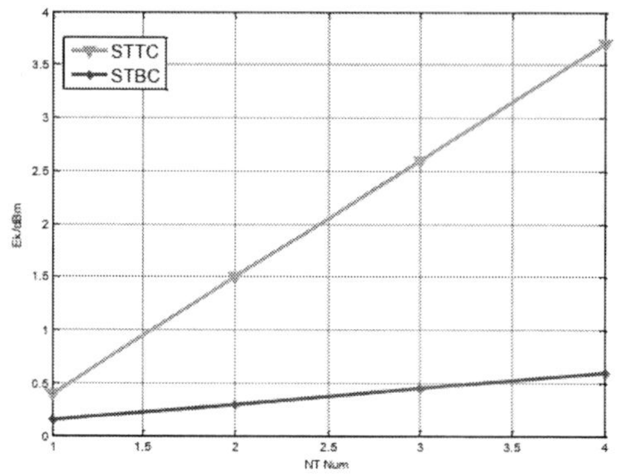

Figure 3. The curve of the energy value harvested by the receiver

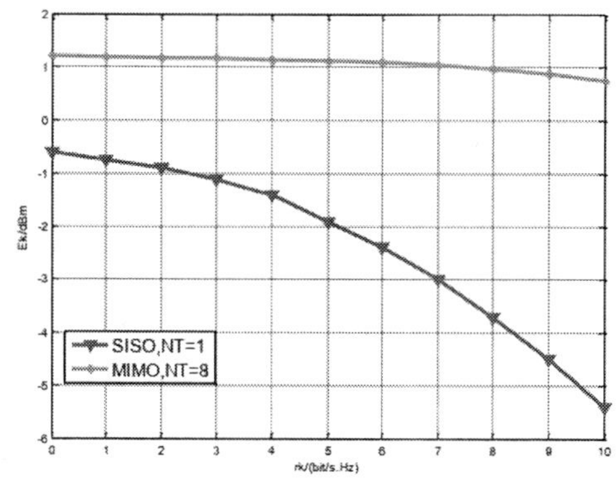

Figure 4. The curve of the relationship between the energy value harvested and the information rate

In summary, the use of MIMO technology and space-time coding in the SWIPT system can reduce the data transmission error rate, and can increase the energy value harvested by the user terminal, thereby improving the performance of the SWIPT system.

4. Conclusion

In this paper, the MIMO SWIPT system with space-time coding is studied. Firstly, the scheme of MIMO SWIPT system used space-time coding is given, and the system model is described in detail. Secondly, the system channel capacity and energy harvesting expression are given based on the space-time coding principle. Then the optimization formula for the user to harvest the maximum energy is analyzed. Finally the performance of the MIMO SWIPT system with different number of transmit antennas is analysed by the space-time block coding simulation. According to the simulation results, the use of space-time block coding in the MIMO SWIPT system can reduce the data transmission error rate, and increase the performance of energy harvesting by the user terminal.

Acknowledgments

This work is financially supported by the Sustained Supported Foundation by National Key Laboratory of Science and Technology on Space Microwave (2018SSFNKLSMT-12, HTKJ2019KL504001).

References

[1] J Tang, K. C. So Daniel, N Zhao. Energy Efficiency Optimization with SWIPT in MIMO Broadcast Channels for Internet of Things[J]. IEEE Internet of Things Journal, 2017:1-4.

[2] K YANG, Q YU, S.P LENG, P ZHANG. Data and energy integrated communication networks[J]. SCIENTIA SINICA Informationis, 2016, 46(5):591-609.
[3] W.J Liu, Sh. Q Han, Ch.Y Yang. Energy Efficiency Comparison of Massive MIMO and Small Cell Network[C]. GlobalSIP 2014: Massive MIMO Communications, 2014:617-621.
[4] R Zhang, C. K Ho. MIMO Broadcasting for Simultaneous Wireless Information and Power Transfer[J]. IEEE TRANSACTIONS ON WIRELESS COMMUNICATIONS, 2013, 12(5):1989-2001.
[5] H. H. M. Tam, H. D. Tuan, A. A. Nasir. Power Splitting for MIMO Energy Harvesting in Multi-User Networks[C]. 2017 International Conference on Recent Advances in Signal Processing, Telecommunications & Computing (SigTelCom), 2017:217-222.
[6] A Zappone, E Jorswieck. A Leshem. Distributed Assignment and Resource Allocation for Energy Efficiency in MIMO Wireless Networks[C]. WSA 2016, March, 2016:165-169.
[7] Y.X Zhu; L.F Wang; K.K Wong; Sh Jin. On Wireless Power Transfer in Two-Tier Massive MIMO HetNets: Energy and Rate Analysis[C]. 24th European Signal Processing Conference (EUSIPCO), 2016:828-832.
[8] G Peron, G Brante, R.D Souza. Physical and MAC Cross-Layer Analysis of Energy-Efficient MIMO Networks[C]. 27th Annual IEEE International Symposium on Personal, Indoor and Mobile Radio Communications, 2016.
[9] N.P Le, N.S Vo, M.T Hoang. Throughput Analysis of Energy Harvesting MIMO Relay Systems over Nakagami-m Fading Channels[C]. 2017 International Conference on Recent Advances in Signal Processing, Telecommunications & Computing,2017:164-169.
[10] B Fang, W Zhong, Z.P Qian. Optimal Precoding for Simultaneous Information and Power Transfer in MIMO Relay Networks[C]. 9th International Conference on Communications and Networking in China, 2014:462-467.

A Centralized Scheduling Algorithm in On-Board Hybrid Optical/Electric Switching

Yi Zhang[1,*], Jun Li[1], Quan Zhou[1], Jingling Li[1], Wei Liang[1] and Tao Cui[1]

[1]National Key Laboratory of Science and Technology on Space Microwave(Xi'an) China Academy of Space Technology, Xi'an, China

[*]lbzytogether@163.com

Abstract: To realize the interconnection of heterogeneous links, it's necessary to develop the on-board hybrid optical/electric switching technology. Because of the separate scheduling strategy in on-board hybrid switch, there is conflict among different ports, and the cell loss rate is greatly reduced. In order to solve this problem, a centralized scheduling algorithm in on-board hybrid optical/electric switching is proposed. According to the parameters and characters of optical data and electric data, the proposed algorithm with the group performance idea can schedule the multi-granularity traffic by analysing the resource utilization of hybrid switch fabric, which has the principle of the highest throughput. This algorithm can increase the matching probability between input and output ports, reduce the inner congestion probability of hybrid optical/electric switch module, and improve the performance of this switch fabric. The analysis and simulation results show that the proposed algorithm has fewer parameters and simple implementation steps, which can effectively avoid congestion, and has the better performance than others in cell loss rate.

1. Introduction

The space information network is a system for acquiring, transmitting and processing spatial information in real time, and the carrier of this network is the space platform, such as GEO/MEO/LEO satellites, stratospheric balloon, and manned or unmanned aircraft, etc.

With the growing demand for spatial information, the spatial information network should not only meet the requirements of high-speed information relay and broadband high-capacity information forwarding, but also should have flexible access and global coverage capabilities. In the space information network, the laser and microwave links will coexist for a long time and have complementary advantages. This network is a heterogeneous network composed of laser-microwave hybrid link.

To realize the intercommunication of heterogeneous links and the exchange of data between heterogeneous nodes, it is necessary to develop the on-board hybrid optical/electric switching technology, which should have the capabilities of service classification, aggregation, convergence and distribution for the different links. Because of the great differences between optical switching and electrical switching[1-3], the use of separate scheduling is a relatively easy control method, but the disadvantages are also obvious: the optical switching unit and the electrical switching unit schedule and manage the services according to the status of the single unit, which can cause conflict between the optical port and the electrical port, and congestion of the hybrid switching unit. The data loss rate is greatly reduced, and the exchange performance has deteriorated.

IOP Conf. Series: Materials Science and Engineering **685** (2019) 012028 doi:10.1088/1757-899X/685/1/012028

To solve this problem,a centralized scheduling algorithm in on-board hybrid optical/electric switching is proposed. According to the characters of optical data and electric data, the algorithm with the group performance idea can schedule the multi-granularity traffic by analysing the resource utilization of hybrid switch fabric, which has the principle of the highest throughput. This algorithm can increase the matching probability between input and output ports, reduce the inner congestion probability of hybrid optical/electric switch module, and improve the performance of this switch fabric.

This paper is arranged as follows: the new algorithm is described in section II, in which the idea of the algorithm is introduced in detail; Section III introduces the implementation method of the proposed algorithm; Then the performance is analysed in section IV and the algorithm is simulated in order to show the better performance in cell loss rate in section V. At last, section VI concludes the paper.

2. Thecentralized scheduling algorithm in on-board hybrid optical/electric switching

2.1. Composition of onboard hybrid optical/electric switching

The module of on-board hybrid optical/electric switching is shown as figure.1, which include input/output procession unit, hybrid switching unit and hybrid scheduling unit.

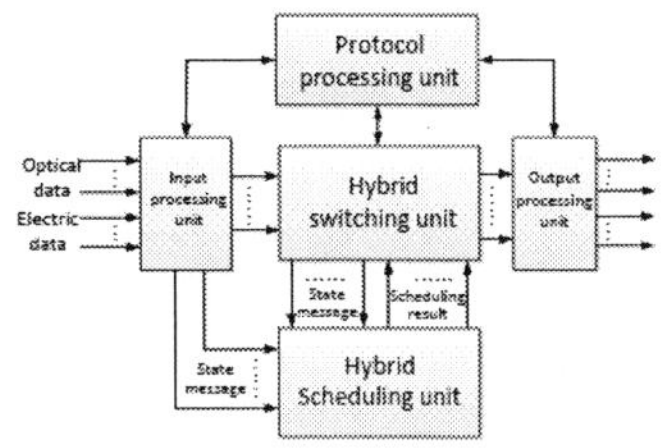

Figure 1. The module of onboard hybrid optical/electric switching

In Figure 1, the hybrid scheduling unit receives state information from the input processing unit and the hybrid switching unit, and computes an internal conflict-free optimal port configuration scheme for the hybrid switching unit. The hybrid switching unit exchanges the mixed data by the matching result. Therefore, the scheduling algorithm directly affects the performance of the hybrid switch.

2.2. Typical optoelectronic hybrid scheduling algorithm

At present, most scheduling algorithms are developed for the scheduling of one single service, such as input/output serial scheduling algorithm [4][5], FC-WFQ for iSCSI Virtualization Switch [6],and so on. Multi-service scheduling algorithms are mainly focused on the switching mechanism [7], variable-length messages [8], multicast [9], such as hierarchical smooth round-robin scheduling algorithm-mFGSR, hierarchical hybrid scheduling algorithm-MSHF, and so on.

The traditional optoelectronic hybrid switch[10] adopts an independent scheduling management for different services, as shown in Figure 2. The optical service and the electrical packet scheduling algorithm operate independently. Since the scheduling unit does not focus on the traffic characteristics of various services and the resource utilization of the hybrid switching unit, conflicts of various services at the output port will occur, which results in the congestion inside the switching unit. The data loss rate is greatly increased, which affects switching performance.

IOP Conf. Series: Materials Science and Engineering **685** (2019) 012028 doi:10.1088/1757-899X/685/1/012028

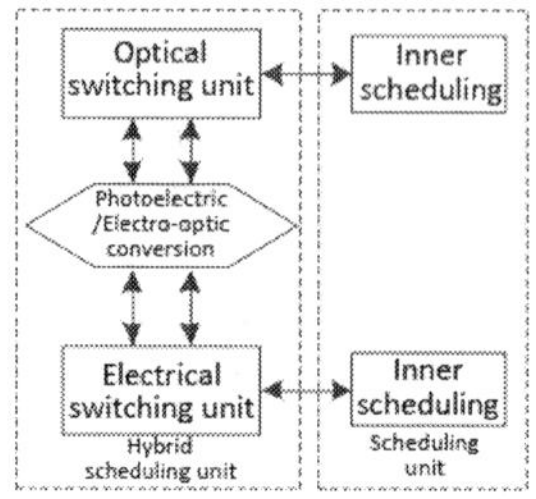

Figure 2. Traditional scheduling strategy in hybrid optical/electric switching

2.3. The centralized scheduling algorithm

To solve the above problems, a centralized scheduling algorithm for on-board hybrid switching is proposed in this paper. Combing the parameters which include the incoming photoelectric service, the burst length and the QoS requirements of each service, and the use of resources such as the buffer and the service tag in the hybrid switching, the matching result between each input port and the output port is calculated and the data among multi-granular ports achieve aggregation and distribution. Figure 3 shows general idea of the proposed algorithm.

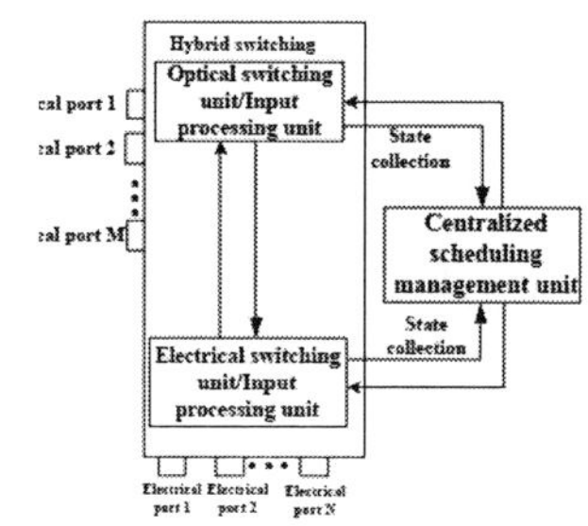

Figure 3. The centralized scheduling algorithm in on-board hybrid optical/electric switching

First, the centralized scheduling management unit collects information such as service parameters of each input port, utilization of exchange resources, and current status of the hybrid switch.

Secondly, the centralized scheduling management unit schedules the optoelectronic service from the high priority to the low priority according to the QoS requirements of different services, and allocates optical and electrical switching resources(such as electrical caches, optical tags, etc.) according to the principle of highest throughput. Considering the convergence of the optical service and the electrical packet service, one optical service may be switched to the multi-channel packet port, and the multi-channel packet service may be switched to the same optical service. Therefore, the optical switching unit and the packet switching unit resources are uniformly managed by the centralized scheduling control unit in this paper.

Then, according to the arrival of the electrical packet/optical service, the cache usage of each port, and the current state of the switch, the internal path of the exchange is calculated, then the output port of the light load is selected for various services in order to satisfying the high priority service, which can avoid conflicts between different services and reduce the data loss rate. In this algorithm, the optical service has a higher priority than the packet service. If the optical service conflicts with the electrical packet service, the optical service should be switched preferentially.

Finally, the matching scheme between each input port and the output port is calculated, and the matching result is sent to the hybrid switching unit.

Due to the constraints of the space environment, the algorithms in the on-board hybrid switch can't be too complicated. The algorithm proposed in this paper requires fewer parameters, simple implementation steps and low complexity, which is suitable for application.

3. Specific implementation method

The main steps of the centralized scheduling algorithm in onboard hybrid optical/electric switching proposed by this paper are as follows:

Step 1, collect various types of state information according to data characteristics.

Step 2, calculate the internal path for various services.

- If the service is optical, it needs to be split into X packets and switched to the electrical output port K(optical->electrical)

a) In the hybrid switching matrix A (i, j), if i in equation (1) exists, the corresponding input port is selected to the receive optical service data, and is transferred to c); if not, it is transferred to b);

$$A(i,K) = 0((M+1) \leq i \leq (M+N)) \tag{1}$$

b) Compare the port buffer states B1, B2, ··· BN of the electrical packet switching unit,and select the electrical input port with less buffer capacity to receive the optical service data;

c) Encapsulate optical service data into internal packet data format and send it to the corresponding electrical input port;

d) If the current status of the output port is idle, go to the third step; if the current status of the output port is busy, the data is cached, and when the output port is idle, the data is output preferentially, and the process goes to the third step;

- If Y-way electrical services need to be aggregated into optical port H:(electrical->optical)

a) In the hybrid switching matrix A(i, j), if the conditional expression referred to (2) satisfied, one of the corresponding optical input ports is selected to receive the optical data, then proceeds to c); if the condition is not met, the process proceeds to b)

$$A(i,H) = 0(1 \leq i \leq M) \tag{2}$$

b) Buffer the electrical data. If i that satisfies Equation (2) exists, goes to c); if not, goes to b);

c) Converge electrical data into optical data;

d) Output the optical data to the optical switching unit and go to the third step;

- If the data is optical, it needs to be output from optical port J:(optical->optical)

a) In the hybrid switching matrix A (i, j), if equation (3) is established, go to step c); if equation (4) is satisfied, go to step b);

$$A(i,J) = 0(1 \leq i \leq M) \tag{3}$$

$$A(i,J) = 1(1 \leq i \leq M) \tag{4}$$

b) Delay the input optical data and wait, if the formula (3) is satisfied, go to step c);

c) If no other optical data is output to port J, the data is directly output and goes to the third step. If there are multiple input port data to be output to port J, compare the data with other data priority of P. If the priority value of other data is higher, go to b). If the business priority value of the data is the highest, go to the third step;

- If the data is electrical, it needs to be output from electrical port J1:(electrical->electrical)

a) In the hybrid switching matrix A(i, j), if the equation(5) is satisfied, the data is directly output to the third step; if the equation (6) is satisfied, go to the step b);

$$A(i,J1) = 0((M+1) \leq i \leq (M+N)) \tag{5}$$

$$A(i,J1) = 1((M+1) \leq i \leq (M+N)) \tag{6}$$

b)When multiple input data are output to port J1, and the number of services to be outputted in port J1 is the least among all the output ports, go to d); if there isn't a port J1, then go to c);

c) Poll all electrical output ports according to the number of services to be output. When polling to output port J1 which met the condition in a), then go to d);

d) When the service priority is the highest among all competing services, the service is output to

port J1, then go to the third step; when the service priority is not the highest and the polling pointer does not point to the input port, the service is cached until the round robin pointer points to this input port, then the service is output to port J1 and go to the third step.
Step 3, switch the services to the corresponding ports through the hybrid switching unit.

4. Performance analysis

An efficient and fair scheduling algorithm should match as many ports as possible in order to ensure the QoS requirements of various services, reduce the internal congestion probability of the switching unit, and improve the performance of the switching unit, such as data loss rate and resource utilization. Because it is difficult to compare the performance of scheduling algorithms through mathematical models, this paper first compares the matching probability of input and output ports in different scheduling algorithms, and then analyzes the performance of centralized scheduling algorithms from the aspects of QoS guarantee, data loss rate, and resource utilization in hybrid switching units.

4.1 Input/output port matching probability

First, we analyse and compare the matching probabilities of the traditional scheduling algorithms with the ports in the centralized scheduling algorithm proposed in this paper.

Definition1: Conditional events and conditional probabilities.

Under the condition that one event occurs, another event occurs, called a conditional event; the probability that a condition event occurs is called a conditional probability. The probability of conditional event B|A is defined as

$$P[B \mid A] = \frac{P[A \cap B]}{P[A]}, P[A] \neq 0 \tag{7}$$

Definition 2: Complete event group

Any two events in the event group are mutually exclusive and have a relationship

$$S = A_1 \cup A_2 \cup \cdots \cup A_m \tag{8}$$

Then, the event group is called a complete event group. The relationship of any event B is:

$$P[B] = \sum_{i=1}^{m} P[B \cap A_i] = \sum_{i=1}^{m} P[A_i]P[B|A_i], P[A_i] > 0 \tag{9}$$

(1) If the output ports of every input service are different, the number of matches obtained by the two scheduling algorithms is the same.

(2) If the output ports of multiple services are the same, assuming that the matching probability of input port is P[B], event A1 indicates that the selected output port of the input port is not matched, and event A2 indicates that the selected output port of the input port has been matched, and A1 and A2 form a complete event group.

In traditional scheduling algorithms,

$$P[B] = P[B \cap A_1]P[B \cap A_2] \tag{10}$$

When the centralized scheduling algorithm matches the input and output ports, the optical service and the electrical service are uniformly considered to avoid output port conflicts. Therefore

$$P[B] = P[B \cap A_1] \tag{11}$$

Because of $P[B \cap A_2] < 1$, the matching probability of the input port in the centralized scheduling algorithm is higher than the traditional scheduling algorithm.

4.2 Comparison of performance indicators

4.2.1. QoS guarantee. In a hybrid switch, the centralized scheduling unit uniformly manages optical/electrical services. According to the QoS requirements and the status of the hybrid switch, the optimal internal path is calculated for the higher priority service to meet the QoS requirements of different services.

4.2.2. Data loss rate. With the separate scheduling control mechanism in different switching unit, if the optical and electrical data are switched, it is easy to match the multiple input data to the same output port with a large amount of traffic, which causes the congestion in the output port. The proposed algorithm selects the internal path for various services with the group decision-making idea. Therefore, the proposed algorithm can effectively reduce the data loss rate and improve the performance of the hybrid switch.

4.2.3. Resource utilization of hybrid switch

When the centralized scheduling algorithm adapts the optoelectronic services, the scheduling and switch of various services are not performed separately. Instead, the parameters of the optical/electrical switch service are considered together by the centralized scheduling management unit. The input/output ports with less traffic are selected to complete the optoelectronic service adaptation, by the highest throughput rate. With this adaptation method, the utilization rate of the hybrid switching resource is improved.

In summary, the centralized scheduling algorithm has certain advantages over the existing scheduling methods. The algorithm can realize integrated scheduling of hybrid switches with fewer parameters and improve the performance of data loss rate in switching units. Therefore, it is suitable for use in on-board hybrid switching systems.

5. Simulation comparison

This section verifies the performance of the centralized scheduling algorithm in terms of data loss rate by constructing a simulation model, and compares the data loss rate of the traditional scheduling method and the centralized scheduling algorithm under different service burst lengths.

5.1 Simulation model

In the simulation, the ON/OFF model is used to simulate the arrival process of data. According to the characteristics of multi-beam onboard switching, each beam should correspond to one port of the switch. This section chooses to simulate and compare the 8-port switch.

In the ON/OFF model, it is assumed that the burst state length and the quiescent state length have geometrical distributions of the parameters Pb and Pe . Therefore, the probability distribution function of the ON state can be obtained, that is, the probability that a burst lasts for x time slot is:

$$p(x) = Pb \times (1 - Pb)^{x-1}, x = 1,2,3, \ . \ . \ . \tag{12}$$

The probability distribution function of the OFF state, that is, the probability of a stationary continuous x time slot is:

$$p(x) = Pe \times (1 - Pe)^{x}, x = 0,1,2, \ . \ . \ . \tag{13}$$

Here x can be 0, indicating that one burst can follow another burst without a stationary period, that is equivalent to two different stream multiplexing, and the two bursts have different destination ports.

In the actual simulation, the parameters required by the service source are the average burst length M and the traffic volume L, and it can be calculated that

$$Pb = 1/M, \quad Pe = L/(M - M \times L) \tag{14}$$

In the simulation, the input ports are independent, so that it is possible to generate multiple services of multiple input ports simultaneously to one output port, and some output ports are idle.

5.2 Simulation result

In this section, 100000 time slots are simulated. The data loss rates under the two cases of centralized scheduling algorithm and existing scheduling algorithm are simulated. Figure 4 compares the data loss rates of two algorithms under different loads and different buffer capacities.

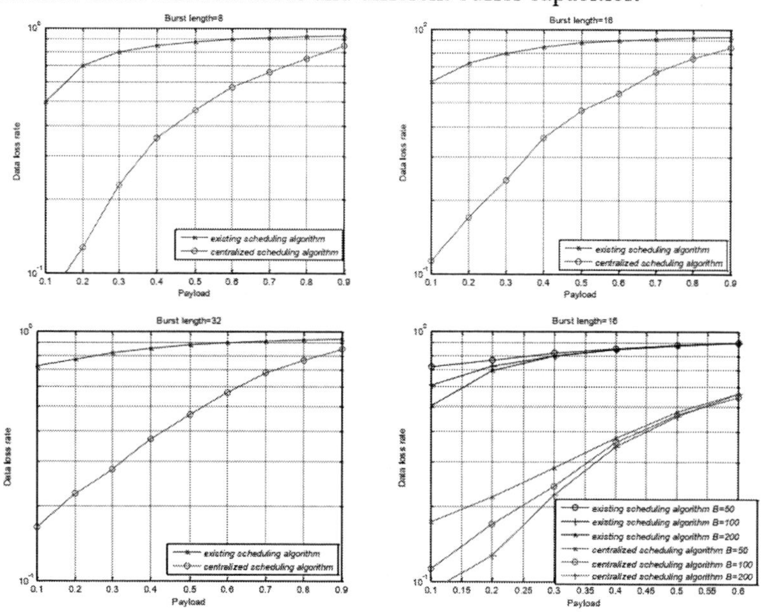

Figure 4. Compare of cell loss ratio between the two scheduling algorithms

The simulation results show that the centralized scheduling algorithm proposed in this paper can significantly reduce the data loss rate of onboard hybrid switches. This is because the existing scheduling algorithm separately manages the optical service and the electrical service, and multiple services are scheduled to the same output port at the same time. Under the existing algorithm, the switching unit is congested and a large amount of data is lost. On the other hand, the centralized scheduling algorithm uses the group decision-making idea to schedule all services, which largely avoids the conflict of output ports, thus effectively reducing the data loss rate.

6. Conclusion

This paper proposes a centralized scheduling algorithm in on-board hybrid switch. According to the different characteristics of optoelectronic services, combined with the utilization of hybrid switching resources, the algorithm centrally dispatches optical and electrical services, realizes the aggregation and distribution among multi-granularity data ports, and reduces the internal congestion probability of on-board hybrid switch modules. The simulation results show that the algorithm can effectively reduce the data loss rate and improve the performance of hybrid switch. At the same time, the implementation of the algorithm requires fewer parameters and the implementation steps are simple. The low complexity of the proposed algorithm can meet the needs of spatial information network, which used laser and microwave links to carry multi-granularity service and achieve data aggregation and distribution function. Therefore, this algorithm can be used in on-board environments.

Acknowledgment

This work is supported by the Supported Foundation by National Key Lab. Of Science and Technology on Space microwave (2018SSFNKLSMT-12).

IOP Conf. Series: Materials Science and Engineering **685** (2019) 012028 doi:10.1088/1757-899X/685/1/012028

References

[1] Liu Kai, Yan Jian, GaoXiaolin,etc."Fully Distributed Fault Tolerant Scheduling for Onboard Clos-network Switching," Journal of Electronics & Information Technology, 2016,38(6), pp.1377-1384.

[2] Xiaoting WANG. "Novel High Performance Scheduling Algorithms for Crosspoint Buffered Crossbar Switches,"IEICE TRANSACTIONS on Information and Systems, 2015, E98-D(12), pp.2091-2104.

[3] JihwanP.Choi, Seok-Ho Chang, Vincent W.S. Chan."Cross-Layer Routing and Scheduling for Onboard Processing Satellites with Phased Array Antenna," IEEE transactions on wireless communications, 2017,16(1), pp. 180-191.

[4] Narayanan Prasanth, KannanBalasubramanian."Performance analysis of buffered crossbar switch scheduling algorithms," International Journal of Information and Computer Security, 2015, 7(1), pp.49-63.

[5] Zhang Yi, Zhou Quan, Li Jun, etc. "The Simulation and Realization of Input-buffer Scheduling Algorithm in Satellite Switching System,"Space Electronic Technology, 2015, 12(1), pp.97-103.

[6] Han Yong, Yao Jianmin, CaiShaobin, etc. "Packet Forwarding Scheduling Algorithm FC-WFQ for iSCSI Virtualization Switch,"Acta Electronica Sinica, 2013,41(3), pp.587-592.

[7] Zhou Zhiqiang, XuZhanqi, Zhang Xiaolei, etc. "Survey on high performance switching architecture,"Application Research of Computers, 2015,32(4), pp.961-966.

[8] Zhen Zheng, LufengQiao,QinghuaChen,etc."A concurrent round-robin-based variable-length packets dispatching scheme for satellite onboard clos-network switches,"2015 8th International Congress on Image and Signal Processing(CISP 2015), 2015,pp.1510-1514.

[9] Yang Fan, XuZhanqi, Li Danwu,etc. "Multicast scheduling algorithm with a dynamic weight for the input buffered Crossbar,"Journal of Xidian University(Natural Science), 2016, 43(6), pp.80-85.

[10] Liu, L.,Zhang, Z.,Yang, Y."In-Order Packet Scheduling in Optical Switch With Wavelength Division Multiplexing and Electronic Buffer ,"IEEE Transactions on communication, 2014, 62(6), pp.1983-1994.

IOP Conf. Series: Materials Science and Engineering **685** (2019) 012029 doi:10.1088/1757-899X/685/1/012029

Analysis of IGS Data Based on Military Action Prediction

Qiang Fu[1,2,3], Guowei Shi[1,2], Wentao Fu[2,3*], Yuanfa Ji[1,2,3]and Xiyan Sun[2,3]

[1] College of Information and Communication, Guilin University of Electronic Technology, Guilin 541004, China

[2] Guangxi Key Laboratory of Precision Navigation Technology and Application, Guilin University of Electronic Technology, Guilin, 541004, China

[3] National & Local Joint Engineering Research Center of Satellite Navigation and Location Service,Guilin University of Electronic Technology, Guilin 541004, China

[*]fuwentaoaa@163.com

Abstract. In order to verify that the analysis of satellite navigation signal observation data can be used as a means of predicting military operations, this paper combines the recent Iranian shooting down of UAV event, using RTKLIB software to analyze the parameter of visible satellite numbers, SNR(signal-to-noise ratio), and accuracy factor, multipath error, positioning accuracy and other indicators in order to avoid errors in the analysis results of a single type of receiver data analysis, further IGS in the vicinity of Iran and around the world according to the type of receiver, antenna type, and firmware version used. Observe the data for analysis. The analysis shows that the signal power of the GPS L2 frequency is enhanced in the United States from 15:00 on June 20, 2019 to 9:00 on June 21, 2019. The enhancement range is global. The analysis system of this paper can as an effective means of accurately predicting and judging military operations.

1. Introduction

At about 9:00 am on June 20, 2019, Iran, a US "Global Hawk" drone was shot down at the entrance to the Strait of Hormuz. According to the media "New York Times" reported on June 20, the US President On the same day, Rump approved the air strike Iran plan, which is scheduled to be launched at 7:00 pm on the 20th, and at 4 am on the 21st, Iran time. Combined with the US military attack on Syria on April 14, 2018 (the local time in Syria), the GPS L2 frequency signal was boosted[1,2] this paper, IGS uses the continuous observation data of IGS in the tracking and monitoring stations around Iran to analyze the GPS signals in five aspects: visible satellite number, accuracy factor, multipath error, SNR, and positioning accuracy, and further according to the received receiver. The global IGS monitoring stations are classified according to the model, antenna type and firmware version, and the observation data of some monitoring stations are analyzed.

2. Introduction to basic software and data sources

The analysis software used in this paper is RTKLIB, and it is also used by Matlab for auxiliary analysis. The data source is the real-time monitoring data of IGS monitoring station. The following introduces RTKLIB and IGS.

2.1. RTKLIB software introduction

RTKLIB is an open source package developed by Takashima I of Tokyo Ocean University, Japan. It consists of a portable library and multiple application tool libraries [3], its main functions include: support for standard and precision positioning algorithms for multiple GNSS systems, support for multiple GNSS standard formats and protocols, support for multiple GNSS real-time and post-processing positioning modes, support for multiple GNSS receiver-specific data protocol formats, and Support a variety of external communication methods.

2.2. IGS introduction

International GNSS Service (IGS) is a permanent service organization established by the International Association of Geodesy in 1992. It consists of more than 400 observatories around the world which are composed of global network, regional network and local network [4].

The GPS data of the global network consists of a daily measurement and navigation data file generated by the receiver-independent exchange format (RENIX), which is compressed by UNIX and transmitted to the corresponding data center. The naming convention for RENIX observation data files is: ssssddd.yyt. ssss is 4 characters indicating the name of the monitoring station. ddd is the annual product date of the first data record and f is the serial number of the file generated during the UTC time of day. ("0" means that the file contains all the data of the day) yy is the year represented by a 2-digit integer, t represents the type of the file, O represents the original observation, N represents the ephemeris, M represents the meteorological data, and G represents the GLONASS ephemeris, H represents the navigation message of the synchronous satellite GPS payload [5].

3. Data processing and analysis of monitoring stations in and around Iran

After viewing the official website of IGS, there are 4 IGS monitoring stations in and around Iran, namely hamd, tehn, isba, bhr4. Since the hamd monitoring station does not have data on GPS L2 frequencies, it is not analyzed. This paper first downloads the data of four monitoring stations from the CDDIS server. These monitoring stations are the tehn monitoring station in Iran and the isba and bhr4 monitoring stations around Iran. The UTC time of the downloaded data is June 20, 2019 and June 21, 2019 (Day of Year (DOY) is 171, 172).

3.1. Visible satellite number and dilution of precision

The satellite can be seen as the most fundamental condition for GPS receivers. When the number of visible satellites received by the receiver is reduced, less than 4, it will not meet the minimum requirements for positioning [6]. In satellite navigation, the concept of dilution of precision (DOP) is generally used to represent the amplification factor of the error, which is independent of the strength of the signal, the positioning algorithm used by the receiver, and the positioning performance.

Processing data of three monitoring stations, tehn, isba and bhr4, for two days (DOY 171, 172), the number of satellites and the value of DOP are shown in Figure 1 to 3.

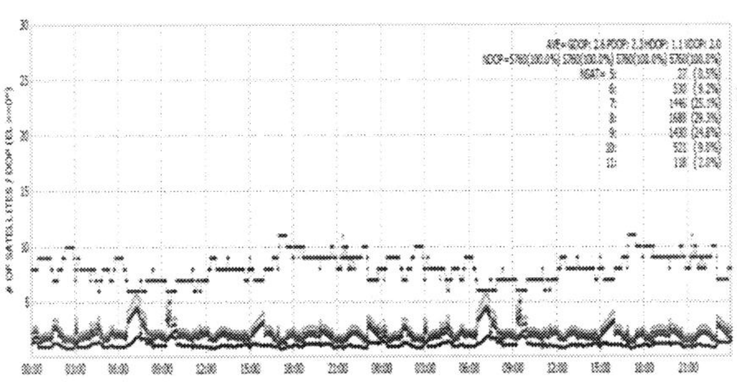

Figure 1. Two-day GPS visible satellite number and DOP value at tehn monitoring station.

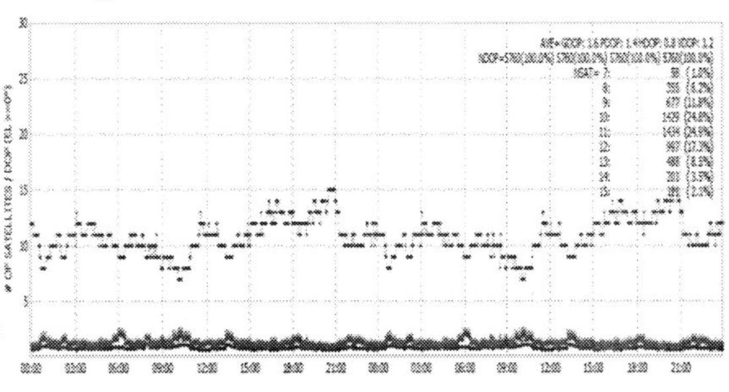

Figure 2. Two-day GPS visible satellite number and DOP value at isba monitoring station.

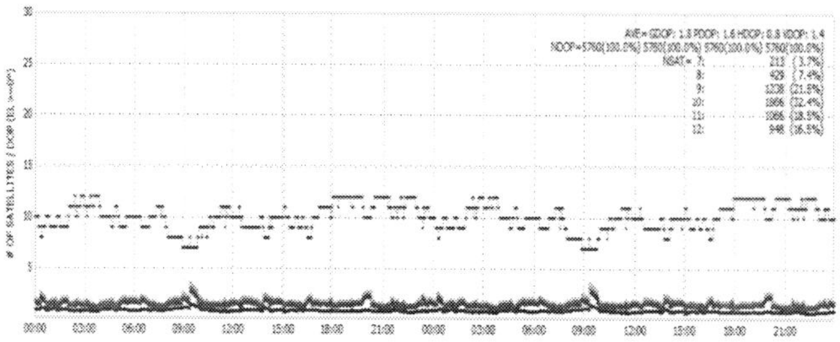

Figure 3. Two-day GPS visible satellite number and DOP value at bhr4 monitoring station.

As can be seen from Figure 1 to 3, the number of satellites observed at the tehn monitoring station in Iran is about 8 or so, the average values of GDOP, PDOP, HDOP, and VDOP are 2.6, 2.3, 1.1, and 2.0 respectively. The number of satellites observed by the isba and bhr4 monitoring stations is about 10, and the respective DOP values are below 2.0. There is no obvious abnormality in the observed data.

3.2. Multipath error

Since the GPS signal passes through various reflectors, such as the ionosphere, mountains, trees, buildings, etc. From the satellite transmitting end to the receiving end of the GPS receiver antenna, the direct wave and the reflected wave exist simultaneously, which causes multipath effect. Multipath effect is an important source of error in GPS measurement and an important factor affecting positioning accuracy [7].

Processing data of three monitoring stations, tehn, isba and bhr4, for two days (DOY 171, 172), The pseudo-code multipath effect error of the GPS L2 frequency signal is shown in Figure 4 to 6.

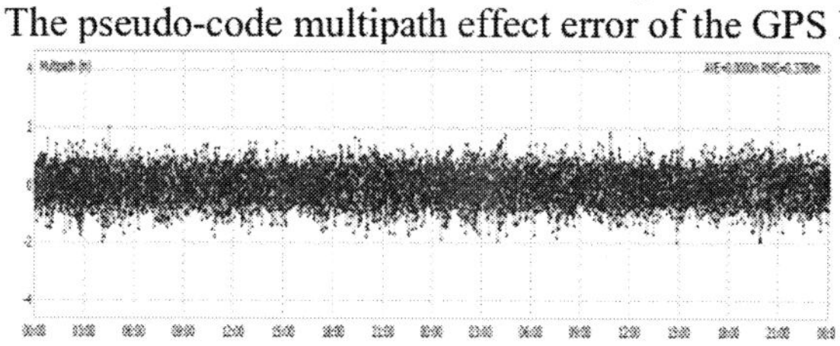

Figure 4. Two-day pseudorange multipath at tehn monitoring station.

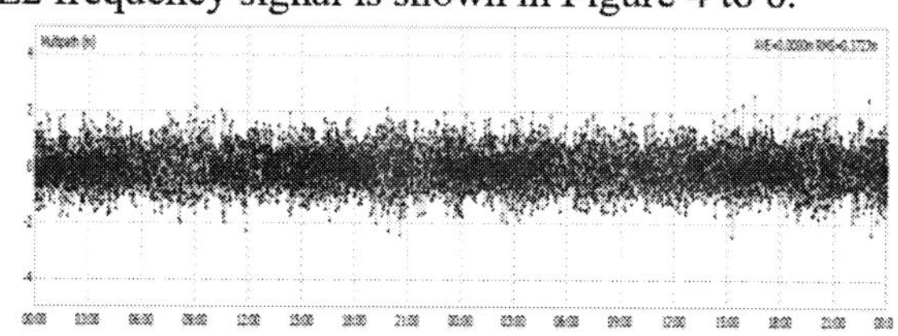

Figure 5. Two-day pseudorange multipath at isba monitoring station.

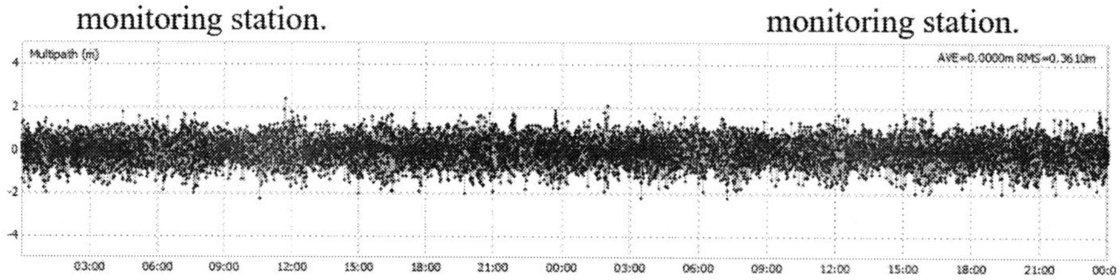

Figure 6. Two-day pseudorange multipath at bhr4 monitoring station.

It can be seen from Figure 4 to 6 that the multipath errors of the three monitoring stations did not change significantly during the two days, and the overall root mean square was around 0.3.

3.3. SNR

The SNR is the ratio of the carrier signal strength to the noise intensity. The larger the value, the stronger the received signal power [8]. Processing data of three monitoring stations, tehn, isba and bhr4, for two days (DOY 171, 172), The SNR variation of GPS L2 signal is shown in Figure 7 to 9.

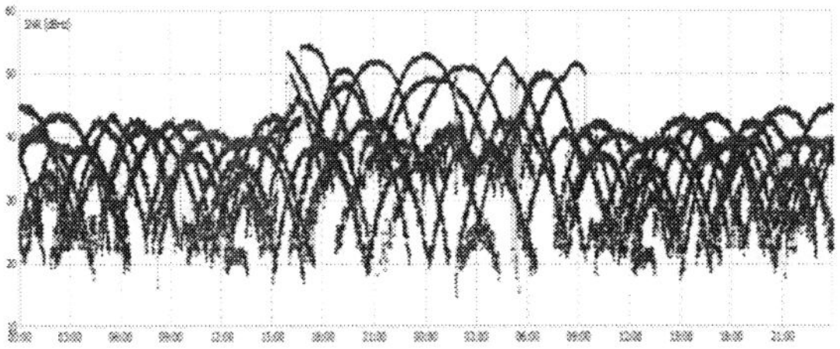

Figure 7. Change of SNR of GPS L2 frequency point signal in tehn monitoring station in two days.

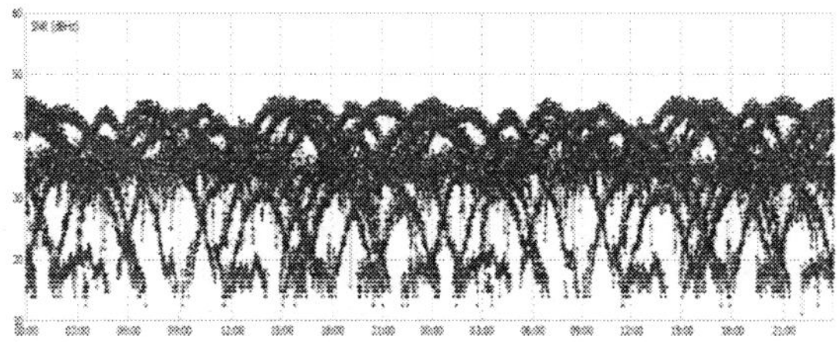

Figure 8. Change of SNR of GPS L2 frequency point signal in isba monitoring station in two days.

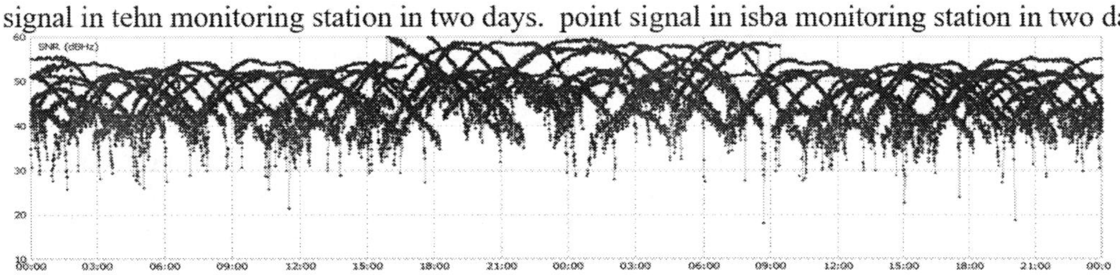

Figure 9. Change of SNR of GPS L2 frequency point signal in bhr4 monitoring station in two days.

It can be seen from Figure 7 to 9 that the signal power of the GPS L2 frequency of the tehn monitoring station and the bhr4 monitoring station is significantly enhanced from 15:00 pm DOY 171 to 9:00 am DOY 172, while that at isba monitoring station did not.

3.4. Positioning accuracy

Positioning accuracy is an important indicator in satellite navigation and positioning systems. The results of single-point positioning of the observed data of three monitoring stations, tehn, isba and bhr4, on DOY 171 and 172 days are shown in Figure 10 to 12.

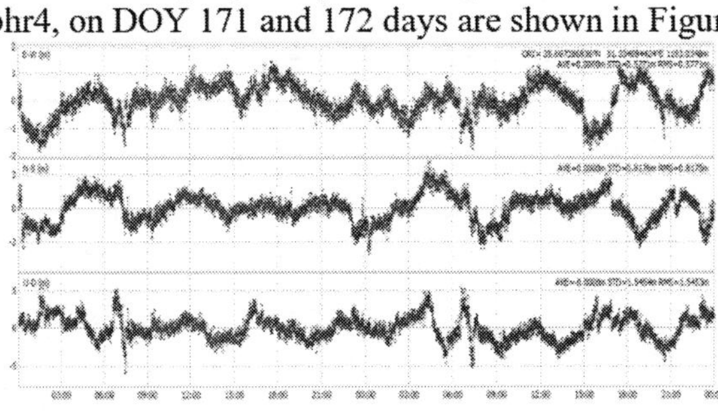

Figure 10. The result of single point positioning of the tehn monitoring station.

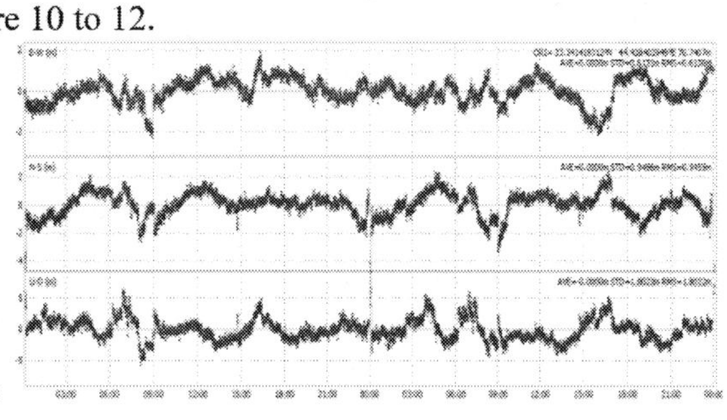

Figure 11. The result of single point positioning of the isba monitoring station.

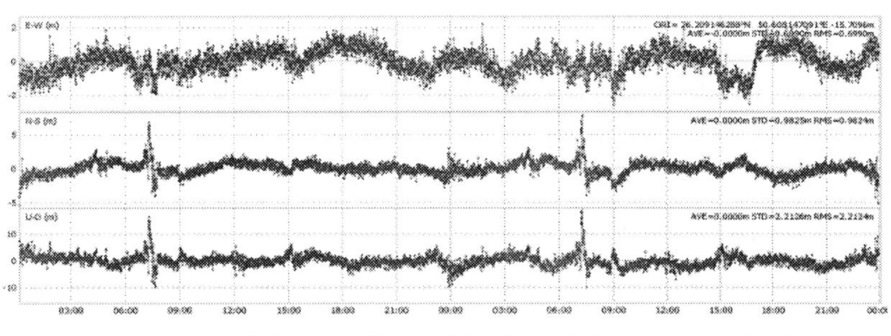

Figure 12. The result of single point positioning of the bhr4 monitoring station.

It can be seen from Figure 10 to 12 that the standard deviation and root mean square of the three monitoring stations in the east and north directions are below 1 m, the standard deviation and root mean square of the vertical direction of the tehn station and the station site are below 2m, the standard deviation and root mean square of the vertical direction of the bhr4 station are 2.2126m and 2.2124m, respectively. Overall, the positioning errors of the three stations are in line with the accuracy requirements of single point positioning, From the local details, there is no significant improvement in the positioning accuracy of the three stations from 15:00 pm DOY 171 to 9:00 am DOY 172.

4. SNR analysis of other monitoring stations in the world

There are many factors affecting the SNR of the receiver. For example, receivers produced by different manufacturers, different firmware versions in the same model receiver, different antenna types, etc., will affect the SNR measured by the receiver. In order to further analyze whether GPS has enhanced the power of L2 signal, and whether the enhancement is regional or global, in this paper, more than 400 monitoring stations on the CDDIS server are classified according to the receiver model, receiver firmware version and antenna type. The monitoring data of some receivers in DOY 171 and 172 are further analyzed.

First, classify the receiver using the model TRIMBLE NETR9 in the same way as the tehn monitoring station, Then randomly select the tabl monitoring station in Letwood, USA, which uses different firmware versions from the tehn monitoring station but the antenna type is the same, for data analysis, the SNR results obtained by the analysis are shown in Figure. 13. Since there is no monitoring station with the same firmware version as the tehn monitoring station after classification, the str2 monitoring station in Canberra, Australia, which is different from the tehn monitoring station firmware version and antenna type, is randomly selected for analysis, the SNR results obtained by the analysis are shown in Figure 13.

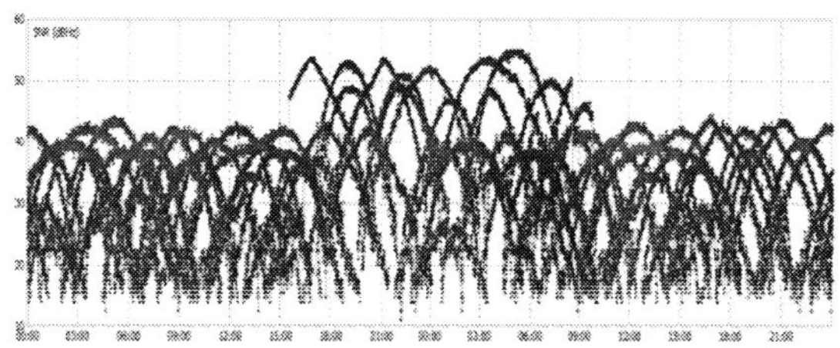

Figure 13. Change of SNR of GPS L2 frequency point signal in tabl monitoring station in two days.

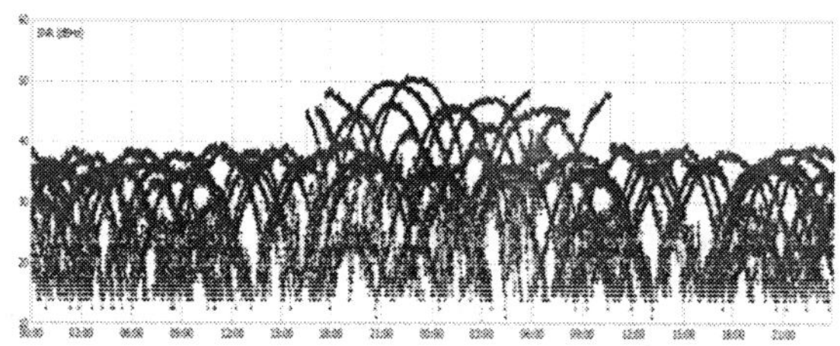

Figure 14. Change of SNR of GPS L2 frequency point signal in str2 monitoring station in two days.

It can be seen from Figure 13 and 14 that there are obvious power boost periods for both monitoring stations. The power enhancement time is from DOH 171 15:00 pm to DOY 172 9:00 am.

Then, according to the same as the isba monitoring station, the model is TRIMBLE NETR5. Since all the stations use different firmware versions and antenna types than the isba monitoring station, therefore, the data of the aspa monitoring station in Pago Pago, USA and the bjco monitoring station in Cotonou, Benin, were randomly selected for analysis, the SNR of the two monitoring stations are shown in Figure 15 and 16, respectively.

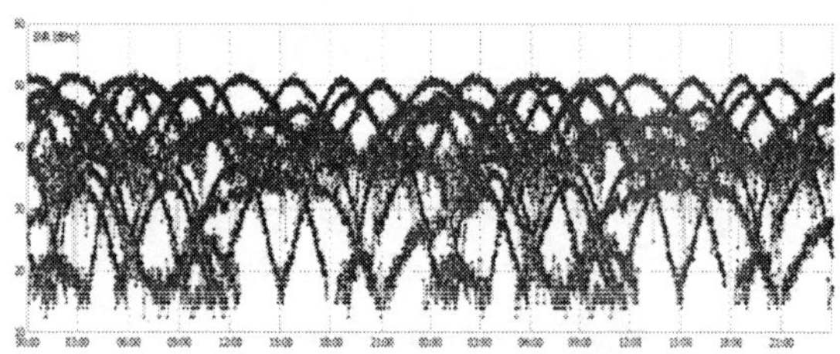

Figure 15. Change of SNR of GPS L2 frequency point signal in aspa monitoring station in two days.

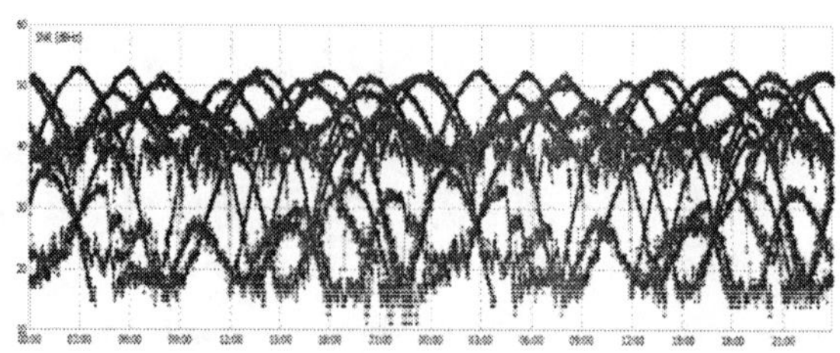

Figure 16. Change of SNR of GPS L2 frequency point signal in bjco monitoring station in two days.

IOP Conf. Series: Materials Science and Engineering **685** (2019) 012029 doi:10.1088/1757-899X/685/1/012029

It can be seen from Figure 15 and Figure 16 that neither of the two monitoring stations has a period of significant power enhancement.

Next, according to the same type of receiver as the ITT 3750300, the same as the brt4 monitoring station. Since all the stations after the classification use the same firmware version and the same antenna type, so randomly select the data from the oka1 monitoring station in Sachem, England for analysis.

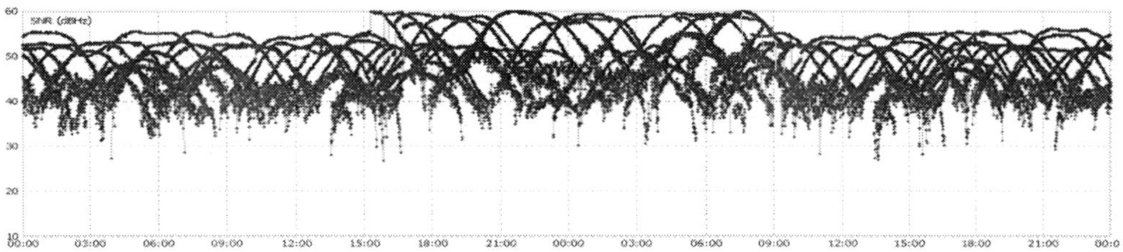

Figure 17. Change of SNR of GPS L2 frequency point signal in oka1 monitoring station in two days.

It can be seen from Figure 17 that significant power enhancement occurs between 15:00 PM at DOY 171 and around 10:00 AM at DOY 172.

Finally, the receiver models used in the above monitoring stations are selected differently, and the data of the monitoring stations using JAVAD receivers are used for further analysis. Firstly, the data of two monitoring stations with firmware versions of 3.7.6 but different antenna types were selected for analysis, the types of antennas used by the two stations are JAVRINGANT_DMand TRM29659.00, respectively, bogt monitoring station in Goboda, Colombia, and RABT monitoring station in Rabat, Morocco, the SNR of the two monitoring stations are shown in Figure 18,19, respectively. Then select the data analysis of the met3 monitoring station in Kilconum, Finland, whose antenna type is also JAVRINGANT_DM but firmware version 3.7.5. the SNR results obtained by the analysis are shown in Figure 20.

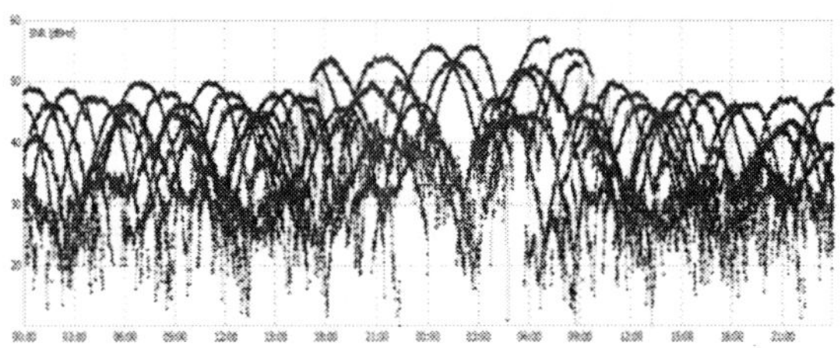

Figure 18. Change of SNR of GPS L2 frequency point signal in bogt monitoring station in two days.

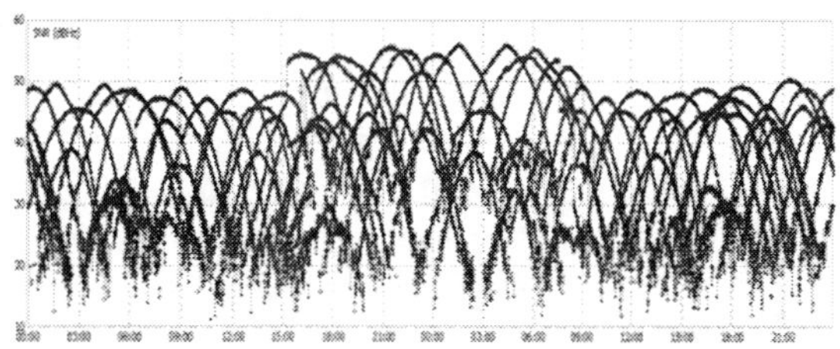

Figure 19. Change of SNR of GPS L2 frequency point signal in tabt monitoring station in two days.

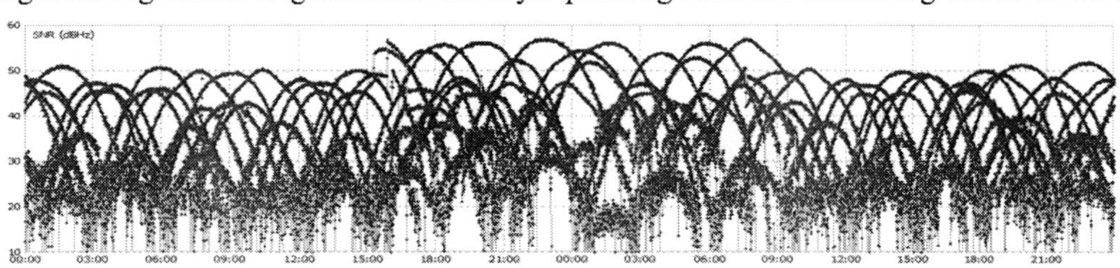

Figure 20. Change of SNR of GPS L2 frequency point signal in met3 monitoring station in two days.

As can be seen from Figure 18 to 20, power boosting occurs in receivers that also use JAVAD receivers but have different firmware versions and different antenna types, the power boost period is from DOY 171 15:00 pm to DOY 172 10:00 am.

As can be seen from the analysis in Figure 13 to 20, in addition to the fact that several monitoring stations using the TRIMBLE NETR5 receiver have not monitored the power boost, the remaining stations around the world have monitored the power boost, since the stations with no power boost monitoring are all using the TRIMBLE NETR5 receiver, the reason may be that the TRIMBLE NETR5 receiver uses automatic gain control. Therefore, the above analysis shows that the power enhancement range of GPS L2 frequency signal is global.

5. Conclusion

This article uses RTKLIB software to analyze two days of data from some IGS stations around the world, these two days are June 20, 2019 and June 21, 2019 (DOY 171, 172). From the above analysis, it can be seen that the power of the signal of the GPS L2 frequency has increased from 15:00 on the 20th to 9:00 on the morning of the 21st, the visible satellite number, multipath effect, DOP value, and positioning error are all within the normal range, and there is no obvious abnormality. According to the analysis of this real case, it can be seen from the media report that the time when President Trump approved the military strike against Iran is located within the power enhancement period, combined with the US military attack on Syria on April 14, 2018 (Syrian local time), the signal of the GPS L2 frequency is also enhanced, it shows that the evaluation and analysis of real-time data of IGS can be used as an important and effective means to predict and judge military operations.

Acknowledgments

This work was supported by the National Key R&D Program of China (2018YFB0505103), the National Natural Science Foundation of China (61561016, 61861008),also the funding of Science and Technology Major Project of Guangxi (AC16380014, AA17202048, AA17202033), the Natural Science Foundation of Guangxi(2018JJA170090). This work was also funded by Guilin Science and Technology Bureau Project (20170216), and the Innovation Project of GUET Graduate Education (2018YJCX19).

References

[1] ZHANG K, ZENG F L, OUYANG X F and ZHAO Y, 2019, Analysis of GPS Interference Effect Based on RTKLIB and IGS Real-time Data, *Journal of Projectiles,Rockets,Missiles and Guidance*, 1-9

[2] LIU M M, JIAO W H and JIA X L, Analysis of GPS Positioning in W artime Syria, *Geomatics Science and Engineering*, 2018, 4:19-25

[3] GITHUB:An Open Source Program Package for GNSS Positioning[EB/OL].[2015-03-28]. http://www.rtklib.com/rtklib.htm

[4] MA Y, YUAN W L, XU Y B and SUN H B, Accuracy of GPS Satellite Orbits and Its Effects on Single Point Positioning, *Geography and Geo-Information Science*, 2013, **29**(06):26-29+129

[5] Wener Gurtner, 2007, RINEX: The Receiver Independent Exchange Format Version 2.10. Astronomical Institute, University of Berne

[6] LIU Z W and JIANG P, Analysis on multi-GNSS satellites visibility and global temporal-spatial variation, *Engineering of Surveying and Mapping*, 2018, **27**(09):26-29+34

[7] MENG W and YANG K W, The Analyze on Multipath Effect Influence GPS Measure Precision, *Digital Communication World*, 2018(02):7+12.

[8] LI X G, CHENG P F, CHENG J Y and WAN J, Quality Analysis of GNSS Data, *Bulletin of Surveying and Mapping*, 2017(03):1-4+8.

IOP Conf. Series: Materials Science and Engineering **685** (2019) 012030 doi:10.1088/1757-899X/685/1/012030

Application of Deep Convolutional Neural Network in Remote Protection of Transmission Lines

Wenjie Zheng[1,*], Chengqi Li[1], Bo Yang[2] and Xuebin Lv[2]

[1]State Grid Shandong Electric Power Research Institute, Jinan, 250003, China
[2]State Grid Shandong Electric Power Company, Jinan, 250001, China

[*]zhwjchww@163.com

Abstract. In order to alleviate the pressure of line operation and maintenance, it is necessary to use advanced image recognition methods to identify hidden line hazards. In this paper, we extract features from hidden targets and perform target recognition using deep convolutional neural network. Firstly, we collect a large number of photos with hidden dangers for transmission line. Secondly, we classify those samples according to the detection targets. Thirdly, the deep convolutional neural network model is trained and tuned by using the sample images of various detection objects. When deployed on the server, identification of hidden dangers such as cranes, tower cranes, excavators, and mountain fires reaches acceptable accuracy.

1. Introduction

Destruction caused by hidden dangers including construction machines, foreign objects on transmission lines, mountain fires and smoke are the main causes for line trips. In order to improve the capability to protect overhead transmission lines, an intelligent identification and early warning system for hidden dangers of transmission lines is proposed. To further reduce the workload of artificial identification of hidden dangers, we try to identify the types of hidden dangers automatically. The traditional target detection method is mainly divided into four steps: pre-processing, feature selection, target recognition and target location, whose research focuses on feature extraction and feature classification. Therefore, researchers have proposed various forms of features and classifiers.

In order to extract better features, Hinton use deep neural networks to automatically learn high-level features from large amounts of data in 2006 [1]. Compared with pre-designed features, learning features are more abundant and expressive. OverFeat mainly uses a multi-scale sliding window for classification, positioning and detection [2]. However, OverFeat uses a sliding window to generate candidate regions, which is essentially exhaustive and has a high time complexity. In 2014, Girshick et al. proposed the R-CNN model [3], which introduced a search algorithm instead of the traditional sliding window idea for target location. However, the process of R-CNN training model is very complicated, and the detection speed is very slow, which is a method that is not dominant in time and space. SPP-Net designed a spatial pyramid pooling layer in convolutional neural networks [4], which can extract feature vectors of the same length from different size feature maps. However, the fine-tuning weights in SPP-NET only update the fully-connected layer behind the network, and the previous convolutional layer is not updated. It is not advisable for deeper networks to adopt this method. Girshick proposed the Fast R-CNN model based on R-CNN and SPP-NET [5]. Compared with the R-CNN model, the feature is extracted for each candidate region. The Fast R-CNN only extracts feature once for the detected image. The target detection and recognition accuracy are already high, but the speed of the model detection is affected by the process of extracting the candidate region. YOLO integrates target area prediction and target category prediction into a single neural network model to achieve rapid target detection and recognition with high accuracy [6]. However, each grid of YOLO only predicts two boxes, one category, which causes the model to predict the accuracy of adjacent targets. YOLO can quickly identify targets in an image, but accurately targeting a target is difficult. SSD takes into account both the accuracy and real-time requirements [7], but the shape of the Default Box and the size of the grid in the SSD are fixed in advance, resulting in the model not being good enough for the extraction of small targets in a particular picture. Ren et al. proposed the Faster R-CNN model [8]. Since RPN and Fast R-CNN share the feature extraction of the convolutional neural network, only one feature of the image to be detected is extracted, which speeds up the target detection.

This paper adopts deep convolutional neural network-based scheme to extract features and target recognition from hidden targets. On the basis of the accumulated image of a large number of transmission line hidden danger samples, the samples are classified according to the detection target, and the deep convolutional neural network model is trained and tuned by using the sample images of various detection objects, and the model is deployed in the system to achieve accurate identification of hidden dangers such as cranes, tower cranes, excavators, and mountain fires.

2. Labelling hidden danger targets and establishing sample library

In order to meet the normative nature of the hidden danger target label, this project requires the marking file recording format to be in accordance with the .xml conforming to the VOC2007 data set format. The marked content is a hidden target in transmission line corridors that is easy to cause failure of the transmission line, which are sown in figure 1.

Figure 1. Sample Instance.

In order to improve the generalization ability of the model and avoid over-fitting, a large amount of training data needs to be marked. In this system, data is marked in a semi-automatic manner. That is, in the initial stage of system establishment, firstly, a batch of data is manually marked, and then the data is used to train the detection model, then the unlabelled picture is detected by the trained detection model to obtain the annotation information, finally a part of the mislabelled image is manually assisted. In the later stage, the result of the model detection can be directly used as the annotation information to realize the automatic labelling of data, which can effectively expand the transmission hidden danger sample image database.

In this project, according to the above marking requirements, after the sample is sorted and manually labelled, a training sample library is formed. The number of hidden target targets is shown in the following table 1.

Table 1. Labelled hidden danger number.

Primary classification	Secondary classification	Quantity of sample photos	Quantity of hidden dangers
Crane	Crane	45000	69000
Tower crane	Tower crane	36000	43500
Construction machinery	Bulldozer	31500	37500
	Excavator	27000	33000
Foreign objects on transmission line	Kite	300	300
	Plastic film	345	345
	Dustproof net	285	285
Mountain fire	Mountain fire	18000	21000
Smoke	Smoke	30000	39000

3. Identification of hidden dangers for transmission lines

It has been determined above that the target detection in the transmission application scenario uses the Faster R-CNN recognition algorithm, and the CNN model needs to be trained using the labelled samples.

3.1. Data preprocessing

There are many changes in the picture of the transmission hidden danger samples. Since photos taken by cameras installed on transmission towers have a large time span and complicated weather conditions, it is necessary to perform pre-processing on those photos in order to obtain a better recognition effect. The main operations include mean removal, normalization, and data enhancement.

1) Mean removal. The average of all training set images is subtracted from each image, after which the data of each dimension of the input data is cantered to 0, by which we can enhance the robustness of image recognition to illumination, weather changes, etc.

2) Normalization. After mean removal, the data in each dimension is divided by the standard deviation of the data in this dimension to ensure that all data is normalized to between -1 and 1. Normalization ensures that the data in each dimension is in a range of changes.

3) Data enhancement. Among the specified types of hidden dangers, excavators and cranes are easily misjudged as trucks, buses and so on, but those misjudgements can be eliminated by increase data diversity using other car databases. In addition, each picture in the training set is shifted by 5 pixels in the up, down, left, and right directions, and then all the pictures are horizontally flipped to obtain an expanded training set.

3.2. Training of Faster R-CNN model

Faster R-CNN combines the RPN network and the Fast R-CNN network, that is, the candidate area acquired by the RPN is directly connected to the ROI pooling layer. Target identification of hidden dangers using deep neural networks consists of two main phases including training and testing.

In the training phase, the hidden danger target appearing in the image is first marked, and the sample database is established. Then the error back propagation (BP) algorithm is used to train the network, and the network weight coefficient is updated with random gradient descent method, this process iterates many times until the model converges.

In the test phase, the convolutional neural network is used to learn the advanced features in the image, and then the RPN deep neural network is used to generate candidate regions for hidden targets such as cranes, tower cranes, excavators and mountain fires. Finally, SoftMax classification and Bbox regression are performed, giving the type and location information of each hidden target.

The specific steps of the loss function design, the Faster R-CNN training process and the BP training algorithm are described in detail below.

1) Loss function. RPN consists of two branches including a classification layer and a regression layer. The classification layer gives a two-category label, which means whether the area contains the target. The regression layer produces the position of the target. Current area is considered to contain the targets when its IOU with some ground-truth box is greater than 0.7, and it is considered to be a negative sample when its IOU with any ground-truth box is less than 0.3. The loss function is expressed as equation:

$$L(\{p_i\}\{t_i\}) = \frac{1}{N_{cls}}\sum_i L_{cls}(p_i, p_i^*) + \lambda \frac{1}{N_{reg}}\sum_i p_i^* L_{reg}(t_i, t_i^*) \quad (1)$$

Among them, N_{cls} and N_{reg} weighs classification and regression. Classification loss is two classification, while regression losses is non-category related. RPN is a full-convolution process, full connection of subsequent vectors is achieved by 1*1 convolution.

2) Back propagation training. The BP algorithm is the basic algorithm of neural network optimization, whose learning process includes two stages of forward propagation and back propagation.

3) Faster R-CNN training process. Training of faster R-CNN continues based on trained CNN models such as VGG16 and Resnet152, and the actual training process is shown in figure 2.

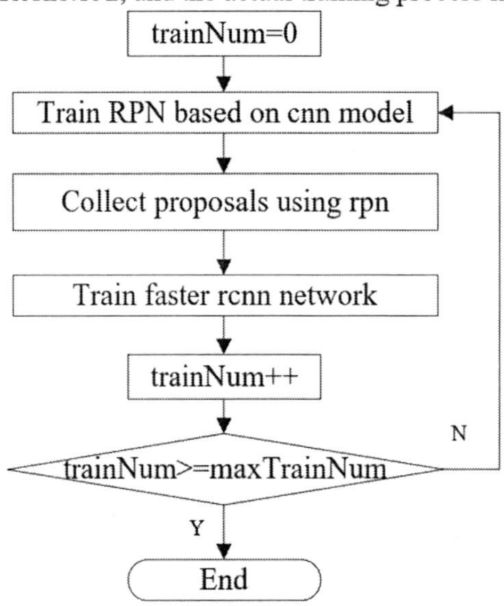

Figure 2. Training process of Faster R-CNN.

The training process is similar to an iterative process and iterated twice here.

4. Experiments

Since TensorFlow is a clear and efficient open source deep learning framework that can be used for computing on a variety of platforms, this paper uses TensorFlow for training.

In the actual training process, the verification set are generally used to check the effect of the current saved model after training N epoch ends, so that the model or parameter problem can be discovered in time. For example, when the model encounters divergence on the verification set, mean Average Precision does not grow, or grows slowly, it is necessary to terminate the training in time, re-adjust or adjust the model without waiting for the training to end. When the effect on the verification set is much worse than that on the training set, it is necessary to consider whether the model is over-fitting, and it is necessary to adjust the hyper-parameter in time to ensure the final training effect of the model, which is shown in table 2.

Table 2. Values of hyperparameters after adjusted.

Hyperparameters	Description	Value
Learning rate	Learning rate determines the speed of weight update	0.01
Weight decay	Influence on the loss function caused by adjusting model complexity	0.05
Iterations	The number of times the entire training set is input to the neural network for training	400000
Learning rate decay	Reduced learning rate every n iteration	0.1 times
Weight_initialization	Initialize the weight of each network layer	Evenly distributed

In order to test the performance of the trained final model, it is necessary to test the trained model using the test set. The test set is completely independent of the training set, and there is no intersection between the two, and the test set data types and quantities are shown in table 3.

Table 3. Tag number of hidden dangers in test set.

Hidden danger	Crane	Tower crane	Excavator	Bulldozer	Kite & Plastic film & Dustproof net	Smoke	Fire
Tag number	1626	864	1173	275	800	756	315

The test results of various hidden dangers are shown in table 4.

Table 4. Missed rate and false rate.

Target	Crane	Tower crane	Excavator	Bulldozer	Kite & Plastic film & Dustproof net	Smoke	Fire
Missed rate	6.31%	7.91%	7.82%	9.23%	12.45%	4.37%	4.61%
False rate	9.93%	5.14%	9.78%	9.67%	14.20%	10.32%	9.61%

The test results in the same scenario under the test set are shown in figure 3.

Figure 3. Detection results at different times in the same scene.

It can be seen from figure 3 that the algorithm can obtain better hidden danger target detection results at different times in the same scene, indicating that the algorithm is robust to complex conditions such as illumination changes and weather changes.

The test results in various scenario under the test set are shown in figure 4.

Figure 4. Detection results in various scenes.

It can be seen from figure 4 that the algorithm can obtain good detection results in different scenarios, indicating that the system has good generalization ability.

5. Conclusions

In order to take full advantage of artificial intelligence technology and achieve better protection for transmission lines, this paper uses Faster R-CNN to extract features from hidden targets and realize target recognition. On the basis of collecting a large number of hidden dangers on transmission lines, the samples are classified according to the detection target, and the deep convolutional neural network model is trained and tuned by using the sample images of various detection objects. When the model is deployed on the test set, test results show that the algorithm is robust to complex conditions such as illumination changes, weather changes and various scenarios, indicating that our algorithm has good generalization ability.

Acknowledgment

This paper was supported by State Grid Science and Technology Project (520626180067).

References

[1] Hinton G E and Salakhutdinov R 2006 Reducing the Dimensionality of Data with Neural Networks *Science* vol 313 pp 504-507

[2] Sermanet P, Eigen D, Zhang X, Mathieu M, Fergus R and Lecun Y 2013 Overfeat: integrated recognition, localization and detection using convolutional networks *Eprint Arxiv:1312.6229*

[3] Girshick R, Donahue J, Darrell T and Malik J 2014 Rich feature hierarchies for accurate object detection and semantic segmentation *Pro. of the IEEE Computer Society Conf. on Computer Vision and Pattern Recognition, CVPR 2014* pp 580-587

[4] K He, X Zhang, S Ren and J Sun 2015 Spatial Pyramid Pooling in Deep Convolutional Networks for Visual Recognition *IEEE Transactions on Pattern Analysis and Machine Intelligence* vol 37 pp 1904-1916

[5] Girshick R 2015 Fast R-CNN *Proc. of the IEEE Int. Conf. on Computer Vision, ICCV 2015* pp 1440-1448

[6] Redmon J, Divvala S, Girshick R and Farhadi A 2016 You Only Look Once: Unified, Real-Time Object Detection 2016 *Pro. of the IEEE Computer Society Conf. on Computer Vision and Pattern Recognition, CVPR 2016* vol 2016-December pp 779-788

[7] Liu W, Anguelov D and Erhan D, et al 2015 SSD: Single shot multibox detector Lecture *Notes in Computer Science* vol 9905 LNCS pp 21-37

[8] Shaoqing R, Kaiming H, Ross G, and Jian S 2017 Faster R-CNN: Towards Real-Time Object Detection with Region Proposal Networks *IEEE Trans. on Pattern Analysis and Machine Intelligence* vol 39 pp 1137-1149

IOP Conf. Series: Materials Science and Engineering **685** (2019) 012031 doi:10.1088/1757-899X/685/1/012031

CFD Simulation Methods for Rotor Hovering Based on N-S Equation

Zhou Cheng-Long*, Chen Ming, Xu An-An and Wang Fang

School of Aeronautic Science and Engineering, Beihang University, Xueyuan Rd, 37th Haidian Dist, Beijing 100191, Peoples R China.

*E-mail: 278427938@qq.com

Abstract. CFD simulation of the hovering condition of a single-rotor helicopter was carried out by three methods: moving reference frame (MRF), sliding mesh and overset mesh, respectively. The parameters such as velocity field, pressure field and pressure coefficient of the helicopter rotor and its surrounding flow field are obtained, and the simulation results are compared with the experimental data of wind tunnel test. The three methods are compared from the differences between the model and the grid, the accuracy of the calculation results, and the calculation convergence time. The characteristics, application and other meaningful results of each method are obtained.

1. Introduction

The main rotor of the helicopter generates lift and propulsion, so it is the most important part of the helicopter. Its aerodynamic performance not only has a great impact on the flight performance of the helicopter, but also has an important impact on the characteristics of vibration and noise. The design of the main rotor is largely influenced by the requirements of its hovering state, so it is of great significance for the aerodynamic performance of the helicopter hovering state[1].

At present, the research methods of rotor aerodynamic performance mainly include experimental methods and numerical simulation methods. The numerical simulation method effectively overcomes the shortcomings of high cost and long period of the experimental method. Therefore, this method has received more attention. The numerical simulation method of helicopter rotor is mainly divided into vortex method[2, 3] and computational fluid dynamics method[4]. The vortex method is relatively mature, and has good effects in the simulation and prediction of the aerodynamic characteristics of the conventional blade, and the calculation efficiency is high. However, since the vortex method is an approximation based on the potential flow hypothesis, it is difficult to accurately simulate the aerodynamic characteristics and flow field of modern helicopter rotors, especially the detailed characteristics of the flow field near the tip of the blade. The CFD method based on the N-S equation can capture the induction of the wake including the interaction of the tip vortex and the continuous blade as part of the entire flow field without specifying any wake model, and the calculation results can meet the accuracy requirements of the new blade flow field simulation, so the CFD method has become a new research hotspot. In particular, this method has unique advantages in the analysis of new blades or blades with strange shapes[5].

In the hovering condition of a single-rotor helicopter, there is no asymmetry of the pulling force and the moment caused by the asymmetry of the airflow on the rotating plane compared to the forward flight, so there is no flapping motion, shimmy motion and periodic variable pitch motion of the blade.

Therefore, a variety of methods can be used for this condition, including momentum source method, MRF method, sliding mesh method, and overset mesh method. The momentum source method is based on the disc assumption, and there is no precise three-dimensional model of the blade, so it is difficult to consider the transport of the vorticity and the viscosity of the gas. Compared with the momentum source method, the MRF, sliding mesh and overset mesh are more detailed and more close to the actual situation. Therefore, they have been widely used in rotor flow field simulation in recent years[6].

This paper mainly discusses the problems such as calculation accuracy and computational efficiency of MRF, sliding mesh and overset mesh for helicopter under hovering conditions. The calculation results of different methods are compared with the wind tunnel test data to obtain the characteristics and application scope of each method.

2. Grid system

2.1. Mesh generation of MRF and sliding mesh

The MRF method is mainly used for the steady state calculation of CFD. In this paper, the rotating reference frame system is mainly used. The solution provided by this method represents the time-averaged behavior of the fluid, not the time-accurate behavior. When using the MRF technique, the flow area is generally divided into two or more calculation areas, and adjacent calculation domains are connected by one or more interfaces. In the calculation, different regions are assigned different coordinate systems, such as assigning a rotating coordinate system to the region containing the main rotor and assigning a static coordinate system to other flow field regions. When using this method, a constant grid flux determined by the appropriate conservation equation will be generated. The mesh flux is calculated based on the properties of the reference frame, rather than from the local motion of the mesh cell nodes. The MRF method does not change the position of the grid cell node.

It should be noted that the MRF can provide results with practical physical meaning only when the fluid in the surrounding volume is axisymmetric. The MRF cannot be used when any large component of the external flow rate is perpendicular to the axis of rotation. The main reason is that when any component of the surrounding fluid is perpendicular to the axis of rotation, a non-physical result proportional to the vertical component will result.

The sliding mesh method is used for the unsteady state calculation (transient analysis) of CFD, which implements time-accurate behavior by moving the region mesh nodes. When using the sliding mesh technique, as with the MRF method, the flow area is generally divided into two or more calculation areas, and adjacent calculation domains are connected by one or more interfaces. Unlike the MRF method, the motion domain slips along the interface during the calculation. At each time step, the position of the grid is recalculated and the relative position of the grid points on the interface is determined to allow flux to be delivered at the interface.

From the above analysis, it is found that the single-rotor helicopter CFD simulation using the MRF method or the sliding mesh method creates a separate area for the rotating paddle, which is the interface with the air domain that wraps it. It should be noted that the grids on both sides of the interface do not need to be aligned, but the grid scales on both sides cannot be too different, otherwise it will be difficult to ensure the accuracy of the interface flux transfer. In fact, in the process of simulation, the MRF method and the sliding mesh method can all use the same set of grids. Since the vertical flight field of the rotor has rotational symmetry, it is only necessary to take one blade and its corresponding calculation domain for the simulation. The influence of the other blades can be achieved by using periodic boundary conditions on the circumferential boundary. In this paper, the MRF method and the sliding mesh method adopt a three-dimensional body-fitted structure grid, in which the sliding region (motion reference frame region) grid is shown in Figure 1, and the static region and the sliding region assembly grid diagram are shown in Figure 2.

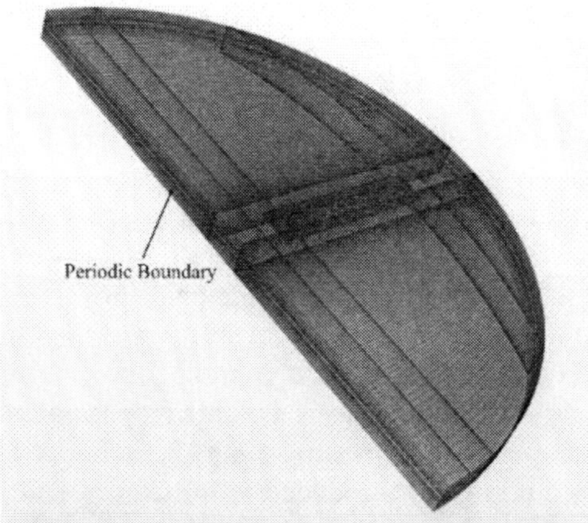

Figure 1. Three-dimensional body-fitted structure grid of the rotating area of the sliding mesh (the rotating frame area of the MRF)

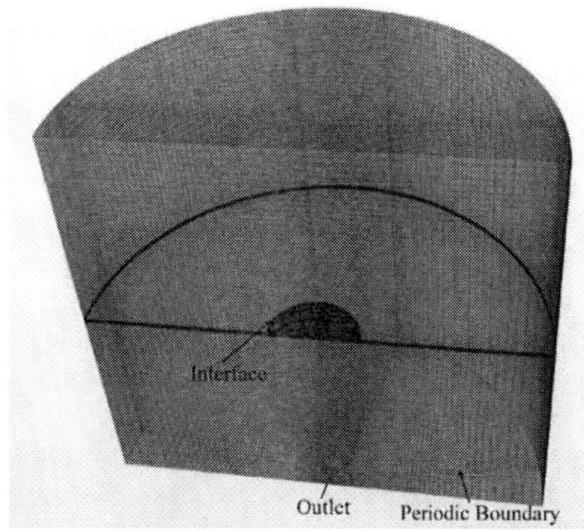

Figure 2. Assembly drawing of the sliding mesh

2.2. Overset mesh method

The overset mesh method can only be used for unsteady computing of CFD, for discrete computing domains with multiple different meshes that overlap each other in any way. Research involving overlapping meshes must have a background region that encloses the entire solution domain, as well as one or more smaller regions that contain the bodies in the domain.

Figure 3 provides an example of how the airfoil curve mesh is located in the background Cartesian mesh. Airfoil grids capture boundary layer, tip vortices and shock waves. The background mesh surrounds the airfoil mesh and takes the solution to the far field. Some background grid points are located inside the airfoil entity area and must be removed from the solution. After deletion, the hole area will remain inside the larger background mesh and create a set of boundary points called hole fringe points. The airfoil mesh inserts the data into the background mesh through the hole fringe points of the background, and the background mesh inserts the data into the airfoil mesh at the airfoil outer boundary points[7].

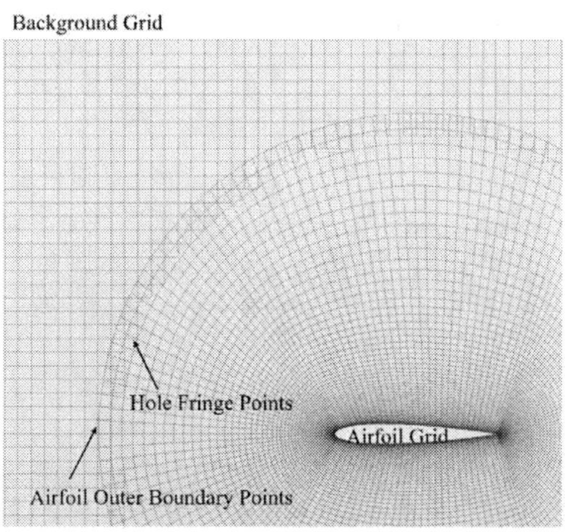

Figure 3. Overset mesh of a two-dimensional airfoil

The advantage of overset mesh is that it allows us to divide several subdomains according to the specific research object, generate meshes for each subdomain separately, and allow the subdomain grids to overlap each other. Therefore, the overset mesh method can simulate the blade boundary layer flow and the rotor wake vortex with high precision, and has satisfactory calculation efficiency. It is generally considered to be a grid method suitable for rotor CFD calculation.

The overset mesh system of this paper is mainly composed of two parts: one is the nearly orthogonal O-H type body structure grid generated around the blade surface, as shown in Figure 4. The second is the background structure grid surrounding the body structure grid. The assembly diagram of the background grid and the body structure grid is shown in Figure 5. The disadvantage of the O-H mesh compared to the O-O mesh is that the paddle tip mesh transition is not smooth enough. The reason why the H-type grid is used in this paper is that the H-type grid has more applications in capturing the tip shock wave[8]. The background grid is static, used to simulate the change of the entire flow field, and the information is transmitted on the hole boundary with the blade body structure grid, thereby completing the information update of the entire flow field. In order to facilitate the overlap of the two grids and to accurately capture the tip vortex, local encryption of the nearby area swept by the tip is required in the background grid. The size of the grid is as large as possible to the outer boundary grid size of the blade body structure grid. At the same time, it is necessary to moderately expand the far-field boundary of the background grid to reduce the influence of far-field boundaries.

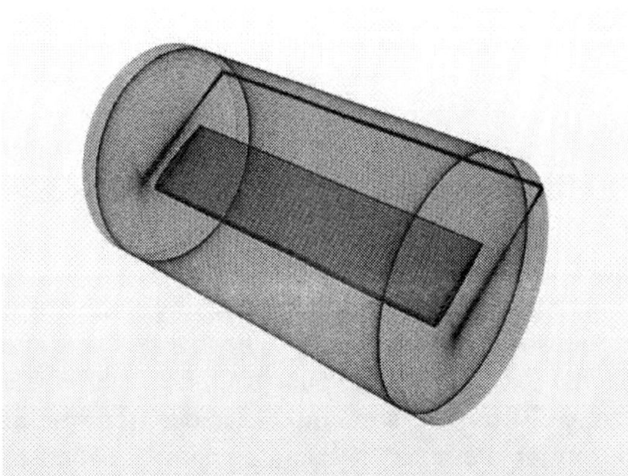

Figure 4. Three-dimensional body-fitted grid with blade rotation domain in the overset mesh system

Figure 5. Assembly drawing of the overset mesh system

Regardless of the grid format, it should be noted that accurately capturing the rotor wake requires multiple grid points in the vortex core to prevent the tip vortex from spreading and maintaining its strength. Observing Figures 1 and 4, it is not difficult to find that the area mesh is encrypted in the vicinity of the vortex core.

3. Governing equations

Considering the continuity equation, the momentum equation and the energy equation, we can obtain the Reynolds average Navier-Stokes equation of the three-dimensional, viscous, unsteady and compressible flow in the inertial coordinate system[9]:

$$\frac{\partial}{\partial t}\iiint_V \vec{W} dV + \iint_S (\vec{F_c} - \vec{F_v}) \cdot \vec{n} dS = 0 \quad (1)$$

Where $\vec{W} = [\rho, \rho u, \rho v, \rho w, \rho e]^T$ is a conserved variable, $\vec{F_c} = (f, g, h)$ is the convective flux, and $\vec{F_v} = (a, b, c)$ is the viscous flux. The specific expression is:

$$f=\begin{bmatrix}\rho u\\ \rho u^2+p\\ \rho uv\\ \rho uw\\ \rho ue+up\end{bmatrix}, g=\begin{bmatrix}\rho v\\ \rho uv\\ \rho v^2+p\\ \rho vw\\ \rho ve+vp\end{bmatrix}, h=\begin{bmatrix}\rho w\\ \rho uw\\ \rho vw\\ \rho w^2+p\\ \rho we+wp\end{bmatrix} \quad (2)$$

$$a=\begin{bmatrix}0\\ \tau_{xx}\\ \tau_{yx}\\ \tau_{zx}\\ u\tau_{xx}+v\tau_{yx}+w\tau_{zx}-q_x\end{bmatrix}$$

$$b=\begin{bmatrix}0\\ \tau_{xy}\\ \tau_{yy}\\ \tau_{zy}\\ u\tau_{xy}+v\tau_{yy}+w\tau_{zy}-q_y\end{bmatrix} \quad (3)$$

$$c=\begin{bmatrix}0\\ \tau_{xz}\\ \tau_{yz}\\ \tau_{zz}\\ u\tau_{xz}+v\tau_{yz}+w\tau_{zz}-q_z\end{bmatrix}$$

In viscous flux, the expression of stress is:

$$\begin{aligned}\tau_{xx}&=\frac{2}{3}\mu\left(2\frac{\partial u}{\partial x}-\frac{\partial v}{\partial y}-\frac{\partial w}{\partial z}\right), \tau_{xy}=\tau_{yx}=\mu\left(\frac{\partial v}{\partial x}+\frac{\partial u}{\partial y}\right), q_x=-K\frac{\partial T}{\partial x}\\ \tau_{yy}&=\frac{2}{3}\mu\left(2\frac{\partial v}{\partial y}-\frac{\partial u}{\partial x}-\frac{\partial w}{\partial z}\right), \tau_{yz}=\tau_{zy}=\mu\left(\frac{\partial w}{\partial y}+\frac{\partial v}{\partial z}\right), q_y=-K\frac{\partial T}{\partial y}\\ \tau_{zz}&=\frac{2}{3}\mu\left(2\frac{\partial w}{\partial z}-\frac{\partial u}{\partial x}-\frac{\partial v}{\partial y}\right), \tau_{xz}=\tau_{zx}=\mu\left(\frac{\partial w}{\partial x}+\frac{\partial u}{\partial z}\right), q_z=-K\frac{\partial T}{\partial z}\end{aligned} \quad (4)$$

In order to obtain a closed system of equations, the relationship between pressure and temperature is introduced as follows:

$$p=\rho(\gamma-1)\left[e-\frac{1}{2}(u^2+v^2+w^2)\right] \quad (5)$$

$$p=\rho RT \quad (6)$$

Where ρ is the density, p is the pressure, e is the total energy of the unit mass gas, μ is the viscosity coefficient, K is the heat transfer coefficient, T is the temperature, q_x, q_y, q_z are the components of the heat flow in three directions, γ is Specific heat capacity, for ideal gas, $\gamma=1.4$, $R=287.3$.

The turbulent viscosity of the flow field was estimated using the Spalart-Allmaras turbulence model[10]. This model is a single equation turbulence model with no accurate transport equations for deriving eddy viscosity coefficients. It is an approximate transport equation obtained from a large amount of experimental data and experience. Because Spalart et al. have rich experience in aeronautical calculation and a large amount of experimental data, the S-A model proposed by them has been widely used in the aviation field, especially in shock capture.

4. Numerical methods and boundary conditions

4.1. Numerical methods

4.1.1. Spatial discretization method. The grid elements of the rotor flow field are spatially discrete based on the finite volume method, and different interpolation formats are formed according to the different discrete methods of the non-viscous flux on the interface of the grid elements. The central format is difficult to simulate the unsteady characteristics of the rotor flow field because of the second-order accuracy in space and the addition of artificial dissipation. Therefore, this paper adopts the format of Roe in upwind style[11]. This format calculates the numerical flux by the approximate Riemann solution method and has strong shock capture capability.

Approximate Riemann solution in Roe format may lead to non-physical understanding. Therefore, the entropy correction method proposed by Harten is used to correct the eigenvalues in Roe format[12].

4.1.2. Time stepping procedures. The slip grid and overlap grid method solves the unsteady state of the rotor. For the solution of the unsteady flow field of the rotor, the time dispersion of the flow control equation adopts the dual time stepping method. This method introduces the concept of physical time and pseudo-time, and pseudo-time advancement in each physical time step is equivalent to solving a steady-state problem. Therefore, measures such as local time step and implicit residual smoothing in the steady solution can be applied to the calculation of the flow field, thus improving the efficiency of the calculation. Jameson[13] explained the method in detail.

4.2. Boundary condition

In order to accurately predict the effects of viscous convection on the flow field and rotor blade parameters, the flow of the turbulent boundary layer in laminar, semi-laminar and fully turbulent zones needs to be considered. The solution is to set a number of grid points in the viscous boundary layer so that the values near the wall satisfy $y+\approx 1$[14]. The innermost layer height of the mesh generated in this paper is less than $1.0*10^{-5}c$, where c is the chord length.

In the hovering state, if the far-field boundary condition is determined based on the static flow outside the calculation box, the flow rate into and out of the box is zero, and the blade is in a closed box atmosphere. When the rotor rotates, the airflow in the box will cycle. Through multiple calculations, it is found that it is difficult to obtain trustworthy calculation results by using such boundary conditions. The reason is that the box size must be very large under such boundary conditions, even beyond the memory limitations of today's supercomputers. Therefore, another far-field boundary condition based on momentum theory is used: when there is a small computational domain around the rotor, a non-zero flow field boundary condition should be specified that allows the airflow to enter and exit the box without violating the conservation law. Strawn and Barth explain in detail the method of determining the boundary conditions[15].

The outflow boundary is in the circle with the radius $R/\sqrt{2}$ directly below the paddle, and the outflow velocity is as follows:

$$V_{out} = M_{tip}\sqrt{C_T} \tag{7}$$

The remaining boundaries are inflow, and the directions are all pointing to the center of the hub. The inflow velocity is as follows:

$$V_{in} = \frac{M_{tip}}{8}\sqrt{C_T}\left(\frac{R}{r_F}\right)^2 \tag{8}$$

Where M_{tip} is the Mach number of rotor tip, C_T is the tension coefficient of the rotor, R is the blade radius, and r_F is the distance from the far field boundary point to the center of the hub.

Srinivasan et al.[16] used this boundary condition to obtain satisfactory results.

5. Examples and discussions

The Caradonna-Tung[17] single-rotor experimental model with experimental data was used as a verification example. The experimental model is a double-leaf, rigid rotor with no distortion. The blade is stretched from the NACA0012 airfoil section with an aspect ratio of 6.0. In this calculation, the pressure coefficient distributions of the upper and lower surfaces of the blade profile at 0.96R in four different hovering states were selected from the experimental data: 1) Mach number at the tip of the blade $M_{tip} = 0.815$, Pitch $\theta = 5°$; 2) Mach number at the tip of the blade $M_{tip} = 0.439$, Pitch $\theta = 8°$; 3) Mach number at the tip of the blade $M_{tip} = 0.877$, Pitch $\theta = 8°$; 4) Mach number at the tip of the blade $M_{tip} = 0.794$, Pitch $\theta = 12°$. The specific parameters of the rotor model are shown in Table 1:

Table 1. Relevant parameters of Caradonna-Tung experimental rotor

Number of blades (N)	2
Rotor radius	1.143m
Plane shape of blade	Rectangle
Airfoil	NACA0012
Chord length (c)	0.1905m
Undercut (R_{cut})	0.243R
Twist	0°

Figure 6 is the pressure contour of working condition 3 calculated by MRF method, and Figure 7 is the velocity contour of working condition 3 calculated by MRF method at 0.96R profile of blade. It can be seen that there is obvious shock wave near the tip under this working condition.

Figure 6. Pressure contours of a blade by MRF method when $M_{tip} = 0.877$, $\theta = 8°$

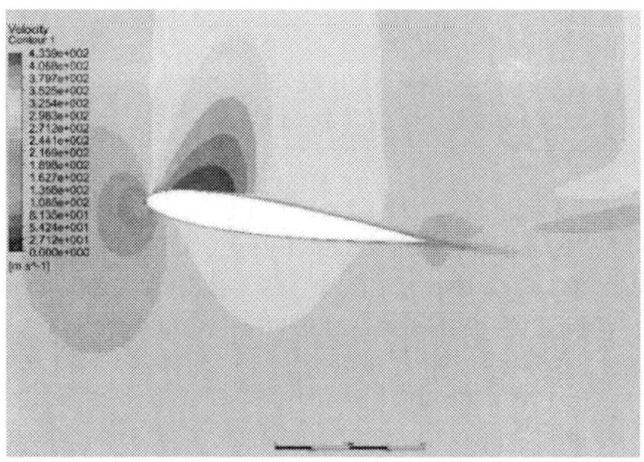

Figure 7. Velocity contours at 0.96R profile of a blade by MRF method when $M_{tip} = 0.877$, $\theta = 8°$

Figure 8 shows the comparison of the surface pressure coefficients obtained by MRF, sliding mesh and overset mesh methods with the experimental data, as follows:

IOP Conf. Series: Materials Science and Engineering **685** (2019) 012031 doi:10.1088/1757-899X/685/1/012031

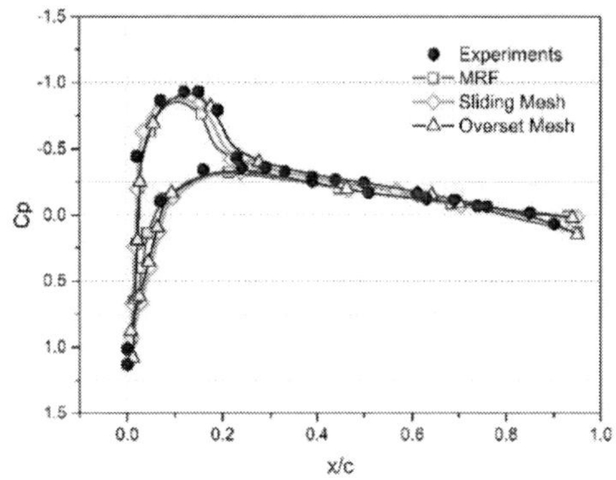

(a) $M_{tip} = 0.815, \theta = 5°$

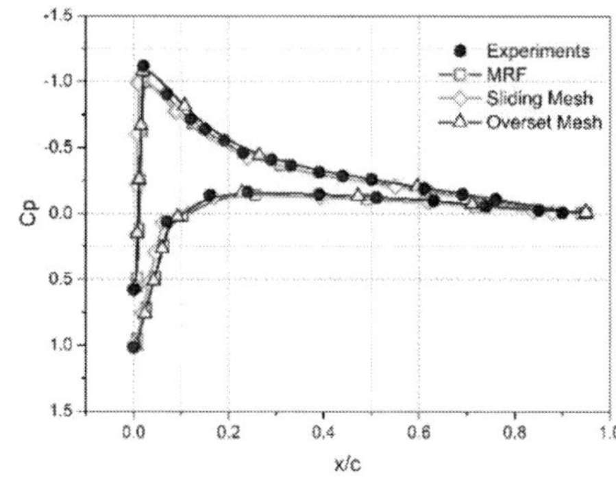

(b) $M_{tip} = 0.439, \theta = 8°$

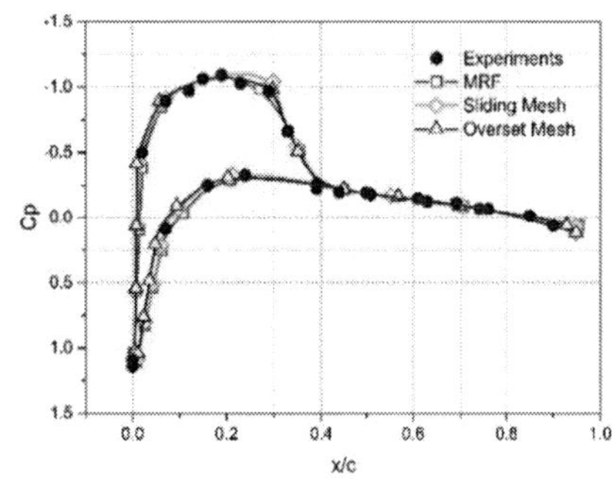

(c) $M_{tip} = 0.877, \theta = 8°$

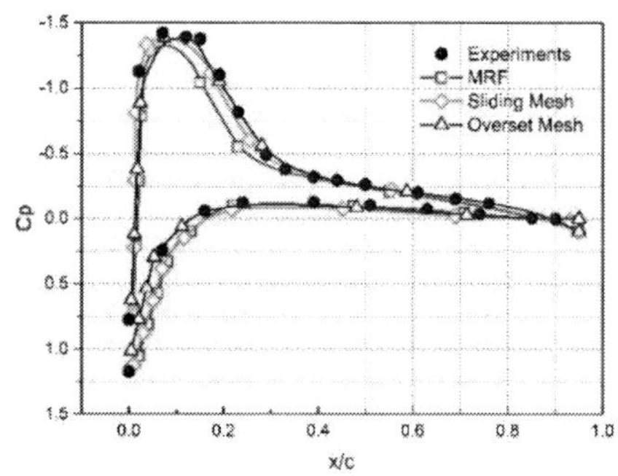

(d) $M_{tip} = 0.794, \theta = 12°$

Figure 8. Surface pressure coefficients of Caradonna-Tung rotors calculated by three methods in hover

It can be seen from the figure 8 that under the four working conditions, all the three methods can obtain reasonable calculation results. Among them, the accuracy of MRF method is worse than that of the other two methods, especially in the acquisition of shocks at propeller tip. However, the time required for convergence of MRF method is much shorter than that of sliding mesh and overset mesh.

Sliding mesh method and overset mesh method can obtain more accurate results than MRF. The convergence time of the two methods is almost the same when the mesh number difference between the two grids is very small. Sliding mesh method needs to re-model and draw meshes in the rotating region when changing the total pitch angle. Therefore, the workload is large and heavy. The overset mesh method can change the pitch angle by simply rotating the mesh in the moving area, and the workload is relatively small.

When calculating the rotor forward flight conditions, due to the complex motion forms such as periodic pitch variation, neither MRF nor sliding mesh method can accurately calculate the rotor forward flight conditions. The overset mesh method can be used to precisely define the forward flight state of the rotor by specifying the angular velocity of the mesh around the variable pitch axis in the rotating region and the initial pitch angle of each blade. Therefore, the overset mesh method has a wider scope of application.

In conclusion, the MRF method is suitable for cases where only the aerodynamic performance of the rotor in hover state is concerned and the accuracy requirement is not high. For example, in the initial stage of engineering design, the reasonable calculation results can be obtained quickly. The sliding mesh method is suitable for further improving the accuracy of calculation on the basis of MRF method. It can directly take the results of MRF method as the initial solution and use the grids of MRF method to calculate. The overset mesh method is applicable to the aerodynamic performance of the rotor in hover, forward flight and other conditions. Compared with MRF method, the calculation amount of overset mesh method is larger, and the calculation results in hover condition are more accurate than that of MRF method.

In addition to the influence of different mesh methods, the aerodynamic simulation of the rotor is also affected by factors such as mesh quality and turbulence model. For unsteady calculations, the size of the time step will also affect the result. In this paper, the analysis of the three methods makes the other influencing factors as consistent as possible, but the rigor is still inevitable.

Acknowledgments
This research was financially supported by the National key research and development project of China (No.2017YFD0701004). Thanks to Professor Chen Ming from the Beihang University, School of Aeronautic Science and Engineering for his valuable comments.

References

[1] Costes M, Renaud T and Rodriguez B 2012 Rotorcraft simulations: a challenge for CFD J. *International Journal of Computational Fluid Dynamics* **26** 383-405
[2] Bagai and Ashish 1995 Rotor free-wake modeling using a pseudo-implicit technique - including comparisons with experimental data J. *Journal of the American Helicopter Society* **40** 29
[3] Bagai A, Leishman J G and Park J 1999 Aerodynamic analysis of a helicopter in steady maneuvering flight using a free-vortex rotor wake model J. *Journal of the American Helicopter Society* **44** 109
[4] Cao Y H, Wu Z L and Huang J S 2014 Numerical simulation of aerodynamic interactions among helicopter rotor, fuselage, engine and body of revolution J. *Science China Technological Sciences* **57** 1206-18
[5] Srinivasan G R, Raghavan V and Duque E P N 1993 Flowfield analysis of modern helicopter rotors in hover by navier-stokes method J. *Journal of the American Helicopter Society* **38** 3
[6] Kang H J and Kwon O J 2002 Unstructured mesh navier-stokes calculations of the flow field of a helicopter rotor in hover J. *Journal of the American Helicopter Society* **47** 90
[7] Ahmad J and Duque E P 1996 Helicopter rotor blade computation in unsteady flows using moving overset grids J. *Journal of Aircraft* **33** 54-60
[8] Baeder J D, Gallman J M and Yu Y H 1997 A computational study of the aeroacoustics of rotors in hover J. *Journal of the American Helicopter Society* **42** 39
[9] Wesseling P 2001 Principles of computational fluid dynamics J. *Springer* **22** 3-9
[10] Spalart P R and Allmaras S R 1994 A one-equation turbulence model for aerodynamic flows J. *Recherche Aerospatiale* **1** 5-21
[11] Roe P L 1997 Approximate Riemann solvers, parameter vectors, and difference schemes J. *Journal of Computational Physics* **135** 250-258
[12] Harten A and Hyman J M 1983 Self adjusting grid methods for one-dimensional hyperbolic conservation laws J. *Journal of Computational Physics* **50** 235-269
[13] Jameson A 1991 Time dependent calculations using multigrid, with applications to unsteady flows past airfoils and wings *10th Computational Fluid Dynamics Conference*
[14] Kang H J and Kwon O J 2002 Unstructured mesh navier-stokes calculations of the flow field of a helicopter rotor in hover J. *Journal of the American Helicopter Society* **47** 90
[15] Strawn R C and Barth T J 1993 A finite-volume Euler solver for computing rotary-wing aerodynamics on unstructured meshes J. *Journal of the American Helicopter Society* **38** 61-67(7)
[16] Srinivasan G R, Raghavan V, Duque E P N and Mccroskey W J 1993 Flowfield analysis of modern helicopter rotors in hover by navier-stokes method J. *Journal of the American Helicopter Society* **38** 3
[17] Caradonna F X and Tung C 1981 Experimental and analytical studies of a model helicopter rotor in hover J. *Vertica* **5** 149-161

AUTHOR INDEX

An-An, Xu 217
Bing, Zhang 35
Bingxiu, Guo 176
Chen, Zhongwei 83
Cheng, Pan 61, 73
Cheng-Long, Zhou 217
Cui, Tao 189, 196
Ding, Zekai 73
Dong, Qing 170
Dong, Shiwei 189
Fan, Jun 148
Fang, Dong 1
Fang, Wang 217
Feng, Jiqiang 127
Fu, Qiang 204
Fu, Wentao 204
Guo, Fushui 148
Guo, Yonggang 141
Hu, Weiwei 98
Huang, Yan 156
Irsyadi, R F 48
Ji, Yuanfa 204
Jiahao, Tang 104
Jianwei, Shi 35
Jun, Zhou 104
Kong, Dalin 1
Kretov, Anatolii 67
Lai, Junyu 83, 91
Lei, Dongyang 98
Li, Benwei 170
Li, Chengkang 83
Li, Chengqi 211
Li, Donglin 1
Li, Jingling 196
Li, Jun 189, 196
Li, Xiaoping 118
Li, Yijun 183
Liang, Wei 196
Liu, Rui 91
Liu, Xiaoyan 61
Liu, Yanming 118
Liu, Zhidong 141
Luo, Lin Qing 133
Lv, Xuebin 211
Ma, Chao 162
Ma, Zhao Xing 133
Mi, Dong 148
Ming, Chen 217
Nirbito, W 41, 48, 54
Ouyang, Zhiping 127
Priatmoko, M R 54
Qi, Chaojie 98
Qian, Renjun 170
Qiang, Zhang 104
Qu, Ying 91
Rammon, N A 41
Rongjun, Deng 35
Shao, Yue 176
Shen, Jun 189
Shi, Guowei 204
Shi, Lei 7
Song, Hanqiang 170
Song, Yuanyuan 183
Sun, Bo 156
Sun, Xiyan 204
Sun, Yingbing 83
Sun, Yufeng 98
Tu, Jianhui 141
Uscategui, J 23
Wan, Qun 183
Wang, Bo 127
Wang, Chao 1, 7
Wang, Jianwei 1, 7
Wang, Jianxiang 141
Wang, Shiwei 141
Wang, Shuquan 17
Wang, Xiaohui 176
Wang, Xiaoming 156
Wang, Yanyan 176
Wang, Yonghua 170
Weijun, Pan 104
Wen, Jiang 112
Xiao, Han 83
Xiao, Sun 35
Xinbing, Su 112
Xinsheng, W 23
Xiwei, Zhao 112
Xu, Chen 127
Yan, Jubin 156
Yang, Bo 211
Yang, Jun 162
Yang, Yuxuan 156
Yingjie, Duan 104
Yuan, Bo 7
Zhai, Kunxian 118
Zhang, Jun 141

Zhang, Luxi 183
Zhang, Wu 73
Zhang, Ying 1, 7
Zhang, Yi 196
Zhang, Yudi 91
Zhao, Shiqi 148
Zhao, Yang 61
Zheng, Wenjie 211
Zhong, Lei 91
Zhou, Chao 162
Zhou, Quan 196
Zhou, Suiping 118
Zhou, Xiaodong 148
Zhu, Lili 141
Zhu, Lingchao 17
Zou, Bo 156